Methods in Molecular Biology

Series Editor
John M. Walker
School of Life and Medical Sciences
University of Hertfordshire
Hatfield, Hertfordshire, UK

For further volumes:
http://www.springer.com/series/7651

For over 35 years, biological scientists have come to rely on the research protocols and methodologies in the critically acclaimed *Methods in Molecular Biology* series. The series was the first to introduce the step-by-step protocols approach that has become the standard in all biomedical protocol publishing. Each protocol is provided in readily-reproducible step-by-step fashion, opening with an introductory overview, a list of the materials and reagents needed to complete the experiment, and followed by a detailed procedure that is supported with a helpful notes section offering tips and tricks of the trade as well as troubleshooting advice. These hallmark features were introduced by series editor Dr. John Walker and constitute the key ingredient in each and every volume of the *Methods in Molecular Biology* series. Tested and trusted, comprehensive and reliable, all protocols from the series are indexed in PubMed.

Alzheimer's Disease

Methods and Protocols

Edited by

Jerold Chun

Translational Neuroscience Initiative, Neuroscience Drug Discovery, Sanford Burnham Prebys Medical Discovery Institute, La Jolla, CA, USA

Humana Press

Editor
Jerold Chun
Translational Neuroscience Initiative
Neuroscience Drug Discovery
Sanford Burnham Prebys Medical Discovery Institute
La Jolla, CA, USA

ISSN 1064-3745 ISSN 1940-6029 (electronic)
Methods in Molecular Biology
ISBN 978-1-0716-2654-2 ISBN 978-1-0716-2655-9 (eBook)
https://doi.org/10.1007/978-1-0716-2655-9

Cover Caption: Volume imaging of neurovascular alterations in the Alzheimer's disease brain. Fibrinogen deposits (red) around areas of vascular tortuosity (green) and Aβ plaques (blue). Image Credit: Mario Merlini and Katerina Akassoglou.

This Humana imprint is published by the registered company Springer Science+Business Media, LLC, part of Springer Nature.
The registered company address is: 1 New York Plaza, New York, NY 10004, U.S.A.

Preface

Alzheimer's disease (AD) is the most common cause of dementia worldwide with enormous medical and societal costs that impact both patients and their families. At the time of writing, AD and related dementias remain without effective disease-modifying therapies, despite intensive research spanning well over a century, since the initial description of AD by Alzheimer and Fischer. The neuropathological identification of what is now recognized as senile plaques, neurofibrillary tangles, and neuronal loss came from careful examination of the human diseased brain itself, as did the identification of amyloid-beta (Aβ) by Glenner and Wong some eight decades later. Indeed, over the last decade, studies on the human brain have identified common comorbid brain diseases distinct from AD, conserved patterns of lesions, neuroanatomical disease progression, and other key features, underscoring the essential importance of analyzing the actual diseased brain, with thoughtful use of experimental systems like human cellular models.

The foundational importance of human brain studies for understanding AD and related neurodegenerative disorders stimulated this collection of modern methodologies. Experts from around the globe have come together to share their expertise through 16 chapters, divided into 5 parts, for studying the diseased human brain: (I) Brain Preparations (through specialized imaging in, and single-cell isolation from, post-mortem brains); (II) Neural Cellular Models (using primary and human induced pluripotent stem cells); (III) Nucleic Acid Analyses (transcriptomic and somatic genomic changes); (IV) Lipid Analyses (via mass spectrometry); and (V) Protein Analyses (particularly examining Aβ and Tau, including their prion forms). Detailed step-by-step protocols designed for clarity have been supplemented with explanatory footnotes and generous use of complementary figures.

Sincere thanks are due to many who enabled this book to materialize. Foremost are the brain donors and their families who selflessly and generously donated brains for scientific study. The availability of these precious samples has been facilitated by brain banks and disease-related organizations, both public and private, and we thank them here as well as specifically within the book's chapters. Thanks go to all of the authors who contributed to the effort during the Covid-19 pandemic: hailing from 12 countries beyond the United States, this meant navigating an unpredictable and constantly changing landscape of health and regulatory challenges during the creation of this volume of *Methods in Molecular Biology*. With sadness, we respectfully offer our condolences to the family of author Professor Svetlana G. Vorsanova (Chapter 10) who was tragically lost to Covid-19. Dr. Laura Wolszon was the formal associate editor of this book, being instrumental in its assembly and completion. Further thanks go to the expert and patient editorial staff at Springer, who enabled the book's completion: John Walker, Patrick Marton, and Anna Rakovsky. We gratefully acknowledge the many funding agencies around the world, particularly the National Institutes of Health and the National Institutes on Aging, for supporting the germane science. Last and by no means least, we thank past and current members of the Chun laboratory, particularly Dr. Gwendolyn Kaeser who helped to get this project off the ground along with Dr. Laura Wolszon, who took on associate editor responsibilities.

We hope that all readers may benefit from this book, towards expanding knowledge and helping patients and their families through new understanding, new treatments, and, one day, curing Alzheimer's disease and related dementias.

La Jolla, CA, USA *Jerold Chun*

LISA M. KEENE • *University of Washington, Department of Laboratory Medicine and Pathology, Seattle, WA, USA*
CAGHAN KIZIL • *German Center for Neurodegenerative Diseases (DZNE) Dresden, Helmholtz Association, Dresden, Germany; Taub Institute for Research on Alzheimer's Disease and the Aging Brain, Columbia University Irving Medical Center, New York, NY, USA; Department of Neurology, Columbia University Irving Medical Center, New York, NY, USA*
MITCHELL K. P. LAI • *Department of Pharmacology, Yong Loo Lin School of Medicine, National University of Singapore, Singapore, Singapore; Memory Aging and Cognition Centre, Department of Psychological Medicine, Yong Loo Lin School of Medicine, National University of Singapore, Singapore, Singapore*
CAITLIN S. LATIMER • *University of Washington, Department of Laboratory Medicine and Pathology, Seattle, WA, USA*
VIRGINIA M. -Y. LEE • *Center for Neurodegenerative Disease Research, Department of Pathology and Laboratory Medicine, Institute on Aging, University of Pennsylvania School of Medicine, Philadelphia, PA, USA*
LIH-FEN LUE • *Human Cell Core for Translational Research, Banner Sun Health Research Institute, Sun City, AZ, USA*
HATICE BÜŞRA LÜLECI • *Department of Bioengineering, Gebze Technical University, Gebze, Kocaeli, Turkey*
CHRISTINE L. MAC DONALD • *University of Washington, Department of Neurological Surgery, Seattle, WA, USA*
IVÁN MANUEL • *Department of Pharmacology, Faculty of Medicine and Nursing, University of the Basque Country (UPV/EHU), Leioa, Spain; Neurodegenerative Diseases, BioCruces Bizkaia Health Research Institute, Barakaldo, Spain*
JONATAN MARTÍNEZ-GARDEAZABAL • *Department of Pharmacology, Faculty of Medicine and Nursing, University of the Basque Country (UPV/EHU), Leioa, Spain*
ERICA J. MELIEF • *University of Washington, Department of Laboratory Medicine and Pathology, Seattle, WA, USA*
BEATRIZ MELLO • *D'Or Institute for Research and Education (IDOR), Rio de Janeiro, Brazil*
MARIO MERLINI • *Gladstone UCSF Center for Neurovascular Brain Immunology, San Francisco, CA, USA; Gladstone Institute of Neurological Disease, San Francisco, CA, USA*
GREGORY E. MERZ • *Institute for Neurodegenerative Diseases, Weill Institute for Neurosciences, University of California San Francisco, San Francisco, CA, USA*
MARTA MORENO-RODRÍGUEZ • *Department of Pharmacology, Faculty of Medicine and Nursing, University of the Basque Country (UPV/EHU), Leioa, Spain*
ALISSA L. NANA • *Department of Neurology, Weill Institute of Neurosciences, University of California, San Francisco, CA, USA*
AMBER L. NOLAN • *University of Washington, Department of Laboratory Medicine and Pathology, Seattle, WA, USA*
KRISTEN DAMS O'CONNOR • *Mount Sinai Hospital, Department of Neurology, New York, NY, USA*
CARTER R. PALMER • *Translational Neuroscience Initiative, Sanford Burnham Prebys Medical Discovery Institute, La Jolla, CA, USA; Biomedical Sciences Program, School of Medicine, La Jolla, CA, USA*
DIEGO PEREZ-RODRIGUEZ • *Department of Clinical and Movement Neurosciences, Queen Square Institute of Neurology, University College London, London, UK*

GEOFFREY PIRES • *Center for Cognitive Neurology, Department of Neurology, New York University Grossman School of Medicine, New York, NY, USA; Alzheimer's and Prion Diseases Team, Paris Brain Institute, CNRS, UMR 7225, INSERM 1127, Sorbonne University UM75, Paris, France*

STANISLAVA POPOVA • *German Center for Neurodegenerative Diseases (DZNE) Dresden, Helmholtz Association, Dresden, Germany; Neuron-D GmbH, Dresden, Germany*

CHRISTOS PROUKAKIS • *Department of Clinical and Movement Neurosciences, Queen Square Institute of Neurology, University College London, London, UK*

STANLEY B. PRUSINER • *Institute for Neurodegenerative Diseases, Weill Institute for Neurosciences, University of California San Francisco, San Francisco, CA, USA; Department of Neurology, Weill Institute for Neurosciences, University of California San Francisco, San Francisco, CA, USA; Department of Biochemistry and Biophysics, University of California San Francisco, San Francisco, CA, USA*

STEVENS REHEN • *D'Or Institute for Research and Education (IDOR), Rio de Janeiro, Brazil; Department of Genetics, Institute of Biology, Federal University of Rio de Janeiro (UFRJ), Rio de Janeiro, Brazil*

RAFAEL RODRÍGUEZ-PUERTAS • *Department of Pharmacology, Faculty of Medicine and Nursing, University of the Basque Country (UPV/EHU), Leioa, Spain; Neurodegenerative Diseases, BioCruces Bizkaia Health Research Institute, Barakaldo, Spain*

CATHERINE SANTUCCI • *Department of Clinical and Movement Neurosciences, Queen Square Institute of Neurology, University College London, London, UK*

AIMEE M. SCHANTZ • *University of Washington, Department of Laboratory Medicine and Pathology, Seattle, WA, USA*

WILLIAM W. SEELEY • *Department of Neurology, Weill Institute of Neurosciences, University of California, San Francisco, CA, USA*

TOHID SIDDIQUI • *German Center for Neurodegenerative Diseases (DZNE) Dresden, Helmholtz Association, Dresden, Germany*

EVA ŠIMONČIČOVÁ • *Division of Medical Sciences, University of Victoria, Victoria, BC, Canada; Neuroscience Graduate Program, University of Victoria, Victoria, Canada*

ELIF G. SOZMEN • *Gladstone UCSF Center for Neurovascular Brain Immunology, San Francisco, CA, USA; Gladstone Institute of Neurological Disease, San Francisco, CA, USA; Department of Neurology, Weill Institute of Neurosciences, University of California, San Francisco, CA, USA*

MARIE-KIM ST-PIERRE • *Axe neurosciences, Centre de recherche du CHU de Québec-Université Laval, Québec, QC, Canada; Département de médecine moléculaire, Faculté de médecine, Université Laval, Québec, QC, Canada; Division of Medical Sciences, University of Victoria, Victoria, BC, Canada*

KESHAV S. SUBRAMANIAN • *Gladstone Institute of Neurological Disease, San Francisco, CA, USA*

TREVOR M. SYTSMA • *University of Washington, Department of Laboratory Medicine and Pathology, Seattle, WA, USA*

MADHAV THAMBISETTY • *Clinical and Translational Neuroscience Section, Laboratory of Behavioral Neuroscience, National Institute on Aging (NIA), National Institutes of Health (NIH), Baltimore, MD, USA*

FEDERICO TORTA • *Precision Medicine Translational Research Programme and Department of Biochemistry, Yong Loo Lin School of Medicine, NUS, Singapore, Singapore; Singapore Lipidomics Incubator, Life Sciences Institute, NUS, Singapore, Singapore*

MARIE-ÈVE TREMBLAY • *Axe neurosciences, Centre de recherche du CHU de Québec-Université Laval, Québec, QC, Canada; Département de médecine moléculaire, Faculté de médecine, Université Laval, Québec, QC, Canada; Division of Medical Sciences, University of Victoria, Victoria, BC, Canada; Department of Biochemistry and Molecular Biology, University of British Columbia, Vancouver, BC, Canada; Department of Neurology and Neurosurgery, McGill University, Montréal, QC, Canada; Centre for Advanced Materials and Related Technology (CAMTEC), University of Victoria, Victoria, Canada*

BEATRIX UEBERHEIDE • *Center for Cognitive Neurology, Department of Neurology, New York University Grossman School of Medicine, New York, NY, USA; Proteomics Laboratory, Division of Advanced Research Technologies, New York University Grossman School of Medicine, New York, NY, USA; Department of Biochemistry and Molecular Pharmacology, New York University Grossman School of Medicine, New York, NY, USA*

DILARA UZUNER • *Department of Bioengineering, Gebze Technical University, Gebze, Kocaeli, Turkey*

BART VANDERBORGHT • *D'Or Institute for Research and Education (IDOR), Rio de Janeiro, Brazil*

GABRIELA VITÓRIA • *D'Or Institute for Research and Education (IDOR), Rio de Janeiro, Brazil*

SVETLANA G. VORSANOVA • *Yurov's Laboratory of Molecular Genetics and Cytogenomics of the Brain, Mental Health Research Center, Moscow, Russia; Vorsanova's Laboratory of Molecular Cytogenetics of Neuropsychiatric Diseases, Veltischev Research and Clinical Institute for Pediatrics of the Pirogov Russian National Research Medical University, Moscow, Russia*

DOUGLAS G. WALKER • *Molecular Neuroscience Research Center, Shiga University of Medical Science, Shiga, Otsu, Japan*

MARKUS R. WENK • *Precision Medicine Translational Research Programme and Department of Biochemistry, Yong Loo Lin School of Medicine, NUS, Singapore, Singapore; Singapore Lipidomics Incubator, Life Sciences Institute, NUS, Singapore, Singapore*

CARSTEN WERNER • *Leibniz-Institut für Polymerforschung Dresden e.V., Max Bergmann Center of Biomaterials Dresden, Dresden, Germany*

ANGELA M. WILSON • *University of Washington, Department of Laboratory Medicine and Pathology, Seattle, WA, USA*

THOMAS WISNIEWSKI • *Center for Cognitive Neurology, Department of Neurology, New York University Grossman School of Medicine, New York, NY, USA; Departments of Pathology and Psychiatry, New York University Grossman School of Medicine, New York, NY, USA*

HONG XU • *Center for Neurodegenerative Disease Research, Department of Pathology and Laboratory Medicine, Institute on Aging, University of Pennsylvania School of Medicine, Philadelphia, PA, USA*

JESSICA E. YOUNG • *Department of Laboratory Medicine and Pathology, University of Washington, Seattle, WA, USA; Institute for Stem Cell and Regenerative Medicine, University of Washington, Seattle, WA, USA*

YURI B. YUROV • *Yurov's Laboratory of Molecular Genetics and Cytogenomics of the Brain, Mental Health Research Center, Moscow, Russia; Vorsanova's Laboratory of Molecular Cytogenetics of Neuropsychiatric Diseases, Veltischev Research and Clinical Institute for Pediatrics of the Pirogov Russian National Research Medical University, Moscow, Russia*

Part I

Human Brain Preparations

Chapter 1

Protocol for the Systematic Fixation, Circuit-Based Sampling, and Qualitative and Quantitative Neuropathological Analysis of Human Brain Tissue

Caitlin S. Latimer, Erica J. Melief, Jeanelle Ariza-Torres, Kim Howard, Amanda R. Keen, Lisa M. Keene, Aimee M. Schantz, Trevor M. Sytsma, Angela M. Wilson, Thomas J. Grabowski, Martin Darvas, Kristen Dams O'Connor, Amber L. Nolan, Brian L. Edlow, Christine L. Mac Donald, and C. Dirk Keene

Abstract

Human brain tissue has long been a critical resource for neuroanatomy and neuropathology, but with the advent of advanced imaging and molecular sequencing techniques, it has become possible to use human brain tissue to study, in great detail, the structural, molecular, and even functional underpinnings of human brain disease. In the century following the first description of Alzheimer's disease (AD), numerous technological advances applied to human tissue have enabled novel diagnostic approaches using diverse physical and molecular biomarkers, and many drug therapies have been tested in clinical trials (Schachter and Davis, Dialogues Clin Neurosci 2:91-100, 2000). The methods for brain procurement and tissue stabilization have remained somewhat consistently focused on formalin fixation and freezing. Although these methods have enabled research protocols of multiple modalities, new, more advanced technologies demand improved methodologies for the procurement, characterization, stabilization, and preparation of both normal and diseased human brain tissues. Here, we describe our current protocols for the procurement and characterization of fixed brain tissue, to enable systematic and precisely targeted diagnoses, and describe the novel, quantitative molecular, and neuroanatomical studies that broadly expand the use of formalin-fixed, paraffin-embedded (FFPE) tissue that will further our understanding of the mechanisms underlying human neuropathologies.

Key words Human brain tissue, Brain tissue preparation, Brain tissue fixation, Postmortem brain tissue, Diagnostics, Alzheimer's disease, Traumatic brain injury, Histology, Whole-slide imaging, Quantitative neuropathology, Ex vivo MRI analysis, Postmortem

1 Introduction

Since the end of the nineteenth century, human brain tissue has been collected for the explicit purpose of better understanding diseases of the central nervous system [1, 2, 3]. In the United

Jerold Chun (ed.), *Alzheimer's Disease: Methods and Protocols*,
Methods in Molecular Biology, vol. 2561, https://doi.org/10.1007/978-1-0716-2655-9_1,

States, a more systematic approach to brain autopsies for research began in the 1960s [4]. Today, there are numerous brain banks worldwide that collect tissue from individuals suffering from a wide range of neurological diseases, as well as from "normal" controls. Studies using these banked human brain tissues have enabled advances in how neuropathologies are diagnosed and have revealed important clues to the underlying pathophysiology, thereby guiding the development of experimental models whose pathologies parallel many of those observed in human tissue.

Much of the characterization of neurological injury and disease in humans has been performed on FFPE tissue sections for which numerous histological protocols have been optimized. Fixation enables detailed observation of the cytoarchitecture, as revealed by immunohistochemical techniques focused on the visualization of specific antigens throughout the human brain. Such direct visualization has significantly advanced our understanding of diseases involving protein misfolding and aggregation, identifying many pathologic proteins underlying disease and characterizing their effects on the molecular biology of the brain.

Although formalin and other fixatives are optimal for preserving the tissue's cytoarchitecture, fresh-frozen tissue remains the material of choice for most biochemical assays; thus, brain banks often freeze some portion of each brain specifically for these studies. However, newer technologies are making many of these biochemical assays feasible in fixed tissue, a boon for neuroscientists because the vast majority of banked human brain is fixed, rather than frozen.

The methods used to study the brain and the cells that comprise it continue to evolve and so must the protocols for acquiring and preserving human tissue. The boundaries of what can be achieved in the realm of molecular biology continue to expand; thus, brain biorepositories must be ready to respond to changing needs of the scientists utilizing this valuable resource to advance the field.

The protocols currently used by most brain banks have been developed over decades in response to the changing demands of scientific research. Although there is no universal brain-banking protocol, certain basic standards have been established that enable the ethical and efficient acquisition of high-quality tissue that is well-characterized, sampled, and preserved. It is imperative that brain tissue acquisition is performed only in instances where the individual, or their legal next-of-kin, provides informed consent during their lifetime, or whose legal next-of kin consents to donation after the passing of the donor.

The value of brain donation is markedly increased if the individual was well-characterized clinically. This includes giving potential donors batteries of cognitive and behavioral tests and conducting imaging and biofluid biomarker studies. To help ensure cases are quickly and appropriately identified and confirmed, it is most helpful to have a rigorous infrastructure in place that includes accessible (but sufficiently safeguarded) donor information and consent and well-established lines of communication, allowing brain removal and tissue preservation to proceed as quickly as possible. Similarly, accurate evaluation of the postmortem tissue is crucial to the integrity and success of the downstream research; indeed, many brain banks employ neuropathologists or similarly trained individuals to provide neuropathologic diagnoses based on the most up-to-date diagnostic and consensus criteria.

Tissue must be sampled and preserved appropriately for current and future uses, which may necessitate both fixed and frozen tissue or, more recently, tissues preserved in ways that permit establishment of future cell and organoid models of disease. There are a multitude of approaches used for collecting, stabilizing, and sampling the brain, for both diagnosis and research, the success of which is highly dependent on the training and background of the individual(s) performing the dissections. Although the brain conforms to a general structural pattern, there is still a remarkable amount of individual variation that can affect the accuracy and precision of sampling. Additionally, the focus and structure of individual brain banks may differ in determining the brain regions to be sampled in order to prioritize particular needs; although this provides focus for limited resources, it may also limit the usefulness of the tissue in future endeavors.

Human brain donation is one of the most valuable contributions an individual can make to science. It is the burden of brain biorepositories to honor this important gift by continuing to adapt methodologies to enable the greatest diversity of research approaches and the highest impact of scientific investigation. Biobanking procedures, particularly for human brain, must be nimble and versatile enough to provide investigators with the highest-quality tissue for use with the most cutting-edge technologies, while still preserving some historical approaches to provide consistency across time. Given how quickly technologies advance, it is now more important than ever to reevaluate constantly tissue procurement, storage, and sampling techniques.

In this chapter, we describe our current protocols for tissue collection, characterization, sampling, and storage, focusing on the acquisition and preparation of fixed tissue that will be appropriate for the rapidly evolving scientific approaches to understanding human neurological disease, while maintaining continuity with historical collection procedures. We outline our protocols for acquiring donations in a time-sensitive manner, tissue procurement and processing, ex vivo MRI analysis, volumetric analyses, imaging- and pathology-guided sampling, histological staining and diagnosis, and quantitative neuropathological analysis using solution-phase immunoassays and immunohistochemical image analysis.

2 Materials

2.1 Brain Removal and Fresh-Tissue Handling

1. Appropriate personal protective equipment (PPE), including:
   (a) Nitrile gloves, sterile.
   (b) Latex gloves, sterile.
   (c) Gown, sterile, disposable, or sterile, disposable arm sleeves, with a disposable apron.
   (d) Hairnet.
   (e) Surgical or N95 masks and power air-purifying respirators (PAPRs).
   (f) Face shield.
2. Razor blades, sterile.
3. Forceps, sterile.
4. pH strips.
5. Bone saw.
6. Scalpel, sterile.
7. Forceps.
8. *Neutral-buffered formalin* (NBF), 10%: for 22.7 L, dissolve 130 g of sodium phosphate (Na_2HPO_4) dibasic and 80 g sodium phosphate monobasic in 2 L ddH_2O, add 2 L of 37% formaldehyde (in fume hood), and then add 18 L ddH_2O.
9. Syringe, 10 cc, sterile.
10. 18-gauge, 8.89 cm needle, sterile.
11. Cryoprotectant solution: 10% sucrose in ddH_2O, sterile.
12. 10 screw-top cryovials, 1.5 mL, sterile.
13. Cryovials, 3 mL, filled with cryoprotectant and kept at 4 °F (-16 °C).
14. Isopropanol.
15. *Mr. Frosty* isopentane freezing container, or equivalent, appropriate for 3 mL cryovials.

2.2 Postmortem Ex Vivo MRI

1. Bouffant cap, 65.8 cm.
2. Agarose powder, 150 g.
3. ddH_2O, 6 L.
4. Hot plate with magnetic stirrer (capable of heating to at least 200 °C).
5. Stir bar.
6. Wooden spoon.
7. Metal rod or equivalent, for keeping the brain submerged.

8. Scaffolding materials for tissue embedding bin:
   (a) Two acrylic sheets, 18.73 cm × 22.7 × 1.27 cm with 0.64 cm holes drilled every 1.27 cm along the 18.73 cm edge, 2.54 cm away from the edge.
   (b) Nine acrylic rods for scaffolding, 26.8 cm long × 0.64 cm in diameter.
9. Rectangular plastic container, approximately 2.4 L (large enough to hold the brain affixed to the scaffold).
10. Twine.
11. 3 T MRI scanner.
12. *FreeSurfer* image analysis software (https://surfer.nmr.mgh.harvard.edu/fswiki/DownloadAndInstall).

2.3 Fixed-Tissue Preparation

1. Rotary deli slicer.
2. Scalpel and blade.
3. Digital camera.
4. Leica tissue processor ASP6025 S, or equivalent.
5. Leica processor tissue cassettes, 1.5 cm × 2 cm.
6. Tissue punch, 1.5 cm × 2 cm (customized to match the size of the processor cassettes).
7. HistoCore Arcadia paraffin-embedding center, consisting of the Arcadia H-heated embedding workstation and the Arcadia C cold plate.
8. Four metal base molds for the embedder: 0.7 cm × 0.7 cm, 1.5 cm × 1.5 cm, 2.4 cm × 2.4 cm, and 3.0 cm × 2.4 cm.
9. Curved forceps.
10. Paraffin.
11. 70%, 95%, and 100% ethanol solutions in ddH_2O.
12. Xylene or xylene substitute.

2.4 Standard Neuropathological Diagnostic Workup

1. Microtome (Leica, or equivalent).
2. High-profile microtome blades.
3. Water bath.
4. Xylene.
5. Ethanol de- or rehydration series, 100%, 95%, 70%, and 50% in ddH_2O.
6. *Tris-buffered saline with Tween* (TBST); for 20 L: 700.8 g NaCl, 484.4 g Tris-HCL, in ddH_2O; adjust pH to 7.2 with NaOH or HCl, then add 20 mL Tween-20 (a nonionic detergent).

7. Optional: NxGen Decloaking Chamber™ and 1× Diva Decloaker solution, or equivalent heat-induced epitope retrieval (HIER) system.
8. Thirty-two (32) plastic staining jars, to hold 200 mL of solution.
9. Two 60 mL glass Coplin jars.
10. Hydrochloric acid (HCl), to acid-wash glassware when called for.
11. *Luxol fast blue (LFB) staining solution* (for 4.02 L): add 4.0 g Luxol fast blue to 4.0 L of 95% ethanol, then add 20 mL of 10% acetic acid.
12. *Lithium carbonate solution* (for 4.0 L): add 2.0 g lithium carbonate (Li_2CO_3) to 4.0 L ddH_2O and mix.
13. Bluing solution (StatLab Diagnostics, or equivalent).
14. Clearview™ hematoxylin (StatLab Diagnostics, or equivalent).
15. Eosin Y with phloxine (StatLab Diagnostics, or equivalent).
16. 70% Ethanol.
17. Two 100 mL glass beakers.
18. Silver nitrate, 20% in ddH_2O (100 mL).
19. Sodium thiosulfate, 2% in ddH_2O (100 mL).
20. Ammonium hydroxide (NH_4OH) solution, fresh.
21. Dilute NH_4OH: 4 drops NH_4OH solution in 100 mL ddH_2O.
22. Glass funnel.
23. Whatman grade 1 filter paper for funnel.
24. Plastic transfer pipet, fine-tipped, sterile.
25. Micropipettor (200 μL) and appropriate tips.
26. *(Aged) developer solution* (for 120 mL): 20 mL of 37% formaldehyde, 100 mL dH2O, 1 drop of concentrated nitric acid (HNO_3), 0.5 g citric acid. *This solution must age for 1 week prior to use.*
27. Microscope for slide-viewing.
28. EcoMount (BioCare, or equivalent).
29. Glass slides.
30. Coverslips, 24 mm × 50 mm.
31. Slide boxes.
32. BioCare intelliPATH™ (semi-automated staining device), or equivalent.

2.5 Digital Neuropathology

1. Slide scanner (Aperio AT2 Leica, or equivalent).
2. Cloud-based file storage system (e.g., Amazon Web Services, or equivalent).
3. *HALO* image analysis platform (Indica Labs).

2.6 Multiplex Liquid-Phase Extracted Protein Quantification

1. Pierce™ BCA protein assay kit, or equivalent.
2. LUMINEX laser capture system, or equivalent.
3. Isopropanol rehydration series: 100%, 96%, 70%, and 50%, in ddH_2O, enough to cover the slides.
4. Polypropylene tubes, 0.6 mL.
5. Four sets (for each sample) of 1.5 mL microcentrifuge tubes, labeled: tissue, pellet, RIPA (radioimmunoprecipitation assay) and GuHCl (guanidine hydrochloride).
6. *Extraction buffer* (for 100 μL per extracted sample): immediately before use, add 6 μL β-mercaptoethanol to 94 μL Qiagen Extraction Buffer EXB Plus, immediately before use, and vortex.
7. Heptane.
8. Methanol (MeOH).
9. β-mercaptoethanol.
10. Qproteome FFPE tissue kit, or equivalent.
11. Chloroform.
12. *RIPA buffer:* sodium deoxycholate, 12 mM, NP-40, 1%, sodium dodecyl sulfate, 0.1%, NaCl, 150 mM, Tris-HCl, 50 mM, pH 7.0, cOmplete™ Mini protease inhibitor cocktail, 1 tablet (or equivalent), PhosSTOP ™ phosphatase inhibitor, 1 tablet (or equivalent).
13. *GuHCl solution*: GuHCl, 5 M, cOmplete™ mini protease inhibitor cocktail, 1 tablet (or equivalent), PhosSTOP™ phosphatase inhibitor, 1 tablet (or equivalent).
14. MILLIPLEX Human A-Beta TAU panel (HNABTMAG-68 K-04).
15. Bio-Rad Bio-Plex® 200 system.
16. xMAP® Sheath Concentrate pack (20×).
17. Bleach.

3 Methods

3.1 Infrastructure and Workflow (See Note 1)

1. Compile database (*see* **Note 2**), to include: subject's legal name and date of birth, the name and contact information for legal next-of-kin (LNOK), and signed consent forms.
2. Pre-arrange a transportation service and ensure that it has access to the paperwork for the consented subjects on file.
3. Discuss consent with the subject and the LNOK, and have the appropriate party sign the consent forms (*see* **Notes 3** and **4**).
4. Scan consent forms and upload to the *REDCap* database, and send copies to the transportation service.
5. Provide LNOK with the contact information for the transport company and instructions for next steps (*see* **Note 5**).
6. Periodically update database with current contact information for the subject and LNOK.
7. At the time of death, the family or caregiver should alert the transport company, which verifies consent with the LNOK and alerts the research team.
8. The transport company should bring the subject to the brain-recovery facility as soon as possible.

3.2 Decedent Arrival, Preparation, and Brain Removal

All personnel coming into contact with human remains should wear appropriate personal-protection equipment.

1. Confirm identification:
   - (a) Find the subject in the database to confirm their enrollment in a study and verify that consent for brain donation has been finalized.
   - (b) Upon arrival of a decedent to the autopsy suite, confirm that at least two physical identifiers associated with the subject (e.g., wrist or ankle bands, etc.) are consistent with all previously received paperwork.
   - (c) Photograph all identification tags on the decedent.
   - (d) Inspect the body and report/document any significant findings.
2. Remove the brain (*see* **Note 6**):
   - (a) Don the appropriate PPE, as follows: a mask or PAPR over the mouth and nose; a face shield protecting the eyes; two pairs of gloves (one pair of latex/nitrile gloves under a pair of latex gloves); a hairnet; and a gown, or disposable arm sleeves with a disposable apron.
   - (b) Perform nasopharyngeal swab sample collection for SARS-CoV-2 testing, and collect any other relevant specimens, such as blood for serology testing (to detect HIV, HCV, HCB, etc.) and for biobanking, skin for primary cell-culture generation, etc.

(c) Position the decedent to ensure a safe workspace.

(d) Incise and reflect the scalp.

(e) Incise the calvarium from the frontal to occipital regions, and use an oscillating bone saw to add a "shelf" in the posterior temporal area to support the apposition of skull bones, taking care not to damage the underlying brain.

(f) Open the skull and remove the bone flap (cranium), and transition to sterile technique to incise and reflect the dura (if not attached to the bone flap).

(g) Collect cerebrospinal fluid (CSF) and measure the pH:

  (i) Insert the 10-cc syringe cannula obliquely through an accessible sulcus in the inferior parietal lobe to access the occipital horn of the lateral ventricle, and extract 10 mL CSF.

  (ii) Place a single drop of CSF on a pH strip and record the value.

  (iii) Aliquot 1 mL of CSF into each of ten screw-top vials, snap-freeze in liquid nitrogen, and store at −80 °C.

(h) Sample and store leptomeninges:

  (i) Using sterile technique (autoclaved or disposable sterile forceps and scalpel), remove three 2 × 2 cm portions of posterior-superior frontal leptomeninges (arachnoid mater), placing each sample in separate, chilled sterile cryovials prefilled with cryoprotectant.

  (ii) Place the cryovials in a Mr. Frosty freezing container (or equivalent) filled to the indicator line with room temperature (RT) 100% isopropanol and store at −80 °C.

(i) Remove the brain, recording the time of removal in the database.

(j) Examine the skull base and skull flap and be sure to note/photograph any lesions, including hemorrhages, fractures, masses, etc., and take their measurements.

(k) Remove the pituitary gland from the sella turcica, bisect it, freeze one half in liquid nitrogen and preserve the other half in formalin.

(l) Remove the dura (if attached to skull flap) and place in 10% NBF (*see* **Note 7**).

(m) Perform gross examination, documenting any notable observations, described below.

3. Weigh the brain and record the weight in the database.
4. Gently clean the brain of obvious blood (using a Kimwipe or similar) and photograph it from all orientations.

5. Perform a gross examination of the brain for any developmental abnormalities or pathologic lesions (i.e., contusions, hemorrhages, infarcts, masses), documenting them liberally with photography, and entering pertinent descriptions (including measurements, as appropriate) in the database.
6. Immediately proceed to Subheading 3.3.

3.3 Rapid Brain Dissection and Processing

1. If frozen tissue sampling is to be performed:
   (a) Divide the brain using a midsagittal cut through the cerebrum (the corpus callosum and other midline tracts, and deep nuclei), cerebellum (vermis), and brainstem (midline), and photograph the medial surfaces of the side to be frozen (*see* **Notes 8**, **9**, and **10**).
   (b) Select one hemisphere ("fixed side") for fixation and place it, medial side down, in a container of 10% NBF.
2. If fixing the entire brain, suspend it in 10% NBF using an extra-large (68.5 cm) bouffant hairnet, making certain the entire brain is covered by the fixative.
3. Continue fixation for at least 2 weeks before further proceeding to **Step 4**, changing the NBF solution at least once every 1–2 weeks (*see* **Note 11**).
4. Just after fixation, but before further processing, perform a gross examination of fixed tissue.
5. Dispose of 10% NBF in appropriate waste container and rinse brains in running tap water for a minimum of 2 h.
6. Weigh brain tissue with the brainstem and cerebellum attached, then remove the brainstem and cerebellum and reweigh them separately.
7. Examine the external surfaces of the brain, noting any gross pathology and assessing the vasculature and degree of cortical atrophy.
8. Photograph the dura, and all surfaces of the brain, cerebellum, and brainstem.

3.4 Postmortem Ex Vivo MRI (See Note 12)

1. Embed the brain in agarose:
   (a) Assemble a scaffold for the brain using two plastic scaffolding plates and 9 acrylic rods (Fig. 1a).
   (b) Using twine, tie the brain and brainstem to the scaffolding to prevent floating when the agarose is poured (cerebellum does not need to be tied because it does not float), as follows:
      (i) Position the brain such that the frontal lobe is almost touching one of the scaffolding plates: for hemibrains, the medial surface should be placed facing up to prevent air bubbles that affect imaging, unless there is

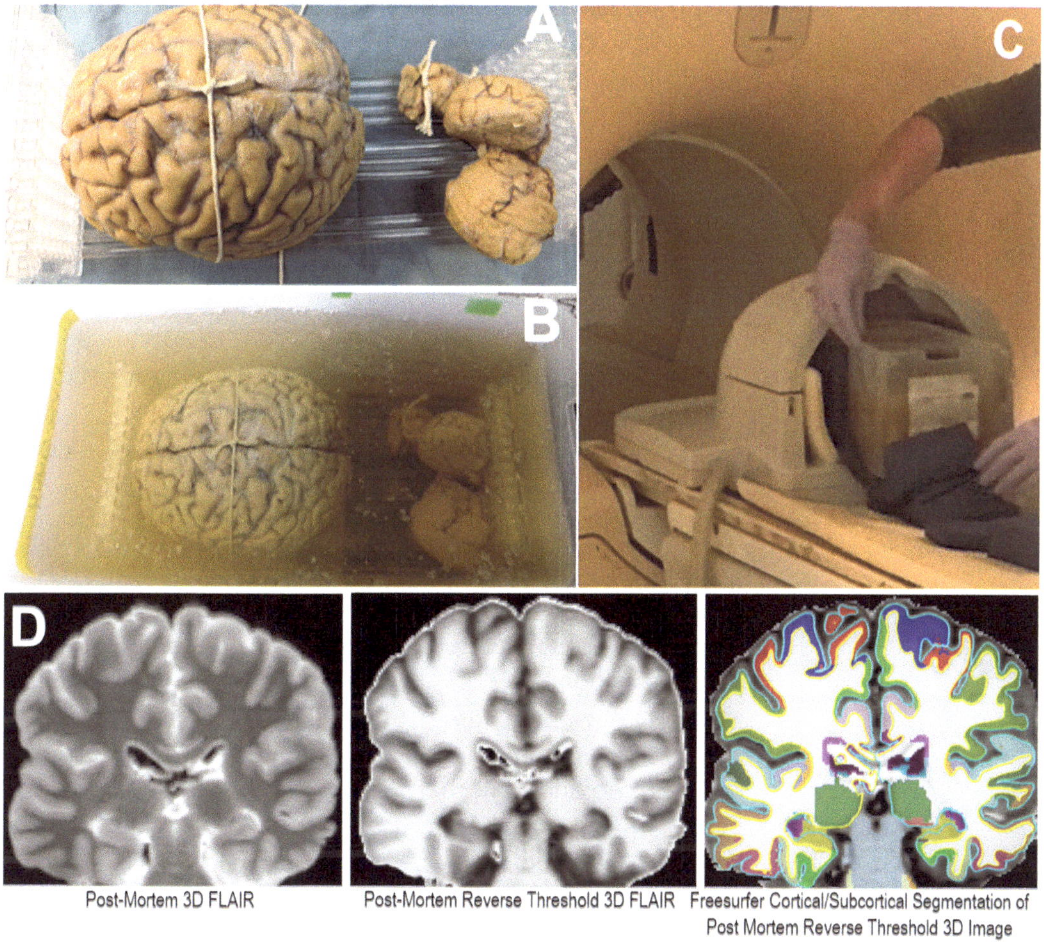

Fig. 1 Brain embedding and imaging. The brain is first oriented on plastic scaffolding (**a**) and placed in a container that is then filled with agarose gel (**b**). After the block solidifies it is imaged in 3 T MRI machine (**c**). The generated images can be manipulated, and standard imaging analysis software leveraged (Post-Mortem 3D FLAIR, Reverse-Threshold 3D FLAIR, FreeSurfer Cortical/Subcortical segmentation) (**d**)

significant pathology on the lateral surface; for whole brains, the brain can be positioned whichever way will ensure more even slicing on the brain slicer (*see* below).

(ii) Position the brainstem such that the inferior portion is closest to the scaffolding plate opposite the one positioned against the frontal lobe of the brain.

(iii) Lay the cerebellum on the scaffolding next to the brainstem, with the vermis positioned against the scaffolding plate. For whole brains, place the left cerebellum closest to the brainstem.

(iv) Place the scaffolded brain into the plastic container, to prepare for agarose embedding.

(c) Make agarose gel by mixing 5 L water and 150 g of agarose powder in a 6 L metal pot.

(d) Place the pot on a hot plate, set the heat to 200 °C, and stir, using a stir bar, on low.

(e) Once boiling, remove the pot from the hot plate and skim the foam off of the top of the liquid, using a wooden spoon.

(f) Once the temperature drops below 32 °C, carefully pour the agarose mixture into the container holding the scaffolded brain (Fig. 1b).

(g) Ensure that the cerebellum stays in place, or reposition it, using a metal rod, after the agarose is poured.

(h) Surround the container of agarose with ice packs until the agarose has solidified and is at RT.

(i) Transport brain(s) to MRI suite.

2. Perform ex vivo MRI scan (Fig. 1c) using the imaging protocol illustrated in Table 1 (*see* **Note 13**).

3. Perform quantitative volumetric postprocessing and segmentation:

   (a) Create a "Pseudo-3DT1" image from the postmortem 800 μm isotropic 3D FLAIR, using signal matrix inversion techniques (Fig. 1d).

   (b) Apply the "Pseudo-3DT1," or reverse signal threshold 3D FLAIR, to the *FreeSurfer* pipeline, which is divided into two primary parts:

      (i) A subcortical/white-matter surface creation and segmentation of the individual structures.

      (ii) Reconstruction of the cortical surface, created by imaging the underlying white-matter surface, followed by parcellating the cortical areas.

Table 1
Postmortem brain MRI protocol

1. 3D FLAIR. 0.8 × 0.8 × 0.8 mm, TE = 294 ms, TR = 4800 ms, 225 images, 28:39 min (min:s)
2. 3D-SWI 0.55 × 0.55 × 0.55 mm, TE = 20 ms, TR = 28 ms, 291 images, 27:05 min (×2, reverse polarity)
3. Multi Echo FLASH10, 15, 20, 25, 30 1 × 1 × 1 mm TE = 2.6 ms, TR = 23 ms, 160 images (5 echoes, 5:23 each)
Total MR acquisition time: 2:21:54 h (h:min:s).

4. Image processing:
   (a) Remove extraneous, non-cortex/non-white-matter tissue, and perform motion correction and alignment with the MNI 305 template.
   (b) Spatially register the brain mask with the subsequent creation of surfaces in native *FreeSurfer* RAS space (RAS: an (x,y,z) coordinate system using *R*ight (right side of brain), *A*nterior, *S*uperior as coordinates), then eliminate topological defects. Then, generate the "pial/cortical" and "white-matter" tessellated surfaces.
   (c) For quality assurance, review QA/QC measures for noise and signal contrast per *FreeSurfer* guidelines, in addition to null values, and fit measures.
   (d) Visually inspect each orthogonal plane of the segmentation, applying control points where needed, to ensure correct surface generation, using *FreeSurfer*, and correct for small, erroneous inclusions of other anatomical features, such as blood vessels, dura, and any white-matter lesions (*see* **Note 14**).
5. Sample any lesions identified at the time of tissue blocking, along with a standard set of brain regions, as described in Subheading 3.5.

3.5 Fixed-Tissue Slicing and Sampling (See Note 15 and Fig. 2)

1. Invert the plastic container onto a flat surface to expel the agarose block, then remove the scaffolding plates and rods.
2. Place the agarose block on the rotary slicer, positioned such that the frontal lobe will be sliced first.
3. Set the slicer to cut at a thickness of 4 mm.
4. Proceed to slice the entire brain from anterior to posterior, catching the 4-mm sections as they are sliced. Stack the slices on top of each other, then flip over the stack of slices so the most anterior slice is at the top of the pile.
5. Starting with the frontal lobe, flip each slice over and remove the surrounding agarose.
6. Lay the slices out from anterior to posterior and take digital photos; any lesions found on the slabs should be indicated with markers and photographed prior to sampling.
7. Perform gross neuropathological examination and modular sampling (*see* Fig. 3 and **Note 16**), by examining the brain slabs for neuropathology, including deviations from normal architecture, developmental anomalies, focal areas of softening, discoloration of cavitary, hemorrhaging, mass lesions, contusions, diffuse or focal atrophy, and other standard diagnostic neuropathologic features.

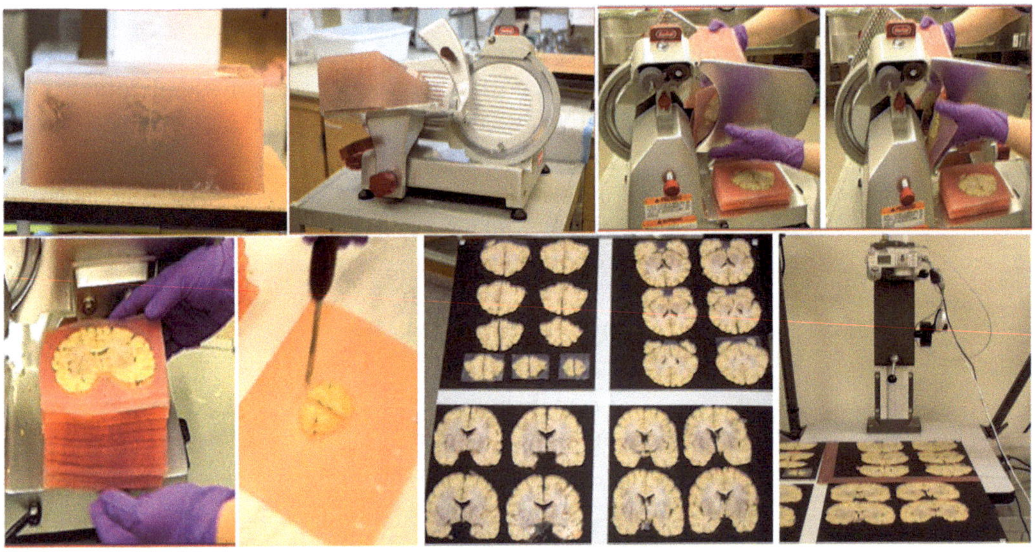

Fig. 2 Fixed brain slicing. The block of agarose with the embedded brain (*top, left panel*) is placed on the rotary slicer and sectioned coronally at 4 mm, from anterior to posterior (*top, right three panels*). The slices are taken off the slicer (*bottom, far-left panel*) and the excess agarose is removed (*bottom, second panel from left*). The slices are arranged and photographed (*bottom, right two panels*) prior to gross assessment and sampling

8. Using the tissue-punch tool, sample areas of interest consistently, across ~80 brain regions, and place in tissue cassettes for subsequent paraffin embedding. Samples should be taken from:
   (a) Standard NIA-AA (National Institute on Aging—Alzheimer's Association) diagnostic regions.
   (b) All brainstem slices.
   (c) Standard neurotrauma (CTE (chronic traumatic encephalopathy) and focal/diffuse TBI (traumatic brain injury)) regions.
   (d) Loci of standard functional networks.
   (e) Any grossly identified lesions.
   (f) MRI-identified lesions.
9. After taking all desired samples in **Step 8**, below, photograph the slices again.
10. Use the Leica automated processor to dehydrate, clear, and infiltrate the tissue.
11. Remove and discard the lid of the processing cassette, then prepare a metal mold (to be used with the HistoCore automated embedder) by adding a layer of hot paraffin to the bottom of the mold. While the wax is still hot, use the curved

Routine Diagnostic Sampling for Neurodegenerative Disease
Bilateral middle frontal gyrus
Bilateral prefrontal white matter
Bilateral supramarginal gyrus
Bilateral superior + middle temporal gyri
Bilateral calcarine cortex
Bilateral anterior cingulate gyri
Bilateral anterior hippocampus
Bilateral mid hippocampus
Bilateral amygdala and olfactory bulbs
Bilateral neostriatum
Bilateral thalamus
Midbrain with substantia nigra
Pons with locus ceruleus
Medulla (mid-section)
Bilateral cerebellum with dentate nucleus
Spinal cord (when available)
Gross lesions

TBI-targeted sampling
Genu of corpus callosum
Splenium of corpus callosum
Body of corpus callosum
Body of corpus callosum with fornix
Bilateral middle cerebellar peduncles
Bilateral posterior cerebellar cortex
Bilateral superior cerebellar peduncle with anterior vermis

CTE-targeted sampling
Bilateral superior frontal gyrus
Bilateral anterior temporal lobe
Bilateral orbital frontal cortex
Bilateral hypothalamus
Bilateral middle frontal sulcus
Bilateral dorsolateral prefrontal cortex
Bilateral caudal orbital frontal cortex

Connectivity-guided sampling
Bilateral precuneus
Bilateral post. superior + middle temporal gyri
Bilateral occipital white matter
Bilateral frontoinsula
Bilateral medial prefrontal cortex
Bilateral hippocampus + parahippocampal sulcus
Bilateral posterior cingulate gyri
Bilateral posterior angular gyrus
Bilateral fusiform gyrus/inferior temporal gyrus
Bilateral parieto-occipital cortex
Bilateral inferior frontal gyrus
Bilateral parietal white matter
Bilateral supplementary motor cortex
Bilateral Broca's area
Bilateral primary motor cortex
Imaging-identified lesions

Fig. 3 Modular sampling approach. Each brain undergoes extensive standardized sampling irrespective of the clinical history or cohort study. The sampling is organized in a modular fashion and slides are generated from the various modules based on diagnostic necessity or research emphasis. The routine sampling for the neurodegenerative disease module (*pink*) generates blocks in accordance with NIA-AA criteria for the assessment of Alzheimer's disease and related dementias [8]. Every case undergoes assessment of these blocks. The chronic traumatic encephalopathy (CTE)-targeted sampling approach (*blue*) is based on CTE consensus criteria and is assessed in cases from cohort studies with an emphasis on CTE pathology, as well as on any case with a suspicious clinical history or routine diagnostic workup. The traumatic brain injury (TBI)-targeted sampling approach (*green*) generates blocks from regions, predominantly large, white-matter tracts that are often secondarily affected by neurotrauma. These blocks are assessed in cases with a history of neurotrauma. The connectivity-guided sampling approach (*orange*) targets brain regions known to participate in specific circuits that have been identified through functional connectivity studies; these are assessed in a research-specific context. Finally, one should assess any blocks (from any module) that contain lesions identified by gross examination or imaging, even if those blocks are not usually examined in detail. Blocks not assessed can be stored safely and used in the future, should the need arise

forceps to remove the tissue from the processing cassette, and place it into the liquid wax in the metal mold. Add more wax to cover the tissue. Invert the bottom of the processing cassette and place over the tissue in the mold, add enough additional wax to reach the top of the mold, then place the mold on the cold plate to solidify the wax. Repeat this process for each tissue sample.

12. Store any remaining tissue in NBF in air-tight containers, for future use.

3.6 Standard Neuropathological Diagnostic Workup (see Notes 17, 18, 19)

We utilize recommended consensus criteria [5] for the evaluation of neurodegenerative diseases, in terms of brain regions to sample and staining protocols to use, as summarized in Table 2.

1. Using a microtome, cut 5 μm thick sections from the paraffin-embedded tissue blocks, forming a ribbon of tissue.
2. Transfer the ribbon to a 2–7 °C water bath.
3. Using a glass slide, pick up the tissue from the water bath.
4. Bake at 16 °C for 10 min to 24 h, to adhere tissue to the slide and melt the paraffin.
5. To remove paraffin, immerse slides sequentially in a series of 3 vessels containing xylene, for 3 min each. Change to fresh xylene every 120–150 slides.
6. Rehydrate in a graded ethanol series (3× at 100%, 2× at 95%, and 1× with ddH_2O) for 3 min per wash. Change to fresh ethanol every 120–150 slides.
7. Wash with TBST twice for 3 min each.
8. Perform immunohistochemical (IHC) staining on a subset of slides (a list of antibodies commonly used as markers for neurodegenerative diseases is provided in Table 3):
   (a) Use heat-induced epitope retrieval (HIER) at 110 °C for 15 min to increase antigen-antibody binding and thereby enhance the IHC signal (*see* **Note 17**).
   (b) Cool the slides for 20 min at RT.
   (c) Wash the slides twice with TBST for 5 min each.
   (d) Place the slides in a BioCare intelliPATH™ (or equivalent) semiautomated staining device, following the manufacturer's protocols, and using reagents recommended for the antibody of choice.
   (e) Once IHC staining is complete, remove the slides and wash for 3 min in TBST, then dehydrate in an ethanol series (70% once, 96% once, then 3 times in fresh 100% ethanol) for 3 min each, then immerse in xylene sequentially in a series of 3 containers, for 3 min each. Change the ethanol and xylene solutions every 120–150 slides.
   (f) Apply mounting medium and a coverslip to each slide.
   (g) Let dry, then store in slide boxes at RT.
9. Stain slides with hematoxylin and eosin (H&E) and Luxol fast blue (LFB) (protocol is for 120 slides):
   (a) Prepare 19 plastic slide-staining jars by adding 200 mL of the appropriate staining solution to each, as follows: make 6 jars of xylene, 4 jars of 95% EtOH, 1 jar LFB staining solution, 1 jar lithium carbonate solution, 1 jar 70% EtOH, 1 jar Clearview™ hematoxylin, 1 jar bluing solution, 1 jar Eosin Y with phloxine solution, and 3 jars of tap water.

Table 2
Standard diagnostic stains for typical baseline neuropathological workup, including assessment for typical neurodegenerative disease neuropathology

Brain Region	Stain	Laterality	Purpose
Middle frontal gyrus	H&E/LFB	Bilateral	Vascular pathology; other lesions
	6E10	Unilateral	Thal staging for amyloid plaques; CAA
	AT8	Bilateral	Braak staging of NFTs; CTE lesions; ARTAG; FTLD-tau
	pTDP-43	Unilateral	LATE staging; assessment for FTLD-TDP pathology
	a-syn	Unilateral	LBD staging; glial synuclein pathology (MSA)
	Bielschowsky	Unilateral	Neuritic plaque density (CERAD score)
Frontal white matter	H&E/LFB	Bilateral	Vascular pathology; other lesions
Inferior parietal lobule	H&E/LFB	Bilateral	Vascular pathology; other lesions
	6E10	Unilateral	Thal staging for amyloid plaques; CAA
	AT8	Bilateral	Braak staging of NFTs; CTE lesions; ARTAG; FTLD-tau
	Bielschowsky	Unilateral	Neuritic plaque density (CERAD score)
Superior/middle temporal gyri	H&E/LFB	Bilateral	Vascular pathology; other lesions
	6E10	Unilateral	Thal staging for amyloid plaques; CAA
	AT8	Bilateral	Braak staging of NFTs; CTE lesions; ARTAG; FTLD-tau
	pTDP-43	Unilateral	LATE staging; assessment for FTLD-TDP pathology
Occipital cortex with V1	H&E/LFB	Bilateral	Vascular pathology; other lesions
	6E10	Unilateral	Thal staging for amyloid plaques; CAA
	AT8	Bilateral	Braak staging of NFTs; CTE lesions; ARTAG; FTLD-tau
Anterior cingulate gyrus	H&E/LFB	Bilateral	Vascular pathology; other lesions
	AT8	Bilateral	CTE lesions; ARTAG; FTLD-tau
	α-syn	Unilateral	LBD staging; glial synuclein pathology (MSA)
Anterior hippocampus	H&E/LFB	Bilateral	Vascular pathology; other lesions
Mid-hippocampus	H&E/LFB	Bilateral	Vascular pathology; other lesions
	6E10	Unilateral	Thal staging for amyloid plaques
	AT8	Bilateral	Braak staging of NFTs; CTE lesions; ARTAG; FTLD-tau
	pTDP-43	Unilateral	LATE staging; assessment for FTLD-TDP pathology
Amygdala	H&E/LFB	Bilateral	Vascular pathology; other lesions
	AT8	Bilateral	ARTAG
	pTDP-43	Unilateral	LATE staging
	a-syn	Unilateral	LBD staging
Neostriatum	H&E/LFB	Bilateral	Vascular pathology; other lesions
	6E10	Unilateral	Thal staging for amyloid plaques
Thalamus	H&E/LFB	Bilateral	Vascular pathology; other lesions

(continued)

Table 2
(continued)

Brain Region	Stain	Laterality	Purpose
Midbrain	H&E/LFB	Bilateral	Vascular pathology; neuron loss; other lesions
	6E10	Unilateral	Thal staging for amyloid plaques
	a-syn	Unilateral	LBD staging; glial synuclein pathology (MSA)
Pons	H&E/LFB	Bilateral	Vascular pathology; neuron loss; Lewy bodies; other lesions
Medulla	H&E/LFB	Bilateral	Vascular pathology; Lewy bodies; other lesions
Cerebellum	H&E/LFB	Bilateral	Vascular pathology; Purkinje cell loss; other lesions;
	6E10	Unilateral	Thal staging for amyloid plaques; CAA
	AT8	Bilateral	CTE lesions; ARTAG; FTLD-tau

ARTAG aging-related tau astrogliopathy, *CAA* cerebral amyloid angiopathy, *CTE* chronic traumatic encephalopathy, *FTLD* frontotemporal lobar degeneration, *LATE* limbic predominant age-related TDP-43 encephalopathy, *LBD* Lewy body disease, *MSA* multiple system atrophy

(b) Immerse slides in 3 successive jars of xylene, for 3 min each.

(c) Immerse slides in 2 successive jars of 95% EtOH, for 3 min each.

(d) Immerse slides in jar with LFB solution, and leave overnight at 60 °C.

(e) Remove from the oven and let the solution come to RT.

(f) Rinse slides with ddH_2O.

(g) To start the reaction, quickly dip the slides, 2 times, in the lithium carbonate solution.

(h) Quickly dip the slides, 2 times, in 70% EtOH, to begin the color differentiation.

(i) Immerse slides in 3 successive jars of tap water, quickly dipping them 10 times in each jar, to remove the EtOH.

(j) Examine a few slides under a microscope to check for over-differentiation (background staining). Only the myelin should be stained blue, not the background tissue.

(k) Repeat **Steps 9.g–j** until the background is clear.

(l) Rinse slides with ddH_2O.

(m) Immerse slides in Clearview™ hematoxylin for 3 min.

(n) Wash slides in tap water.

(o) Dip slides 20 times in bluing solution.

(p) Wash slides in running tap water.

Table 3
Partial list of antibodies and antigen retrieval protocols available for more common neuropathologic evaluations

Primary antibody	Clone/host	Supplier	Catalogue #	Dilution	Pretreatment
Alpha synuclein	LB509/mouse	Invitrogen	180215	1:200	Proteinase-K (10 min)
Beta amyloid	6 E10/mouse	BioLegend	803003	1:1500	Diva decloaker (110 °C- 15 min)
FUS/TLS	3A10B5/ mouse	Abcam		1:3000	Diva decloaker (110 °C- 15 min)
Anti-Huntington's protein	mEM48/ mouse	Millipore	MAB5374	1:3000	Diva decloaker (110 °C- 15 min)
Notch 3/ N3ECD	1 E4/mouse	Millipore	MABC594		Proteinase-K (5 min)
SQSTM1/p62	2C11/mouse	Abnova	H00008878-M01	N/A	ER 1 (10 min)
PHF-tau	AT8/mouse	Thermo Fisher	MN1020	1:1000	Diva decloaker (110 °C- 15 min) or ER 1 10 min
Anti-prion protein	3F4/mouse	Millipore	MAB1562	1:50	ER1 (20 min)
TAU RD3	8E6/C11/ mouse	Millipore	05-803	1:2500	Diva decloaker (110 °C- 15 min)
TAU RD4	1E1/A6/ mouse	Millipore	05-804	1:400	Diva decloaker (110 °C- 15 min)
phosTDP-43 (Ser409/ Ser410)	ID3/rat	BioLegend	829901	1:2000	Diva decloaker (110 °C- 15 min)
TDP-43 (TARDBP)	Rabbit	Proteintech Group	10782-2-AP	1000	ER 2 (20 min)
Ubiquitin	Ubi-1/mouse	Millipore	MAB1510	500	ER 1 (20 min)

(q) Immerse slides in Eosin Y with phloxine solution for 3 min.

(r) Wash slides in tap water.

(s) Dehydrate slides by passing them through 2 jars of fresh 95% EtOH followed by 3 jars of fresh xylene.

(t) Mount slides with EcoMount (or equivalent) and glass coverslips.

10. Perform a modified Bielschowsky silver stain (for 6 slides with 8 μm tissue sections; adjust volumes for more slides):
    (a) *Use HCl to acid-wash all glassware before staining.*
    (b) Prepare 13 plastic staining jars as follows: 6 jars with 200 mL xylene; 4 jars with 200 mL 95% EtOH; 1 jar with 200 mL ddH_2O; 1 jar with 100 mL *dilute* NH_4OH; and 1 jar with 100 mL of 2% thiosulfate.
    (c) Prepare 2 glass Coplin jars: one with 50 mL of 20% silver nitrate solution, and one empty jar to receive filtrate.
    (d) Place slides in 3 successive staining jars containing xylene, for 3 min each.
    (e) Place slides in 2 successive staining jars containing 95% EtOH, for 3 min each.
    (f) Place 20 mL of NH_4OH in a 100 mL beaker and allow it to evaporate at RT in a fume hood while you perform the next step.
    (g) Immerse the slides in the Coplin jar containing 20% silver nitrate, for 20 min.
    (h) Transfer slides to the staining jar containing 200 mL ddH_2O, then refresh the water 3 times. Save the silver nitrate, which will contain precipitants, for the next step (**Step 10.i**).
    (i) Pour the silver nitrate into a 100 mL beaker (be sure it is acid-washed).
    (j) To titrate the silver nitrate solution with NH_4OH, use the fine-tipped plastic pipet to add the evaporated NH_4OH, one drop at a time and swirling between drops, until the precipitant disappears (the solution goes from cloudy to clear).
    (k) Filter this solution through the glass funnel (fitted with filter paper) into the empty Coplin jar.
    (l) Place the slides into the filtered silver nitrate for 20 min; this solution will be used again in **Step 10.n**.
    (m) Transfer the slides into the staining jar containing dilute NH_4OH while you carry out the next step.
    (n) Add 120 μL of ***aged*** developer to the filtered, titrated silver nitrate in the Coplin jar.
    (o) Place the slides into this solution. The tissue will gradually take on a tan/gold background: use a microscope to check for this after 3 min, and then every 1 min afterward, for a total of 5–8 min. Stop when the background is a gold/ light brown. Do not overdevelop.
    (p) Wash the slides under running H_2O to clear silver deposits from the tissue, then rinse well with ddH_2O.

11. Perform microscopic examination of the stained slides (*see* **Notes 18** and **19**).
12. Generate a research diagnostic report that lists the neurodegenerative disease processes present, along with any incidental findings.
13. Record all neuropathologic data points in the database.

3.7 Digital Neuropathology (See Note 20)

1. Clean slides with 70% ethanol and Kimwipes to remove any particles that could cast shadows on the underlying tissue sections.
2. If possible, scan the slides at not more than 20× magnification, which is suitable for most manual histochemical and IHC imaging and analysis [6, 7], and requires less time and storage space than 40× digital images.
3. Carry out quality-control observations, such as checking for false positives in control slides, and for IHC artifacts (e.g., non-tissue- or cell-specific DAB stains, bubbles, or unstained areas) and features that might impede digital analysis (e.g., dark shadows, folds, or tears).
4. Set up cloud-based servers for storage of exported images, creating individual user accounts to control access, to maintain HIPAA (Health Insurance Portability and Accountability Act (of 1996), to protect patient privacy) compliance and to facilitate analysis.
5. Begin visualization and analysis by annotating brain regions and objects of interest.
6. Use appropriate image analysis software (such as HALO (https://indicalab.com/halo/)) to identify, extract, and quantify neuropathologic features (the image analysis modules that we use are described in Table 4).

3.8 Multiplexed Liquid-Phase Extracted Protein Quantification in Fast-Frozen Paraffin-Embedded (FFPE) Tissue (See Note 21)

1. Deparaffinize the slides (mounted with 15 μm thick FFPE tissue) by immersing in 3 sequential baths of 100% xylene, 3 min per bath.
2. Rehydrate with an isopropanol series, 3 min for each bath: 100% (twice), 96% (once), 70% (once), and 50% (once).
3. Scrape off the FFPE tissue into a 1.5 mL labeled microcentrifuge tube, 3 sections per tube.
4. Pipette 0.5 mL heptane into each tube and close tubes tightly.
5. Vortex for 10 s and incubate for 1 h at RT.
6. Add 25 μL MeOH into each tube, close tubes tightly, and vortex for 10 s.
7. Centrifuge the tubes at 9000 ×*g* for 2 min. Tissue will form a pellet at the bottom of the tube.
8. Carefully remove the supernatant, using a pipette (do not decant), and discard.

Table 4
HALO modules used and the type of data obtained with each

Module	Type of data obtained by ROI	Units
Area quantification (up to 3 markers)	Percentage of area by marker	percentage
Object colocalization (up to 2 markers)	Total ROI area	mm^2
	Total object 1 and 2 count	μm^2
	Total object 1 and 2 area	qty
	Object 1 and 2 percent area	percentage
	Total object colocalized area	qty
	Percent of object 1 and 2 colocalized	percentage
	Area of positive stain by object	μm^2
	Colocalized area by object	μm^2
	Percent of colocalized area of each marker by object	percentage
Multiplex IHC (up to 5 markers)	Total cells by marker	qty
	Area of positive stain in nucleus by marker and by object	μm^2
	Area of positive stain in cytoplasm by marker and by object	μm^2
	Area of positive stain in membrane by marker and by object	μm^2
Microglia module	Total number of microglia cells	qty
	Number of inactivated microglia cells	qty
	Number of activated microglia cells	qty
	Average process area	μm^2
	Average process length	μm^2
	Nucleus area (average and single-cell)	μm^2

9. Air-dry the pellet for 5 min.
10. Pipette 100 μL of extraction buffer from the Qproteome kit into each tube, and vortex for 10 s. Seal the tubes with sealing clips.
11. Incubate on ice for 5 min, then vortex again.
12. Incubate the tubes in a ThermoMixer at 100 °C for 20 min. Check the sealing clips to ensure they are tight.
13. Incubate the tubes in the ThermoMixer at 80 °C for 2 h at 750 rpm.
14. After incubation, remove the sealing clips from the tubes, and centrifuge the tubes for 15 min at 14,000 ×*g*, at 4 °C.
15. Transfer the supernatant (which will contain the extracted proteins) to a different 1.5 μm microcentrifuge tube labeled, *Pellet tube.*
16. Add 400 μL MeOH to the supernatant in the Pellet tube. Close the tubes tightly and vortex for 10 s.
17. Centrifuge the tubes at 9000 ×*g* for 10 s.
18. Add 100 μL chloroform to each tube. Close tubes tightly and vortex for 10 s.

connectivity patterns. In addition to these brain regions, more brain sections should be examined following the identification of gross lesions and/or those identified via MRI.

17. There are many ways to accomplish HIER. For convenience, we use NxGen's Diva Decloaking Chamber™ after immersing the slides in 1× Diva Decloaker solution. This protocol works for the vast majority of antibodies we use. For others, enzymatic antigen retrieval is often effective.
18. Often, the diagnostic evaluations require assessment beyond the standard diagnostic workup, in which case the neuropathologist may request additional sampling and/or staining.
19. These histology methods are specific for slides stained immunohistochemically for single protein markers. However, IHC staining of two or more protein markers simultaneously, in a single tissue section, can be used successfully in some instances. Multiplexing protein markers in a single tissue section enables colocalization analyses of multiple antigens, identification of different cell populations within a sample, and optimization of protocols and histological reagents.
20. 20. The value of histologic stains can be increased by quantitative analysis of the pathologic peptide burden [9]. This is necessary for understanding how pathologic burden differs across brain regions and how it correlates with cognitive function and/or biomarker studies, because similarly classified individuals can vary significantly in their manifestation of pathology. Digital pathology utilizes digitized images and computer workstations to analyze and view the whole-slide images (WSIs), enabling sharing, teaching, and primary reporting of diagnostics [7]. Optimal WSI quality and analyses are dependent upon high-quality, reproducible histochemical, and IHC staining of the tissue, so standardization of IHC staining protocols is essential for reproducible results.
21. For this assay, we use a 15 μm thick sections rather than the 5 μm sections that are standard for histology. Thicker sections yield higher amounts of protein in order to ensure being in the optimal range for the assay. A liquid-phase analysis can quantify those oligomers and soluble peptides that may not be evident on a slide but are greatly impacted in disease. For this purpose, we use a LUMINEX-based approach to analyze extracts from FFPE tissue [10]. This approach is sensitive, cost-effective, and can be multiplexed to limit tissue consumption, while enabling correlation with cytoarchitectural features documented in IHC-stained adjacent sections.

References

1. Schachter AS, Davis KL (2000) Alzheimer's disease. Dialogues Clin Neurosci 2(2):91–100
2. Davies J, Everall IP, Lantos PL (1993) The contemporary AIDS database and brain bank--lessons from the past. J Neural Transm Suppl 39:77–85
3. Cruz-Sánchez FF, Tolosa E (1993) The need of a consensus for brain banking. J Neural Transm Suppl 39:1–4
4. Tourtellotte ME, Jacobs RE (1960) Physiological and serologic comparisons of PPLO from various sources. Ann N Y Acad Sci 79:521–530
5. Montine TJ et al (2012) National Institute on Aging–Alzheimer's Association guidelines for the neuropathologic assessment of Alzheimer's disease: a practical approach. Acta Neuropathol 123(1):1–11
6. Kumar N, Gupta R, Gupta S (2020) Whole slide imaging (WSI) in pathology: current perspectives and future directions. J Digit Imaging 33(4):1034–1040
7. Fraggetta F et al (2018) The importance of eSlide macro images for primary diagnosis with whole slide imaging. J Pathol Inform 9:46
8. Hyman BT et al (2012) National Institute on Aging–Alzheimer's Association guidelines for the neuropathologic assessment of Alzheimer's disease. Alzheimers Dement 8(1):1–13
9. Latimer CS et al (2019) Resistance and resilience to Alzheimer's disease pathology are associated with reduced cortical pTau and absence of limbic-predominant age-related TDP-43 encephalopathy in a community-based cohort. Acta Neuropathol Commun 7(1):91
10. Keene CD et al (2018) Luminex-based quantification of Alzheimer's disease neuropathologic change in formalin-fixed post-mortem human brain tissue. Lab Investig 99:1056

Chapter 2

Extraction and Purification of Single Nuclei from Frozen Human Brain Tissue

Carter R. Palmer and Jerold Chun

Abstract

Resolving the complexity of the human brain at the level of single cells is essential to gaining an understanding of the immense diversity of cell types and functional states in both healthy and diseased brains. To exploit fully the technologies available for such studies, one must extract and isolate pure nuclei from unfixed postmortem tissue while preserving the molecules to be interrogated. Currently, nuclei are necessary substitutes for individual brain cells, since myriad cell types/sub-types constituting the human brain are embedded within the neuropil—a complex milieu of interconnected cells, processes, and synapses—which precludes intact and selective isolation of single brain cells. Here, we describe a protocol for the extraction and purification of intact single nuclei from frozen human brain tissue along with modifications to accommodate numerous downstream analyses, particularly for transcriptomic applications.

Key words Human brain nuclei, Extraction, FANS, Single nucleus, Transcriptomics, Epigenomics, Genomic mosaicism, Cell diversity

1 Introduction

Modern advancements in single-cell technologies have opened the door to analyses of tissues at the resolution of single cells in a high-throughput manner. Notably, analysis of the genome, epigenome, transcriptome, and combined multi-omic analyses have become essential to advancing our understanding of the brain. Although a detailed examination of whole, single cells would offer an ideal system for profiling the transcriptome, it is currently unfeasible because of myriad intermixed cells and connections that comprise the human brain, epitomized by the complex neuropil, preventing the specific isolation of purely intact cells from postmortem tissue. Early studies on genomic mosaicism in the brain demonstrated the feasibility of accessing genomic information from single-cell nuclei, particularly when combined with fluorescence-activated flow cytometry and sorting [1, 2], and currently the field routinely uses

Jerold Chun (ed.), *Alzheimer's Disease: Methods and Protocols*,
Methods in Molecular Biology, vol. 2561, https://doi.org/10.1007/978-1-0716-2655-9_2,

isolated nuclei as a surrogate for whole cells in analyzing components of the genome (DNA) [1–9] and the transcriptome (RNA) [10–12]. For transcriptomic studies, although there are undoubtedly transcripts that localize outside of the nucleus that may also have different expression profiles in whole cells, nuclei have been shown to recapitulate faithfully many of the transcriptomic changes observed in intact cells, enabling cell-type classification comparable to that of whole cells [13–15].

The purity of nuclear isolates is essential for generating optimal data. When using droplet-based or combinatorial barcoding technologies, artifacts produced by nuclear doublets, clumping, exogenous RNA, and damaged nuclei can all lead to inaccurate results and unnecessary sequencing costs. Further, although bioinformatic filtering can reduce artifacts, exogenous RNA and damaged nuclei are more difficult to filter out and can cause erroneous transcript-calling. Accordingly, a pure nuclei preparation, tailored to the specific downstream processing technique, is essential for maximizing the quality and quantity of the information obtained from each experiment and providing consistent and reliable data.

Here, we report how to prepare pure and intact nuclei from - fresh-frozen human brain tissue and provide protocol modifications for various downstream applications. Our protocol details preparation and dissociation of brain tissue and the subsequent purification of nuclei. We have utilized and refined this protocol across numerous applications for the effective isolation of nuclei for single-nucleus transcriptomic [12, 16] and genomic/multi-omic approaches [1, 3, 5, 9, 10, 17].

2 Materials

Prepare all buffers using ultrapure water and chill them to 4 °C prior to use. Extraction and wash buffers should be made fresh for each experiment. All human material and any supplies that touch human material should be treated as biohazardous and disposed of or cleaned appropriately. It is not necessary to perform this protocol under sterile conditions, but every effort should be made to use pure and RNase-free materials.

1. Cryostat, capable of 100 μm sectioning, chilled to −20 °C (e.g., Leica CM3050 S or equivalent).
2. Cutting blades (preferably disposable) for cryostat.
3. Curved metal microdissection forceps for handling tissue.
4. Small histology brush.
5. Centrifuge and corresponding rotors, pre-cooled to 4 °C, capable of accommodating 15 mL conical tubes at a force of at least 1000 × *g*.

3.3 Staining of Nuclei

1. Select the appropriate pre-chilled dye solution for the desired downstream analysis and add 500 μL of cold staining buffer to the pellet (*see* **Note 12**).
2. Resuspend the pellet in the dye solution by gently triturating with a P1000 pipette and a filter tip, approximately 5–10 times.
3. Using the P1000 pipette and the same filter tip, pass the nuclear suspension through a 50 μm filter and into a pre-chilled FACS tube.
4. Tap the filter against the tube to recover any remaining liquid (*see* **Note 13**).
5. Prepare the tube to be used for collecting the FACS-sorted nuclei by adding 200 μL of sorting buffer and vortexing vigorously, to coat the sides with BSA. Store on ice until needed (*see* **Note 14**).

3.4 Fluorescence-Activated Nuclei Sorting (FANS)

1. Set up the FACS instrument according to the manufacturer's specifications (*see* **Note 15**).
2. Utilize the appropriate FACS software to establish two plots: a forward-scatter vs. side-scatter plot, and a linear histogram on the (stain-appropriate) fluorescent channel.
3. Load a tube of nuclear suspension and begin acquiring data.
4. Use the software to adjust the forward and side voltages such that the majority of events (>90%) appear on the forward- vs. side-scatter plot.
5. Create and modify a gate on the forward- vs. side-scatter plot, to gate out the smallest debris (*see* **Note 16** and Fig. 1).
6. Use the software to display this newly gated population on the linear histogram for the fluorescent stain.
7. Adjust the voltage of the fluorescent channel such that the main peak, which should correspond to singlets, is approximately centered in the histogram.
8. Create a gate on the histogram to select singlets (*see* **Note 17** and Fig. 1).
9. Sort the events captured in the singlets gate, corresponding to single nuclei, into tubes previously coated with Sorting buffer. There is no need to collect debris or doublets.
10. At this point, nuclei will likely be too dilute for most downstream processes; concentration by centrifugation and resuspension are often necessary (*see* **Note 18**).

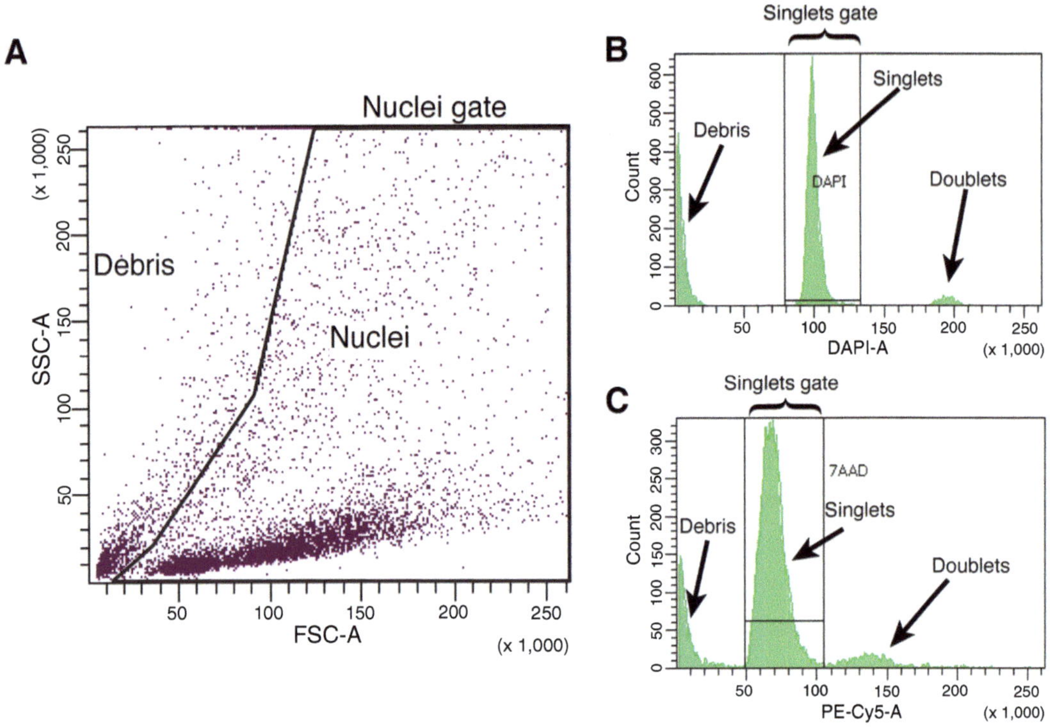

Fig. 1 Representative FANS plots. (**a**) Forward- vs. side-scatter plots to gate out smallest debris; (**b**) histogram of DAPI fluorescence, for selection of single nuclei; and (**c**) histogram of 7-AAD fluorescence, for selection of single nuclei

4 Notes

1. There are various suppliers for RNase inhibitors and their compositions can vary. Notably, many RNase inhibitors are shipped in buffers containing DTT, and some require DTT to function effectively. We commonly utilize a recombinant RNase inhibitor purchased from Takara Bio; however, certain downstream applications recommend the use of specific RNase inhibitors, and we suggest careful selection depending on the final assay technique.
2. Most RNase inhibitors should stably bind to RNases. Accordingly, we have found that a high concentration of RNase inhibitors during the initial dissociation step in NEB is highly beneficial for maintaining RNA integrity throughout nuclear isolation. For transcriptomic work, we use lower levels of RNase inhibitors in subsequent wash steps (with NEWB and Sorting buffer, described below), as we have not seen significant improvement in RNA recovery with high concentrations, and thus do not believe it is worth the cost of maintaining such high levels in these buffers. However, we have observed

benefits from elevated levels of RNase inhibitors throughout the protocol for multi-omic approaches and recommend following application-specific guidelines.

3. Protease inhibitors are included to improve nuclear membrane integrity during dissociation and purification. Broad protease protection is desirable, and the recommended inhibitors offer protection from serine-, cysteine-, and metalloproteases. We utilize the tablets at 0.4× the recommended concentration to save costs without any obvious detriment.
4. Other groups have successfully used a slightly simplified version of this buffer, comprised of 10 mM Tris-HCl, 250 mM sucrose, 25 mM KCl, 5 mM $MgCl_2$, 0.1% Triton X-100, 0.1 mM DTT, and RNase and protease inhibitors [18]. Both versions have yielded high-quality nuclei across multiple processing modalities.
5. Intercalating dyes that bind strongly to DNA have been shown to disrupt chromatin structure, modifying downstream epigenetic profiling. For this reason, we recommend avoiding DAPI (4′,6-diamidino-2-phenylindole) if ATAC-seq (assay for transposase-accessible chromatin, using sequencing) will be performed. We typically use DAPI for transcriptomic methods and 7-AAD (7-aminoactinomycin D) for epigenetic profiling. Dilute DAPI 1:4000 from a 5 mg/mL stock (*see* Subheading 2, **item 24**) in sorting buffer for transcriptomic analyses. Use of 7-AAD is advised for epigenetic profiling at a 1:200 dilution in sorting buffer from a 1 mg/mL stock (*see* Subheading 2, **item 25**).
6. Brain banks are constantly improving isolation, extraction, and freezing protocols. We have found that the best tissue for single-nucleus applications is often characterized by a high RNA integrity number (RIN) [16]. For our work, we use a cutoff of a RIN > 6 for analysis.
7. When blades are in use, always wear cut-proof gloves in addition to standard latex or nitrile gloves.
8. For this protocol, 300 μm of tissue should yield between 200,000 and 500,000 nuclei after sorting. Fixation and antigen staining will drastically decrease this yield and will, in turn, require significantly more tissue. Using large sections of white matter will further result in a significant number of oligodendrocytes being profiled; to enrich for non-oligodendrocyte cell populations, remove white matter regions using a scalpel.
9. Some groups have had success isolating specific layers of the cortex via microdissection for single-nucleus analysis and we recommend this if a subregion of the brain is desired [18]. Depending on the application, quality-control studies

should be pursued to ascertain the integrity of molecular targets; for example, some laser-capture protocols can degrade RNA [12].

10. The number of strokes should be the minimum number necessary to obtain a solution with no visible particulates. Avoid forming bubbles during homogenization. By the end of this step, cells should be lysed, and individual nuclei should remain in suspension. Subsequent wash steps are important to minimize cellular debris and to enable buffer exchange prior to sorting.
11. This centrifugation step successfully pellets nuclei without major damage. Small deviations (+/−30 × *g*) would likely not result in major changes to recovery, but we discourage more extensive modifications.
12. If faster sorting is desired, the pellet can be resuspended in a smaller volume.
13. This second filtration step is strongly recommended as it can significantly diminish the incidence of clogs during sorting.
14. Determining what tube-size to sort into is typically done by estimating the number of nuclei to be sorted. With our system, using the 100 μm nozzle on the FACSAria, 500,000 nuclei generate an approximately 2 mL final sorted volume.
15. We use a BD FACSAria Fusion sorter, with a 100 μm nozzle, but have also had success utilizing a 130 μm nozzle. If working alone, we set up the instrument prior to starting the nuclear isolation, but if working in pairs, one scientist sets up the instrument while the other prepares the nuclei. It is important that sorting starts as soon as possible once the nuclei are prepared.
16. The actual creation of gates in the software will be variable, depending on the software used, and we advise spending time becoming familiar with the software, using resources provided by the manufacturer. We typically use gating on forward- and side-scatter profiles only to eliminate the smallest debris but include all other events to be subsequently gated for the stain used. We have noticed that different regions of the brain have slight variations in the appearance of forward- and side-scatter plots; thus, samples may not match the plots in Fig. 1 exactly.
17. We advise gating from a histogram plot with a linear x-axis for DAPI- or 7-AAD-stained nuclei. It is important to adjust the voltage of the laser to optimize the spacing between debris, the singlets peak, and the doublets peak. The singlets peak should be significantly higher (approximately tenfold) than the doublets peak. An abnormally high doublets peak does not preclude

the effective isolation of single nuclei, but may be indicative of poor filtering, overly aggressive handling of the nuclei, or a need for more BSA in the sorting buffer.

18. When needed, centrifuge the suspension at 820 × *g* at 4 °C for 5 min and aspirate the remaining liquid very carefully, as fewer than ~750,000 nuclei will not form a visible pellet. Resuspend the pellet in the final desired volume of buffer suitable for downstream applications.

Acknowledgments

Federal support was provided by the National Institute on Aging of the National Institutes of Health under award numbers R01AG071465, R01AG065541, and R56AG073965 (J.C.) and the Department of Defense under award number W81XWH-21-10642-01 (J.C.). Nonfederal support was provided by the Shaffer Family Foundation and the Bruce Ford & Anne Smith Bundy Foundation (J.C.). The content is solely the responsibility of the authors and does not necessarily represent the official views of the National Institutes of Health. We would like to thank Dr. Blue Lake and the laboratory of Dr. Kun Zhang for their continued collaboration as well as Dr. Laura Wolszon for her essential feedback. We also thank the donors and their families who shared precious postmortem material for this work and the many brain donation centers responsible for the acquisition, care, and distribution of these materials. Specifically, we would like to thank the University of Washington Biorepository and Integrated Neuropathology Laboratory, the University of Maryland Brain and Tissue Bank, the Goizueta Alzheimer's Disease Research Center (ADRC) at Emory University, the London Neurodegenerative Diseases Brain Bank, the Newcastle Brain Tissue Resource, the University of Miami Brain Endowment Bank, the Maritime Brain Tissue Bank at Dalhousie University, the South West Dementia Brain Bank, and the UCSD ADRC.

References

1. Rehen SK et al (2005) Constitutional aneuploidy in the normal human brain. J Neurosci 25(9):2176–2180
2. Rehen SK et al (2001) Chromosomal variation in neurons of the developing and adult mammalian nervous system. Proc Natl Acad Sci U S A 98(23):13361–13366
3. Bushman DM et al (2015) Genomic mosaicism with increased amyloid precursor protein (APP) gene copy number in single neurons from sporadic Alzheimer's disease brains. elife 4:e05116
4. Kaushal D et al (2003) Alteration of gene expression by chromosome loss in the postnatal mouse brain. J Neurosci 23(13):5599–5606
5. Westra JW et al (2010) Neuronal DNA content variation (DCV) with regional and individual differences in the human brain. J Comp Neurol 518(19):3981–4000
6. Westra JW, Barral S, Chun J (2009) A reevaluation of tetraploidy in the Alzheimer's disease brain. Neurodegener Dis 6(5–6):221–229

7. Rohrback S et al (2018) Genomic mosaicism in the developing and adult brain. Dev Neurobiol 78(11):1026–1048
8. Rohrback S et al (2018) Submegabase copy number variations arise during cerebral cortical neurogenesis as revealed by single-cell whole-genome sequencing. Proc Natl Acad Sci U S A 115(42):10804–10809
9. Lee MH et al (2018) Somatic APP gene recombination in Alzheimer's disease and normal neurons. Nature 563(7733):639–645
10. Lake BB et al (2018) Integrative single-cell analysis of transcriptional and epigenetic states in the human adult brain. Nat Biotechnol 36(1):70–80
11. Lake BB et al (2017) A comparative strategy for single-nucleus and single-cell transcriptomes confirms accuracy in predicted cell-type expression from nuclear RNA. Sci Rep 7(1):6031
12. Lake BB et al (2016) Neuronal subtypes and diversity revealed by single-nucleus RNA sequencing of the human brain. Science 352(6293):1586–1590
13. Bakken TE et al (2018) Single-nucleus and single-cell transcriptomes compared in matched cortical cell types. PLoS One 13(12): e0209648
14. Krishnaswami SR et al (2016) Using single nuclei for RNA-seq to capture the transcriptome of postmortem neurons. Nat Protoc 11(3):499–524
15. Zaghlool A et al (2021) Characterization of the nuclear and cytosolic transcriptomes in human brain tissue reveals new insights into the subcellular distribution of RNA transcripts. Sci Rep 11(1):4076
16. Palmer CR et al (2021) Altered cell and RNA isoform diversity in aging Down syndrome brains. Proc Natl Acad Sci U S A 118(47): e2114326118
17. Bakken TE et al (2021) Comparative cellular analysis of motor cortex in human, marmoset and mouse. Nature 598(7879):111–119
18. Hodge RD et al (2019) Conserved cell types with divergent features in human versus mouse cortex. Nature 573(7772):61–68

Chapter 3

Isolation of Human Microglia from Neuropathologically Diagnosed Cases in the Single-Cell Era

Lih-Fen Lue, Douglas G. Walker, Suet Theng Beh, and Thomas G. Beach

Abstract

This chapter describes the core procedures that we have developed over the last two decades to isolate routinely the microglia from postmortem human brains. The method is suitable for brain slices consisting of both gray and white matter.

The ability to concomitantly isolate vascular cells with glial cells provides the opportunity to investigate multiple cell types originating from the same donor. This represents a novel approach for *-omics* research, with the potential for discovering the shared or distinct molecular features among the glia and vascular cells from the same individual.

Key words Human glia, Single cell suspension, Postmortem, Aging, Neurodegeneration

1 Introduction

Brain microglia are innate immune cells distinct from peripheral macrophages. Over the last three decades, their roles in brain homeostasis, neuroprotection, inflammatory responses, and neurodegeneration have been supported by an abundance of evidence from research using postmortem human brains, animal models, and cultured microglia [1–5]. The momentum of microglial research has been accelerated in recent years owing to the identification of an array of microglia-associated single nucleotide polymorphisms (SNPs) that correlate with increased risk of developing sporadic Alzheimer's disease (AD). In addition, the availability of single-cell RNA-sequencing (RNA-seq) technologies [6–10] has enabled the characterization of microglial transcriptomic profiles across different brain regions, disease stages, pathologies, and cell types, in the human brain [11–15].

Lih-Fen Lue and Douglas G. Walker contributed equally for protocol development.

Jerold Chun (ed.), *Alzheimer's Disease: Methods and Protocols*,
Methods in Molecular Biology, vol. 2561, https://doi.org/10.1007/978-1-0716-2655-9_3,

In addition to transcriptomic research, single-cell isolates have been used in proteomics research. Cell isolates labeled by known cell-type markers are sorted by magnetic-activated cell sorting (MACS) and fluorescence-activated cell sorting (FACS), followed by mass spectrophotometry. Proteomic data analyzed by network co-expression algorithms have led to the discovery of an astrocyte/microglia metabolism module that is differentially affected by AD and aging [16]. Further, cell sorting, coupled with tandem mass tag (TMT)-mass spectrophotometry, have identified disease-associated microglial modules and their pro- and anti-inflammatory subsets [17]. Notably, a comparative study between cell sorting techniques (MACS versus FACS) demonstrated that the choice of cell selection methods can affect the proteomic data outcomes, in that MACS isolates may be contaminated by other cell types and cellular fragments [18]. Furthermore, other isolation procedures appear to cause changes in the patterns of gene expression [19, 20].

In this era of single-cell research, the findings from the transcriptomic and proteomic research models have expanded our ability to understand brain microglia under different biological and pathophysiological conditions, including neurodegenerative disease. It is therefore necessary to validate and extend these findings using human cell experimental systems, as there are likely to be discrepancies between species [21, 22].

Human cells directly isolated from neuropathologically defined tissues have a unique role to play in validation and drug-testing studies. The uniqueness lies in the heterogeneity of the genetic and epigenetic backgrounds associated with individual human cell donors, in contrast to the uniform backgrounds of cell lines or laboratory-bred animals. It follows that, when using postmortem human cells in an experiment, it is necessary to use cells derived from several donors, based on the inclusion and exclusion criteria relevant to the type of study.

We have used cells isolated from postmortem human brains for modeling microglial functions and inflammatory mechanisms. The results have been published in over 20 peer-reviewed articles [5]. In this chapter, we provide a step-by-step procedure that can concomitantly harvest human brain glial cells, endothelial cells, smooth muscle cells, and pericytes. This provides the opportunity to perform studies using a variety of cell types from a single donor. We have omitted the cell sorting from this procedure as we consider this technique to be a part of the downstream application.

2 Materials (Illustrated in Fig. 1)

The quantities listed below assume 50 g of starting material (brain tissue).

Fresh human postmortem brain tissue obtained at short postmortem interval (PMI); the methods described here are based on processing the brain tissues provided by the Brain and Body Donation Program (BBDP) of Banner Sun Health Research Institute and the Arizona Study of Aging and Neurodegenerative Disorders (ASAND). The mean and standard deviation of the PMI of the 25 autopsy cases processed with this methodology were 3.7 ± 0.9 h; this and other properties of the tissues are listed in Table 1.

2.1 Autopsy Tissue Storage

1. *Autopsy tissue storage medium* (sterile): a CO_2-independent Hibernate™-A medium (Thermo Fisher Scientific) or equivalent, 1× B27™ Supplement (Thermo Fisher Scientific, from 50×, stored −20 °C) or other serum-free supplements containing antioxidants, and 1× Antibiotic-Antimycotic (containing streptomycin, amphotericin B, and penicillin; from a 100× solution). Store at 4 °C.
2. Sterile 500 mL screw-top storage bottles.

2.2 Tissue Washing and Dissection

1. *Basal medium* (sterile): Dulbecco's Modified Eagle's Medium with high glucose (*DMEM*), 50 μg/mL gentamicin (from 50 mg/mL stock). Store at 4 °C.
2. Sterile, medium-length, curved, serrated forceps.
3. Three sterile, 300 mL glass beakers.
4. Sterile 150 mm Pyrex glass PETRI dish.
5. *Autopsy tissue storage medium.*
6. 1 pair of sterile, blunt-tipped, angled forceps.
7. 1 pair serrated forceps.
8. 2 pairs of sterile, fine-tipped forceps.
9. 1 pair of sterile, long-bladed, angled scissors.
10. 1 pair of sterile, sharp, curved scissors.
11. 1 sterile scalpel with No. 11 blade.
12. Digital scale disinfected with 70% alcohol.
13. Two sterile, 150 mm Pyrex glass petri dishes.
14. 25 mL sterile, disposable plastic pipettes.
15. Motorized pipette controllers.

Fig. 1 Stepwise isolation of postmortem human microglia. (**a**) Frontal cortical slices collected at autopsy are used for isolating cells. (**b**) After visible blood vessels and meninges are removed, tissues are coarsely cut and

2.3 Digestion

1. *Basal medium.*
2. *Processing medium* (sterile): Hibernate™-A medium, 1× B27™ supplement (from 50×, stored at −20 °C), 50 μg/mL gentamicin. Store at 4 °C.
3. Sterile-filtered 0.5 M EDTA, pH 8.0 (used as 1000× stock).
4. *Digestion medium*: *processing medium* with EDTA (above) added at a final concentration of 0.5 mM.
5. Hanks' Balanced Salt Solution (HBSS), without Ca^{+2} or Mg^{+2} (stored at 4 °C).
6. High-purity papain (from papaya latex), lyophilized powder, ≥10 units/mg protein, stored at 4 to −20 °C.
7. A sterile spatula, to weigh the papain.
8. Sterile 2 mL microtubes, pre-tared, to weigh the papain.
9. A digital scale, disinfected with 70% alcohol.
10. *DNase I glycerol solution:* 10 mg/mL DNase I in a solution of a 1:1 ratio of 0.15 M NaCl and glycerol, stored at −20 °C.
11. 6–8 sterile 50 mL conical centrifuge tubes for washing cells in the filtrate.
12. Sterile, disposable 10 mL and 25 mL plastic pipettes.
13. A rotary, shaking water bath, temperature-controlled (35 °C) and speed-controlled (130 rpm) for tissue digestion.

Fig. 1 (continued) weighed. (**c**) Tissues are further minced into 1 mm^3 cubes. (**d**) Tissues are suspended in medium containing EDTA and papain and placed in a shaking water bath for digestion. (**e**) Twenty minutes into digestion, tissue clumps are broken up by trituration of the tissue suspension after the addition of DNase. (**f**) Cells released from the digested tissues result in a dense and opaque cell suspension. (**g**) The cell suspension is filtered through a large funnel covered with a 120 μm nylon mesh, while the undigested tissues remain behind. (**h**) Enzymes are separated by centrifugation, leaving the loose cell fraction in the bottom of the tube. (**i**) Cell fractions are reconstituted (the 4 tubes on the *left* side), then transferred to high-speed centrifugation containers (on the *right* side) and mixed with Percoll. (**j**) The Percoll gradient centrifugation is performed in a high-speed centrifuge (the conditions illustrated are for a Sorvall RC6 centrifuge). (**k**) After centrifugation, myelin-associated components form a dense opaque upper layer. (**l**) The myelin layer is removed using a suction device attached to a glass Pasteur pipette. (**m**) The remaining content of the container is a solution of dispersed live cells lying over a band of red blood cells. (**n**) The live-cell solution is collected until only 0.5 cm (1–2 mL) remains above the band of red blood cells. (**o**) Loose cell pellets are obtained after washing and centrifugation to remove Percoll. (**p**) The cells are resuspended and filtered through a series of strainers to remove blood vessel fragments. (**q**) Endothelial cells, attached to Biotinylated UEA I-conjugated streptavidin magnetic beads, are removed using a magnet. (**r**) Cells are washed using centrifugation to form cell pellets which can be used for different applications (in vitro experiments, analysis of proteins and mRNA by gene array, RNA sequencing, proteomics, etc.) or stored in liquid nitrogen for later use

Table 1
Demographic and neuropathological characteristics of the 25 autopsy cases in this study

Characteristics	Mean ± SD
Age (years)	85.3 ± 8.8
Tissue PMI (hours)	3.7 ± 0.9
Culture PMI (hours)	8.7 ± 6.3
Sum of amyloid plaque density from 5 brain regions [23]	7.9 ± 6.6
Sum of neurofibrillary tangle density from 5 brain regions [23]	8.6 ± 3.9
Sum of Lewy body density from 10 brain regions [23]	9.7 ± 13.2
RNA yield (ng/μL)	25.3 ± 12.7
RNA quality (RIN: scale of 1–10; 10 is best)	8.7 ± 0.5
Cell yield (million cells per gram of wet tissues)	3.0 ± 1.6
Ratios	Numbers
Males: females APO E 2/3:3/3:3/4:4/4	15:10 3:17:3:2

2.4 Removing Undigested Tissues

1. 165 mL polypropylene 2-piece Buchner funnel (Bel Art, Cat#H14602-000).
2. Autoclavable adhesive tape.
3. Aluminum foil.
4. A 1 L plastic specimen cup in which to autoclave the nylon mesh filter cup (*see* **Item 5**, below), and use as a filtrate collector.
5. A homemade, autoclavable 120 μm nylon-mesh filter cup (*see* **Note 1**).
6. Sterile, disposable 10 mL and 25 mL plastic pipettes.
7. 8–10 sterile 50 mL conical centrifugation tubes.
8. *Processing medium* for washing cells.

2.5 Percoll Gradient Separation

1. Percoll™ density gradient medium (Cytiva), stored at 4 °C.
2. Sterile, 10× phosphate-buffered saline (PBS), stored at room temperature (RT), for making isotonic Percoll solution.
3. *Isotonic Percoll solution*: add 1 volume of 10× PBS to 9 volumes of Percoll.
4. *Processing medium.*
5. Two sterile, 250 mL wide-mouth conical centrifuge bottles with screw-top lids that fit in the centrifugation rotor.
6. Base to hold the conical centrifuge bottles, which are secured to their bases with a vertical strip of tape (Fig. 1i).

7. Rack to hold the centrifuge bottles and their bases.
8. Sterile, disposable 10 mL and 25 mL plastic pipettes.
9. Superspeed (>30,000 × g) refrigeration centrifuge with a fixed-angle rotor sufficient to accommodate 50–250 mL centrifuge bottles.
10. Vacuum suction apparatus to remove myelin layer formed during Percoll gradient separation.
11. Sterile, glass Pasteur pipettes for the above.
12. A 500 mL sterile plastic cup to collect the cell suspension retrieved from the gradient (cell collection cup).
13. 18–20 sterile 50 mL conical centrifuge tubes for washing the cells collected from the gradient.

2.6 Removal of Blood Vessels in Cell Suspension

1. Sterile, disposable cell strainers in different mesh sizes: 20 μm, 40 μm, 70 μm, and 100 m, placed on top of conical centrifuge tubes (below).
2. 6–12 sterile 50 mL conical centrifuge tubes.
3. Sterile, disposable 10 mL plastic pipettes.
4. *Basal medium*.
5. Sterile collagen-I-coated 12-well plates, if using blood vessels to plate for vascular cell culture.

2.7 Removal of Endothelial Cells

1. Orbital rotation mixer to hold 2 mL microtubes and 50 mL centrifuge tubes.
2. CELLection™ Biotin Binder Kit, containing Streptavidin magnetic microbead solution and 2 mg Biotinylated Ulex europaeus agglutinin I (UEA I) in 10 mM HEPES, pH 7.5, 0.15 M NaCl, 0.1 mM $CaCl_2$, and 0.08% sodium azide. Follow the manufacturer's instructions (Vector Laboratories) to prepare the *Biotinylated UEA-I Solution*: 2 mg of Biotinylated UEA I (UEA-I) (above), dissolved in 1 mL sterile water and aliquoted as 50 μL per 1.5 mL Eppendorf microtube, and then stored at −80 °C.
3. PBS with 0.1% Tween-20 (to wash magnetic microbeads).
4. PBS with 0.1% bovine serum albumin (BSA) to resuspend the UEA-conjugated magnetic microbeads, for storage at 4 °C.
5. *Endothelial cell selection solution*: 100 μL *Streptavidin magnetic microbead solution* mixed with 4 μL *Biotinylated UEA-I solution*, prepared as in **Note 2**.
6. *Basal medium*.
7. *DMEM/FBS medium* (sterile) to wet nylon strainers (*see* **Note 3**): DMEM, 0.2% fetal bovine serum (FBS), stored at 4 °C.

8. Sterile disposable cell strainers: 20 μm, 40 μm, 70 μm, and 100 μm pore sizes; pre-treating strainers is recommended (*see* **Note 3**).
9. 15 mL/50 mL magnetic separation racks.
10. 6–8 50 mL sterile, conical centrifuge tubes to receive filtered cell suspension.
11. Sterile, disposable 10 mL plastic pipettes.

2.8 Cell Counting

1. 0.4% sterile-filtered trypan blue solution, suitable for culture.
2. Hemocytometer and coverslip.
3. Sterile 2 mL microtube.
4. 100 μL and 10 μL pipettors with sterile pipette tips.
5. 70% alcohol to disinfect the hemocytometer and coverslip.
6. Phase-contrast microscope with 10× objective.
7. A hand-tally counter.

2.9 Cryoprotection for Ex Vivo Mixed Glial Cells

1. CryoStor^{R} CS10, 100 mL (STEMCELL Technologies).
2. Cryotubes for liquid nitrogen storage: sterile, 2 mL, round-bottomed, self-standing, polypropylene, with internal threads.
3. Sterile pipette tips for 10-, 20-, 200-, and 1000 μL pipettors.
4. Isopropanol, for control of the freezing speed.
5. Freezing containers, filled with isopropanol, that can hold 18 × 1–2 mL cryotubes and provide a 1 °C/min cooling rate, stored at −80 °C.
6. Temperature-monitored −80 °C freezer.
7. Liquid-nitrogen storage tank with specimen box holders.
8. Liquid nitrogen level monitor and alarm.
9. Specimen boxes sustainable in liquid nitrogen tank.

2.10 Plating for Microglial Culture

1. 1–3 (# depends on cell yield) sterile, flat T75 flasks with filter-screw lid.
2. Sterile, disposable 10 mL plastic pipettes.
3. *Microglia growth medium* (10% FBS in *basal medium*).

2.11 Replating Microglia for Experimental Use

1. HBSS.
2. 0.05% trypsin in 0.5 mM EDTA solution or equivalent substitute (such as TrypLE from Thermo Fisher).
3. 0.4% trypan blue solution.
4. Hemacytometer with coverslip.
5. *Low-serum medium* (0.5–2.0% FBS).
6. Sterile, disposable 10 mL plastic pipettes.

7. Sterile 50 mL plastic centrifuge tubes.
8. Sterile 24- or 12-well plastic plates with lids.
9. Inverted phase-contrast microscope.

3 Methods (the Main Steps of the Procedure Are Illustrated in Fig. 1).

3.1 Brain Tissue Preparation

Brain tissue removal and storage are described in previously published reports [23, 24]. The methods described below are the procedures to obtain cells from two, fresh, 1 cm thick frontal-cortical anterior-posterior slices.

3.2 Preparing the Workstation before the Procedure

1. Take the following solutions out of the refrigerator 30 min to 1 h before the procedure: *autopsy tissue storage medium*, *processing medium*, *basal medium*, DMEM, HBSS, 10× PBS, FBS, and Percoll. For the solutions stored at −20 °C, such as the B27™ supplement and *DNase I glycerol solution*, take them out of the freezer and place on ice until use.
2. Gather, on a cart near a biosafety Level II hood: sterile glass Petri dishes; 50 mL centrifuge tubes; Percoll gradient centrifuge bottles; bottle base and rack; autoclaved instrument packages (containing 1 pair of sterile, long-bladed, angled scissors; 1 pair of sterile, blunt-tipped, angled forceps; 1 pair of autoclaved sterile, medium-length, curved, serrated forceps; 2 pairs of sterile, fine-tipped forceps; 1 pair of sterile, sharp, curved scissors; and a small spatula); sterile glass beakers; glass enzymatic digestion container; sterile 120 μm nylon mesh filter cup in a receiving container; cell strainers; pipettes; Percoll centrifugation bottles and supporting bases; and 50 mL centrifugation tubes.
3. Add water to the water-bath shaker until it reaches half of the height of the digestion bottle, and set the temperature to 35 °C.
4. Spray the interior of the hood with 70% ethanol and turn on the UV light for 20 min, to disinfect the surfaces.
5. Turn UV light off before tissue processing begins.

3.3 Brain Tissue Washing and Dissection (the Procedure from Here Onward Is Done inside the Biosafety Hood, Unless Otherwise Specified)

1. Spray the exterior of the brain-sample storage container with 70% ethanol.
2. Open the lid of the storage container.
3. Discard the *autopsy tissue storage medium*.
4. Transfer the brain slices to the first sterile 300 mL glass beaker filled with 100 mL of *basal medium*, using a pair of sterile, blunt-tipped, angled forceps.
5. Wash the brain slices by gently swirling the glass beaker.

6. Remove the medium.
7. Repeat washing steps via two additional transfers of the tissue slices to clean glass beakers, washing each time with fresh *basal medium*.
8. Tare an empty glass Petri dish (sterile) on the digital scale.
9. Transfer brain slices into the Petri dish using the sterile, blunt-tipped, angled forceps, after removing excessive liquid.
10. Weigh the brain slices in order to calculate the volume of the *digestion medium* needed to achieve a 2:1 ratio of medium (in mL) to brain weight (in g) for Subheading 3.4. For example, if you have 50 g of tissue, you'll need 100 mL of *digestion medium*.
11. Pour 5 mL of *processing medium* into the Petri dish to keep the brain slices moist.
12. Examine and record the appearance and texture of the brain slices (*see* **Note 4**).
13. Use the serrated and fine-tipped forceps to remove visible blood vessels from brain tissues.
14. Exchange the medium in the dish with 5 mL fresh *processing medium*.
15. Use a pair of angled, medium-tipped scissors to cut brain slices into longitudinal strips of approximately 1 cm length; then use the serrated forceps to hold each tissue strip with one hand, while using the curved, sharp scissors in the other, to cut the strips into approximately 5 mm^3 tissue cubes.
16. Use a sterile scalpel or razor blade against the bottom of the Petri dish, to mince the cubes into approximately 1 mm^3 pieces.
17. Add 10 mL of *processing medium* to the Petri dish to disperse tissue cubes.
18. Transfer tissue suspension to another sterile, 300 mL glass beaker by using a 10 mL or 25 mL sterile pipette or by pouring tissue suspension into the beaker.
19. Add 2 mL of additional *processing medium* to the beaker for each gram of brain tissue (e.g., add 100 mL *processing medium* to 50 g of tissue).

3.4 Enzymatic Digestion

1. Use the volume of *processing medium* added in Subheading 3.3, **Step 19** to calculate the amount of lyophilized papain required to give a final concentration of 1 mg of papain per 1 mL of the *digestion medium*. Using a small sterile spatula and a pre-tared sterile microtube, weigh out the proper amount of papain into the microtube (e.g., for 100 mL of additional *processing medium*, weigh out 100 mg of papain).

2. Slowly add 1 mL of the calcium- and magnesium-free HBSS to the microtube containing the papain and swirl to disperse the papain powder.
3. Transfer the papain/HBSS solution into the beaker containing the *digestion medium* and tissue pieces, and gently swirl.
4. Transfer the tissue suspension from **Step 3** into the digestion bottle, screw on the lid, and then cover the lid with aluminum foil to prevent splashing in the next step.
5. Place the digestion bottle into the 35 °C water bath and set the speed to 130 rpm.
6. After 15–20 min, remove the bottle from the water bath.
7. Spray the exterior of the bottle with 70% alcohol and wipe dry.
8. Add 100 μL of 10 mg/mL *DNase I glycerol solution* into the tissue suspension. (This represents 1:1000 of the 100 mL *processing medium* added in Subheading 3.3, **Step 19** to suspend 50 g of brain tissue.)
9. Triturate gently with a sterile, 10 mL pipette to break up tissue fragments and cell clumps.
10. Return digestion bottle to the water-bath shaker and continue shaking at 35 °C for an additional 20–25 min.
11. Remove the digestion bottle from the water bath and wipe clean the exterior of the bottle.
12. Triturate again to break up tissue fragments and cell clumps.

3.5 Removal of Undigested Tissue and Enzyme

1. Remove the foil from the 1 L specimen cup that contains the 120 μm nylon mesh filter cup.
2. Wet the surface of the nylon mesh with *basal medium*.
3. Using a sterile, 25 mL pipette, transfer the tissue suspension, 25 mL at a time, to the reservoir of the filter cup to start the filtering process.
4. If the mesh becomes clogged, use the tip of a sterile, 10 mL pipette to move the undigested material away from the center of the filter.
5. Rinse the filter reservoir with *processing medium*, 5–10 mL at a time, and repeat until the filtrate becomes clear.
6. Discard the remaining undigested material.
7. Using a new pipette, divide the cell filtrate into 8–10 of the 50 mL sterile, conical centrifuge tubes, and bring the volume in each tube to 50 mL using *processing medium*.
8. Centrifuge the tubes at 130 × g for 10 min at 10 °C in a swing-bucket rotor centrifuge.

9. Using sterile 10 mL plastic pipettes, remove the liquid above the loose, opaque cell layers settled in the cones of the centrifuge tubes.
10. Combine the cell layers from all centrifuge tubes into a sterile, glass 300 mL beaker.
11. Add *processing medium* to suspend the cells in a final volume of 200 mL.

3.6 Percoll Gradient Separation (See Note 4)

1. Transfer 100 mL of the cell suspension into each of the two 250 mL conical centrifugation bottles, followed by 50 mL of *isotonic Percoll solution*. This results in a 33% (v/v) Percoll concentration in the final mixture.
2. Balance the weights of the two centrifugation bottles on a scale by correcting the difference in the weights of the two bottles with *processing medium*.
3. Centrifuge the mixture to form a gradient in a Sorvall RC6 centrifuge or equivalent at 28,600 × g for 20 min at 10 °C in a fixed-angle rotor with an acceleration setting of 7–9, a deceleration setting of 4 and without braking.
4. Carefully place the centrifugation bottles in the rack without disturbing the three-layer gradients (*see* **Note 5**).
5. With a sterile, glass Pasteur pipette connected to a vacuum suction device, aspirate and discard the upper, opaque myelin layer in the centrifuge bottles.
6. Using another pipette, transfer the cell suspension that lies above the red blood cell layer, formed at the conical portion of the bottle, to a 500 mL sterile collection cup. Stop collecting the suspension when the level is 0.5 cm above the red blood cell layer.
7. Transfer no more than 20 mL of the combined cell suspension to each of the 50 mL sterile centrifuge tubes and add processing medium until the volume reaches 50 mL. Approximately 10–12 tubes will be used.
8. Centrifuge at 500 × g for 10 min at 10 °C in a swing-bucket rotor centrifuge.
9. Remove and discard the liquid above the cell pellets.
10. Add 10–15 mL of *basal medium* to each cell pellet and resuspend the cells by gentle trituration with a 10 mL pipette.
11. Combine all the cell suspensions into a sterile cup and adjust the volume to 300 mL.
12. Distribute 50 mL of the cell suspension into each of 6 conical centrifugation tubes.
13. Centrifuge at 10 °C, 300 × g for 10 min.

14. Remove the liquid and resuspend cell pellets with 5 mL fresh *basal medium* in each tube.
15. Combine all cell suspensions and divide into 2 conical centrifuge tubes, adding *basal medium* to a volume of 50 mL.
16. Centrifuge at 10 °C, 130 × g for 10 min.
17. Remove liquid above the pellets.
18. Resuspend cells in a total of 20 mL of *basal medium*.

3.7 Removal of Blood Vessels in Cell Suspension

1. Remove the screw-top lids of four 50 mL conical centrifugation tubes.
2. Place the cell strainers to fit into the openings of 50 mL sterile, conical centrifuge tubes in the sequential order of 100-, 70-, 40-, and 20 μm strainers.
3. Pour 10 mL of *DMEM/FBS medium* into strainer reservoirs to wet the surfaces and the tubes, and then discard the flow-through.
4. Transfer the cell suspension, 2–3 mL at a time, to the reservoir of the 100 μm strainer, allowing cells to flow into the 50 mL tubes, but straining out fragments of blood vessels.
5. When finished, rinse the screen surface with 3 mL of *basal medium* to release any trapped cells.
6. Remove the strainer from the collection tube. If blood vessels are to be collected for vascular cell culture, keep the cell strainer moist by placing in a Petri dish filled with *basal medium* (*see* **Note 6**).
7. Using the filtrate, repeat **Steps 3–5** with the 70 μm strainer, then the 40 μm strainer and, finally, the 20 μm strainer.
8. In the final conical tube, add *basal medium* to a total volume of 50 mL.
9. Fill another centrifuge tube to 50 mL with water, to balance out the centrifuge, and spin at 10 °C, 130 × g for 10 min.
10. Resuspend the cell pellet by adding 20 mL of *basal medium* into a 50 mL conical centrifuge tube and gently triturating it.

3.8 Removal of Endothelial Cells

1. Add 50 μL of *Endothelial cell selection solution* to the cell suspension.
2. Mix for 10 min at room temperature on a rotatory mixer.
3. Place the cell suspension tube against a magnetic separation rack for 3 min.
4. Carefully transfer the cell suspension to a new 50 mL tube without disturbing the magnetic beads adhered to the tube wall.

5. Add 5 mL of *processing medium* to the tube containing microbeads and mix.
6. Place the tube against the magnetic separation rack for another 3 min.
7. Again, remove the cell suspension to add to the 50 mL tube (*see* **Step 4**), while leaving the magnetic beads undisturbed. The brain microvascular endothelial cells can be collected from the microbeads fraction (*see* **Note** 7).
8. Add *processing medium* to the cell suspension to 50 mL.
9. Centrifuge at 10 °C, 130 × g for 10 min.
10. Resuspend the cell pellet in 20 mL of *processing medium* and place the tube on ice.
11. Transfer 10 μL of the suspension into a 1.5 mL microtube for cell counting (*see* Subheading 3.9).
12. After counting, cells are cryoprotected as a single cell suspension (*see* Subheading 3.10). If cells are to be used for microglia research, mixed cell suspension is plated in culture plates or flasks (*see* Subheading 3.11).

3.9 Cell Counting

1. Add 90 μL of the 0.4% trypan blue solution to the 10 μL of cell suspension in the 1.5 mL microtube from Subheading 3.8, **Step 11**, and mix briefly.
2. Place a coverslip on the hemocytometer slide.
3. Deliver 10 μL of the mixture to the sample chamber of the hemocytometer, under the coverslip.
4. Using the phase-contrast microscope, count the number of cells in each of the four corner squares of the gridded area on the hemocytometer slide, and add the 4 numbers together for "total cell number" in the formula below.
5. Calculate the cell concentration (C: number of cells per ml) and the total cell yield (Y) according to the formula: C = (total cell number/4) × fold of dilution × 10,000; Y = C × volume of cell suspension [for 50 g of starting material, use a 10-fold dilution and a cell suspension of 20 mL]. Cell yield per gram of brain weight is calculated by dividing Y by the total brain weight (in grams) (*see* **Note 8**).

3.10 Cryoprotection of the Isolated Cells

The mixed cell population contains a variable number of different cell types (microglia and astrocytes in every case, oligodendrocytes in most cases and occasionally neuronal cell bodies). Cells that will not be immediately used for experimentation should be cryoprotected and stored in the vapor phase in a liquid nitrogen tank. We monitored the RNA quality of the cell isolates by measuring RNA integrity number (RIN) (Table 1, *see* **Note 9**). One can characterize cell types using quantitative PCR (qPCR) for using cell type-specific genes (*see* **Note 9**).

1. Transfer the cell suspension from Subheading 3.8, **Step 10** into a 50 mL conical centrifuge tube and spin on a table-top centrifuge at 10 °C, 130 × g, for 10 min.
2. Remove the medium with a sterile 10 mL pipette and discard.
3. Resuspend cell pellets in the pre-cooled CryoStor^R CS10 at a concentration of 10 million cells per ml of cryoprotectant (use the cell-counting results to determine how many cryovials are needed).
4. Transfer cells to pre-cooled, 2 mL cryovials and place in a previously cooled freezing container filled with isopropanol.
5. Place the freezing container into a −80 °C freezer for 3–4 h.
6. Remove the cryovials from the freezing container and place inside a specimen box designed for long-term liquid nitrogen storage, and then place the box into the storage position for long-term storage in the vapor phase.

3.11 Plating for Microglial Culture

1. Using a sterile pipette, transfer the appropriate volume of the freshly isolated cell suspension into one or two T75 flasks, such that the cell count is approximately 50–100 million per flask.
2. Add FBS into each flask to obtain a final concentration of 10% FBS in the *basal medium* (this is *microglia growth medium*).
3. Place flasks in the humidified cell-culture incubator at 37 °C, with a 5% CO_2/95% air mix.
4. Remove unattached cells 12–24 h after plating; these cells can be replated for astrocyte culture (*see* Subheading 3.12) or discarded.
5. Add 10–15 mL of *microglia growth medium* to the original flasks and continue to incubate at 37 °C, changing the medium every 7 days.
6. Observe the morphological changes in the following 7–14 days under a phase-contrast microscope (Fig. 2 and *see* **Note 10**). Microglia can be replated for experimental use for up to 14 days (*see* Subheading 3.12).

3.12 Replating Microglia for Experimental Use

1. Change culture medium to *basal medium* (serum-free) or *low-serum medium* and continue to incubate at 37 °C for 24 h prior to trypsinization (next step).
2. Bring 0.05% trypsin containing 0.5 mM EDTA (or equivalent) to room temperature.
3. Remove culture medium from flasks and discard.
4. Add 5 mL of calcium- and magnesium-free HBSS to each flask.
5. Gently rock the flask side to side (manually) for 1 min.
6. Remove and discard the HBSS.

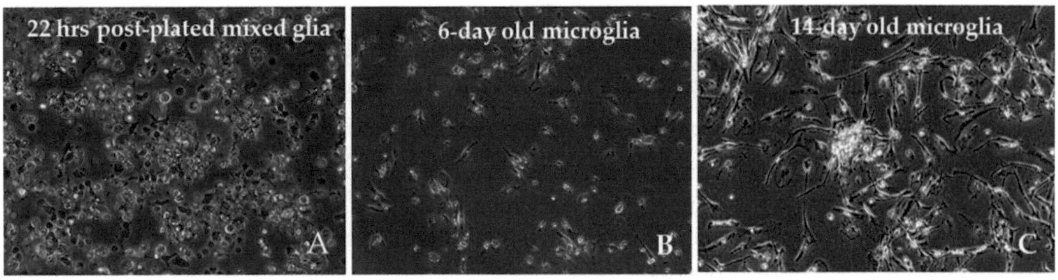

Fig. 2 Morphological changes in cultured human microglia. One application for the isolated, mixed glial cells is to culture microglia for experimentation. Microglia adhere to plastic culture plates within 24 h, and then the non-adherent cells are removed by replacing the medium. The appearance of the cell cultures changes over time, as shown here. (**a**) Cell isolates before non-adherent cells were removed. (**b**) Cultured microglia after 6 days in vitro. (**c**) Microglia, after 14 days in vitro, developed diverse morphologies. The purity of the microglial culture, as determined by observing distinct morphological characteristics, was >95%

7. Add 2 mL of 0.05% trypsin containing 0.5 mM EDTA to each flask, and gently rock the flasks side to side, to distribute solution evenly.
8. Place the flasks back in the 37 °C incubator for 3 min.
9. Take flasks out of the incubator, and gently tap on the side of the flasks while holding them horizontally.
10. Examine the flasks under a phase-contrast microscope to determine whether the cells are rounding up and ready to harvest; if so, go to **Step 11**.
11. Add 8 mL of *microglia growth medium* to the flask.
12. Transfer the cell suspension from the flask to a 50 mL sterile, conical tube and place on ice.
13. Transfer 10 μL of the cell suspension to a 1.5 mL microtube for counting.
14. Count the trypan blue-excluded cells as described in Subheading 3.9.
15. Based on the cell count, determine how many cells per culture well are to be plated for an experiment; aim for 75,000 to 100,000 cells per well in a 12-well plate (3.5 cm^2 area). Transfer appropriate volume of the cell suspension to each well.
16. Incubate microglia for 1–3 days at 37 °C in *microglia growth medium* until using for experimentation.

4 Notes

1. To make the filter cup (and *see* Fig. 1g): place a 10 cm^2 square, 120 μm nylon mesh sheet between the top and bottom units of the 165 mL Buchner funnel. Use autoclavable adhesive tape to

hold the funnel units together and secure the position of the nylon mesh. Place the filter cup into the 1 L plastic specimen cup and cover the opening of the specimen cup with a piece of aluminum foil before autoclaving.

2. Prepare the *Streptavidin magnetic microbead solution* per the manufacturer's instructions for the CELLection™ Biotin Binder Kit:

   Mix 100 μL of *Streptavidin magnetic microbead solution* with 4 μL of *Biotinylated UEA-I solution* for 30 min in a sterile, 2 mL microtube, using a rotation mixer. Place the tube against a magnetic stand for 2 min, then discard the solution while the tube is still on the magnetic stand (so the beads will remain). Release the tube from the stand and add PBS with 0.1% Tween-20 (to wash the beads), resuspend the beads, again place the tube against the magnetic stand for 2 min, and then discard the solution. Repeat the wash step one more time, and then reconstitute the beads in 1 mL PBS with 0.1% BSA. Store at 4 °C.

3. Blood vessels are prone to adhering to hygroscopic surfaces. Pre-rinsing the surfaces of the strainers and the interiors of the filtrate collection tubes, using a medium containing albumin or serum, can reduce the clogging of strainer meshes and prevent vessels from adhering to the tubes.

4. Documentation of the appearance and texture of the brain slices is important if an abnormality is found during cell isolation. For example, a clumped and clot-like red blood cell layer after Percoll-gradient centrifugation could indicate pre-existing micro-bleeding, hemorrhaging and/or long autolysis in the brain tissue. Look for signs of discoloration, hemorrhage, microbleeds, sclerosis, and atrophy; assess the texture in terms of tissue dehydration and firmness by observing how well the tissue slice retains its three-dimensional shape.

5. Percoll gradient centrifugation is used to concentrate myelin-containing components, a step that is important for separating glial cells in the myelin-rich adult brain. When isolating from a small amount of human adult brain tissues, Myelin Removal Beads II can be used instead (Miltenyi Biotec) [25].

   The centrifugation leads to gradient formation. The three layers formed are, from top to bottom: dense myelin and neuronal fragment layer, glial cell layer, and erythrocyte layer.

6. The vessel fragments remaining on the strainer surfaces are a good source for smooth muscle cells and pericytes for cultures. To harvest the vessels, the strainers are flipped upside down, and then cell type-specific medium for pericytes or smooth muscle cells is used to wash the blood vessels into sterile 50 mL conical centrifugation tubes. The vessels can then be plated onto collagen I-coated 12-well plates.

7. The UEA-1-conjugated magnetic microbeads bind to endothelial cells [26], and have been used to isolate endothelial cells from many organs. When using this method, we found that endothelial cells cultured on the extracellular matrix exhibited junction-like features detected by a trans-endothelial electrical resistance (TEER) device.
8. The total cell yield varies by donor. The 25 cell isolations conducted in 2018–2019 yielded an average of 3.0 million cells per gram of brain tissue (Table 1). The cell yields were not affected by the postmortem interval (PMI; the time elapsed before the autopsy), the age of the donors, or the presence of a diagnosed disease. This indicates that antemortem factors might play more important roles in cell yield.
9. The RNA quality (RIN; determined using the Agilent 2100 Bioanalyzer System) of the mixed glial isolates was 8.7 ± 0.5 (mean ± S.D.) for the 25 cases processed in 2017–2019 (5 controls and 20 diseased) (Table 1). The cells from the control cases had slightly higher RINs (9.04 ± 0.31) compared to the disease group (8.55 ± 0.49, $P = 0.044$).

   We used quantitative PCR (qPCR) to analyze the expression of cell-type-specific genes, in order to determine the cell types present in our cultures: allograft inflammatory factor 1 (*AIF1*) for microglia; glial fibrillary acidic protein (*GFAP*) for astrocytes; oligodendrocyte transcription factor 2 (*OLIG2*) for oligodendrocytes; and RNA-Binding Fox-1 Homolog 3 (*RBFOX3*) for neurons. The qPCR results (Table 2) showed the range, mean, median, and standard deviation of the ΔCt values (ΔCt is the Ct value of a specific gene minus the Ct values of β actin, for each sample). Thus, higher positive values of the ΔCt mean lower abundance of gene expression. For example, the microglial gene (*AIF1*) had the lowest average ΔCt; therefore, the highest gene abundance among the four genes tested. *Olig2* gene levels were undetectable (ΔCt values ≥30) in 20% of the cases, and neuronal gene expression (*RBFOX3*) could not be detected in 60% of the cases.

Table 2
Gene expression levels indicated by ΔCt (obtained by subtracting the Ct values of each gene with the Ct values of β actin, for each sample)

Genes	N =	Minimum	Maximum	Mean	Median	Standard deviation
AIF	25	0.612	6.442	2.960	2.733	1.484
GFAP	25	1.546	7.222	5.022	5.207	1.184
RBFOX3	25	2.245	13.992	8.618	8.661	3.085
Oligo2	25	3.290	9.234	6.045	5.596	1.684

We found that cells from higher-yield preparations more frequently exhibited neuronal structures. Morphological examination under a phase-contrast microscope confirmed the occasional presence of large neuronal cell bodies in the mixed cell suspensions, suggesting the procedure enriched the population of glial cells. When microglia are cultured, the neuronal structures, along with other non-adherent cells, are removed with medium change.

10. Microglia in the mixed glia isolates will adhere to the plastic surfaces of the flasks within 24 h. Thus, by differential adherence, other glial cells can be removed, either discarded altogether or plated as astrocyte cell cultures. Adhered microglia will develop a process-bearing morphology (ramified microglia-like) within 7 days. The development of microglial morphology under these culture conditions is shown in Fig. 2.

Acknowledgments

This work was supported by Arizona State DHS grant funding ADHS-14-052688. The operation of the Banner Sun Health Research Institute Brain and Body Donation Program has been supported by the National Institute of Neurological Disorders and Stroke (U24 NS072026 National Brain and Tissue Resource for Parkinson's Disease and Related Disorders), the National Institute on Aging (P30 AG19610 Arizona Alzheimer's Disease Core Center), the Arizona Department of Health Services (Contract 211002, Arizona Alzheimer's Research Center), the Arizona Biomedical Research Commission (Contracts 4001, 0011, 05-901, and 1001 to the Arizona Parkinson's Disease Consortium), the Michael J. Fox Foundation for Parkinson's Research (the Prescott Family Initiative), and Sun Health Foundation.

References

1. Sarlus H, Heneka MT (2017) Microglia in Alzheimer's disease. J Clin Invest 127(9): 3240–3249
2. Colonna M, Butovsky O (2017) Microglia function in the central nervous system during health and neurodegeneration. Annu Rev Immunol 35:441–468
3. Deczkowska A et al (2018) Disease-associated microglia: a universal immune sensor of neurodegeneration. Cell 173(5):1073–1081
4. Bohlen CJ et al (2019) Microglia in brain development, homeostasis, and neurodegeneration. Annu Rev Genet 53:263–288
5. Lue LF, Beach TG, Walker DG (2019) Alzheimer's disease research using human microglia. Cells 8(8):838
6. Efthymiou AG, Goate AM (2017) Late onset Alzheimer's disease genetics implicates microglial pathways in disease risk. Mol Neurodegener 12(1):43
7. Tansey KE, Cameron D, Hill MJ (2018) Genetic risk for Alzheimer's disease is concentrated in specific macrophage and microglial transcriptional networks. Genome Med 10(1):14
8. Hashemiaghdam A, Mroczek M (2020) Microglia heterogeneity and neurodegeneration: the emerging paradigm of the role of immunity in Alzheimer's disease. J Neuroimmunol 341:577185
9. Villegas-Llerena C et al (2016) Microglial genes regulating neuroinflammation in the

progression of Alzheimer's disease. Curr Opin Neurobiol 36:74–81

10. van der Poel M et al (2019) Transcriptional profiling of human microglia reveals grey-white matter heterogeneity and multiple sclerosis-associated changes. Nat Commun 10(1):1139
11. Friedman BA et al (2018) Diverse brain myeloid expression profiles reveal distinct microglial activation states and aspects of Alzheimer's disease not evident in mouse models. Cell Rep 22(3):832–847
12. Mathys H et al (2019) Single-cell transcriptomic analysis of Alzheimer's disease. Nature 570(7761):332–337
13. Olah M et al (2020) Single cell RNA sequencing of human microglia uncovers a subset associated with Alzheimer's disease. Nat Commun 11(1):6129
14. Keren-Shaul H et al (2017) A unique microglia type associated with restricting development of Alzheimer's disease. Cell 169(7): 1276–1290.e17
15. Rangaraju S et al (2018) Identification and therapeutic modulation of a pro-inflammatory subset of disease-associated-microglia in Alzheimer's disease. Mol Neurodegener 13(1):24
16. Johnson ECB et al (2020) Large-scale proteomic analysis of Alzheimer's disease brain and cerebrospinal fluid reveals early changes in energy metabolism associated with microglia and astrocyte activation. Nat Med 26(5): 769–780
17. Rangaraju S et al (2018) Quantitative proteomics of acutely-isolated mouse microglia identifies novel immune Alzheimer's disease-related proteins. Mol Neurodegener 13(1):34–34
18. Rayaprolu S et al (2020) Flow-cytometric microglial sorting coupled with quantitative proteomics identifies moesin as a highly-abundant microglial protein with relevance to Alzheimer's disease. Mol Neurodegener 15(1): 28
19. Serrano GE, et al (2021) Whole-cell dissociated suspension analysis in human brain neurodegenerative disease: a pilot study. medRxiv: 2021.01.08.21249455
20. Volovitz I et al (2016) A non-aggressive, highly efficient, enzymatic method for dissociation of human brain-tumors and brain-tissues to viable single-cells. BMC Neurosci 17(1):30
21. Gibbons HM et al (2011) Valproic acid induces microglial dysfunction, not apoptosis, in human glial cultures. Neurobiol Dis 41(1): 96–103
22. Smith AM, Dragunow M (2014) The human side of microglia. Trends Neurosci 37(3): 125–135
23. Beach TG et al (2015) Arizona study of aging and neurodegenerative disorders and brain and body donation program. Neuropathology 35(4):354–389
24. Beach TG et al (2008) The Sun Health Research Institute Brain Donation Program: description and experience, 1987-2007. Cell Tissue Bank 9(3):229–245
25. Olah M et al (2018) A transcriptomic atlas of aged human microglia. Nat Commun 9(1):539
26. Jackson CJ et al (1990) Binding of human endothelium to Ulex europaeus I-coated Dynabeads: application to the isolation of microvascular endothelium. J Cell Sci 96 (Pt 2):257–262

Chapter 4

Microglia in Human Postmortem Brain Samples: Quantitative Ultrastructural Analysis of Scanning Electron Microscopy Images

Marie-Kim St-Pierre, Eva Šimončičová, Micaël Carrier, and Marie-Ève Tremblay

Abstract

In this protocol, we describe the specific steps required to prepare human postmortem brain samples for ultrastructural microglial analysis. A detailed procedure is provided to improve the ultrastructural quality of the samples, using aldehyde fixatives followed by immunoperoxidase staining of allograft inflammatory factor 1 (AIF1, also known as IBA1), a marker of myeloid cells, and cluster of differentiation *68* (CD68), a marker of phagolysosomal activity. Additionally, we describe an osmium-thiocarbohydrazide-osmium (OTO) post-fixation method that preserves and increases the contrast of cellular membranes in human postmortem brain samples, as well as the steps necessary to acquire scanning electron microscopy (SEM) images of microglial cell bodies. In the last section, we cover the quantitative analysis of various microglial cytoplasmic organelles and their interactions with other parenchymal elements.

Key words Human postmortem brain samples, Scanning electron microscopy, Ultrastructural analysis, Microglia, Phagolysosomal activity, Intercellular relationships

1 Introduction

Evidence of a central role for microglia in the proper development and function of the central nervous system (CNS) has been continuously strengthened through extensive research, both in animal models and in human postmortem brain samples. Investigation of human microglia provides crucial clues about these immune cells that would otherwise be lost in translation from rodent models to human contexts. However, this tissue also presents challenging limitations, such as the availability of an adequate sample size for proper statistical power to account for substantial interindividual differences in both pre- and postmortem conditions. Nevertheless, used in combination with -omics research, super-resolution microscopy [e.g., electron microscopy (EM)] of human postmortem brain

Jerold Chun (ed.), *Alzheimer's Disease: Methods and Protocols*,
Methods in Molecular Biology, vol. 2561, https://doi.org/10.1007/978-1-0716-2655-9_4,

samples provides an indispensable source of in-depth knowledge about microglial structure and function, under both homeostatic and pathological conditions [1, 2].

EM has become a pillar for the study of cellular ultrastructure in human postmortem tissues [3]. Contrasting with light microscopy, EM utilizes an accelerated, focused beam of electrons, having sufficiently short wavelengths to enable imaging at atomic and subatomic levels. Depending on the detector used, scanning EM (SEM) can create an image using high-energy reflected secondary electrons or low-energy surface-emitted ones, determined by the molecular composition of the scanned cellular components [3]. Additionally, specialized software for array tomography or chip mapping enables automatic surface scanning of tissue on a chip, in a rectangular raster pattern, efficiently mapping large sample surfaces. Moreover, three-dimensional reconstruction of ultrastructural data can be achieved by utilizing more advanced SEM-based techniques, such as serial block-face SEM or focused ion beam SEM [1, 3, 4].

The most common challenge a researcher faces when dealing with human postmortem samples is inadequate tissue preservation, as tissue structures are subjected to natural hypoxia-related degeneration and are sensitive to microbial degradation. The postmortem interval (PMI), defined as the delay between the time of death and the fixation of the brain tissue, varies by up to 100 h across EM studies [5]. This highlights the necessity of comparing tissues with similar PMIs, to avoid possible degradation-related bias [6]. Another factor affecting the brain ultrastructure is the pH of the cerebrospinal fluid (CSF), with neutral values found to be more beneficial and values below 6 more deleterious [7].

Whereas in animal research, methods such as in vivo transcardiac perfusion enable a fast and sufficient fixative penetration in brain tissue [8], preservation of human samples is limited to postmortem immersion in fixatives such as buffered formalin, paraformaldehyde, or glutaraldehyde [9]. We describe here an additional post-fixation step that is crucial to protect the samples from disruption during sample preparation and imaging, greatly improving the quality of the images acquired. Formaldehyde (paraformaldehyde) is a frequently used post-fixative, reflecting its ability to form cross-links between proteins [8–12], but to achieve sufficient preservation of subcellular structures for EM analysis, glutaraldehyde or acrolein should be added to the post-fixation solution to provide stronger and irreversible cross-linking of proteins [10, 11].

Structural artifacts of aldehyde fixation in brain samples subject to hypoxia at the time of death (as determined by low pH values in the CSF) include tissue shrinkage, loss of extracellular space, mitochondrial alterations, and docking of synaptic vesicles in presynaptic axon terminals [11, 13]. These anomalies may affect the ultrastructural analysis and should be taken into consideration when interpreting the results. To reduce this occurrence, fixatives should be prepared in a buffer solution with an osmolarity similar to that of the experimental tissue (to prevent osmotic stress), and the samples kept at low temperature (4 °C) to prevent further degradation [3, 11, 13].

This protocol has been optimized for SEM imaging and subsequent ultrastructural analysis of microglial intracellular components and of their extracellular interactions with parenchymal elements (e.g., blood vessels, dendritic spines, axon terminals and myelin sheaths) [14–16]. As artifacts can occur at any time point between fixation and imaging, it is crucial to follow the protocol carefully for reliable quantitative analysis.

Following post-fixation, brain samples undergo an immunoperoxidase staining protocol based on the interaction of a biotinylated secondary antibody with an immunoperoxidase-diaminobenzidine complex [17, 18]. Here, we visualize microglia using allograft inflammatory factor 1 (AIF1) and their phagolysosomal activity using cluster of differentiation 68 (CD68). AIF1 is associated with actin-bundling activity typical for membrane ruffling and phagocytic activity of microglia [19, 20]. Although AIF1 is expressed by non-parenchymal macrophages and infiltrating myeloid cells [21, 22] and is therefore not specific to brain microglia, it is a preferred microglial target across immunohistochemical methods because it allows exceptional visualization of cellular morphology, from the cell body to the finest processes, at least during homeostatic conditions [23, 24]. However, AIF1 may become non-homogeneously distributed during aging, or with comorbid pathology (e.g., Alzheimer's disease or dementia with Lewy bodies), resulting in dystrophic microglial appearances [25, 26]. Further, AIF1 is downregulated in particular microglial subsets, such as disease-associated microglia and dark microglia [14, 16, 27–29].

A further focus of this protocol is on CD68, a transmembrane glycoprotein primarily expressed on the surface of endosomes, secondary and tertiary lysosomes, and the plasma membranes of monocytic cells [30]. CD68 expression is associated with microglial phagolysosomal activity, often present under inflammation-related conditions (e.g., aging, Alzheimer's disease, multiple sclerosis) [31–34].

Osmium-thiocarbohydrazide-osmium (OTO) is utilized to enhance contrast and facilitate delineation of cellular and subcellular structures during EM imaging, as an alternative to the more classical osmium tetroxide protocol [5, 35]. Osmium is highly lipophilic; thus, its deposits are localized primarily within plasma membranes, nuclear envelopes, mitochondrial cristae, lysosomes, membranes of endoplasmic reticulum, Golgi apparatus, and endosomes, as well as in lipid bodies and vesicles [36, 37]. Lipid fixation makes these structures more electron-dense and resistant to dehydration-induced damage, facilitating their visualization under

EM [38]. The three-step OTO protocol intensifies the signal of tissue-deployed osmium, via a secondary osmium tetroxide staining, using thiocarbohydrazide (TCH), a thiamide acting as a linking bridge [35].

During the embedding step (using Durcupan, a hydrophobic resin), samples undergo a series of graded dehydration steps, followed by incubation in propylene oxide. The removal of water leads to an increased fragility of the tissue, requiring careful handling of the samples, but insufficient water removal prevents complete infiltration of the tissue with resin, which precludes proper ultrathin sectioning, and may cause mechanical distortion of tissue structures inside the SEM chamber [11, 39].

The last section of the protocol describes the parenchymal and microglial ultrastructure and provides an example of a standardized quantitative analysis of human postmortem microglial cells that can provide insight into their structures and functions (e.g., their synaptic contacts, vascular interactions and phagocytosis). This analysis protocol is based on previous EM studies of microglia (for more information, see [14–16, 23, 28, 30, 31, 40–42]).

2 Materials

2.1 Post-Fixation

1. *Sodium phosphate-buffered saline* (PBS), *50 mM*: 5.97 g Na_2HPO_4, 1.2 g NaH_2PO_4, 9 g NaCl, diluted in 900 mL MilliQ H_2O, pH adjusted to 7.4, then diluted to 1 L with MilliQ H_2O.
2. *Phosphate buffer* (PB), *100 mM*: 11.74 g Na_2HPO_4, 2.4 g NaH_2PO_4, diluted in 900 mL MilliQ H_2O, pH adjusted to 7.4, then diluted to 1 L with MilliQ H_2O.
3. Paraformaldehyde (PFA), 2% (g/v): 20 g PFA powder in 1 L of 55 °C PB (prepare in fume hood, just before use; *see* Subheading 3.1).
4. Glutaraldehyde, 2%, EM grade: 40 mL of 50% stock reagent diluted to 1 L in 2% PFA at room temperature (RT), prepared in fume hood, just before use; *see* Subheading 3.1).
5. *Cryoprotectant*: 400 mL PBS, 300 mL glycerol, 300 mL ethylene glycol.
6. Hotplate magnetic stirrer.
7. Aluminum foil.
8. Syringe filter, 25 μm pore-size.
9. Razor blade.
10. Petri dish.
11. *Glass* vials, 50 mL.
12. Magnetic stir bar.

13. Orbital shaker.
14. Culture plate, 24-well.
15. Funnel.
16. Beaker, 250 mL.

2.2 Tissue Preparation for Immunohistochemistry

1. PBS.
2. MilliQ H_2O.
3. *Tris–HCl buffer* (TB), *0.5 M*: 78.8 g Tris-HCl in 1 L MilliQ H_2O, pH 8.0.
4. *Tris-buffered saline* (TBS), *50 mM*: 100 mL TB, 9 g NaCl, to 1 L with MilliQ H_2O, pH 7.4.
5. Hydrogen peroxide (H_2O_2), *0.3%*: from 30% stock solution (v/v), diluted in PBS.
6. Sodium borohydride ($NaBH_4$), 0.1% (g/v): diluted in PBS (*see* **Note 1**).
7. Triton X-100, 10%: 1 mL Triton X-100 to 10 mL with MilliQ H_2O (*see* **Note 2**).
8. Fetal bovine serum (FBS).
9. Bovine serum albumin (BSA).
10. *Blocking buffer for AIF1*: 10% FBS, 3% BSA, 0.01% Triton X-100, diluted in TBS.
11. *Blocking buffer for CD68*: 10% FBS, 3% BSA, 0.01% Triton X-100, diluted in TBS.
12. Primary antibody, AIF1: diluted 1:1000 in AIF1 blocking buffer [Wako, or equivalent].
13. Primary antibody, CD68: diluted 1:50 in CD68 blocking buffer [Dako, or equivalent].
14. Secondary antibody, biotinylated goat anti-rabbit: diluted 1: 300 in TBS [Jackson ImmunoResearch, or equivalent].
15. Secondary antibody, biotinylated goat anti-mouse: diluted 1: 300 in TBS with 0.01% Triton X-100 [Jackson ImmunoResearch, or equivalent].
16. Avidin-biotin-complex solution, diluted 1:100 in TBS (*see* **Note 3**).
17. *Diaminobenzidine* (DAB) *solution, 0.05%*: 18 mL MilliQ H_2O, 10 mg DAB, 2 mL TB, 0.015% H_2O_2 (*see* **Note 4**).
18. Culture plate, 24-well.
19. Transfer pipettes.
20. Orbital shaker.

2.3 Post-Fixation in OTO and Embedding

1. PB.
2. MilliQ H_2O.
3. Potassium ferrocyanide, 3% (g/v), diluted in PB.
4. Osmium tetroxide, 4% (*see* **Note 5**).
5. Thiocarbohydrazide (TCH), 0.1%: 0.1 g TCH in 10 mL MilliQ H_2O (*see* **Note 6**).
6. Osmium tetroxide (aq), 2%: 4% solution diluted 1:1 with MilliQ H_2O.
7. Ethanol solutions: 35% (v/v), 50% (v/v), 70% (v/v), 80% (v/v), 90% (v/v), and 100% (v/v), diluted in MilliQ H_2O.
8. Propylene oxide (*see* **Note 7**).
9. *Resin*: 20 g component A, 20 g component B, 0.6 g component C, 0.4 g component D [Durcupan] (*see* **Note 8**).
10. Syringe filter, 0.22 μm.
11. Stable-temperature incubator with gravity convection, set to 55 °C.
12. Fluoropolymer film sheets.
13. Glass vials, 50 mL.
14. Aluminum foil.
15. Aluminum weighing dishes.
16. Transfer pipettes.
17. Fine paintbrushes.
18. Corn oil.
19. Resin block mold.

2.4 Ultramicrotomy and SEM Imaging

1. Superglue.
2. Razor blades.
3. Resin block.
4. Stereoscope.
5. MilliQ H_2O.
6. Xylene.
7. Ultramicrotome with a diamond knife (*see* **Note 9**).
8. Bibulous paper.
9. Pair of ultrafine tweezers.
10. Serological pipette, 1 mL.
11. Syringe, 20 mL.
12. Syringe filter, 0.22 μm.
13. Specimen mounts.

14. Containers for specimen mounts.
15. Silicon wafers.
16. Conductive carbon adhesive disks.
17. Delicate-task wipers.

2.5 Ultrastructural Analysis of Microglia

1. ImageJ software.
2. Microsoft Excel.

3 Methods

3.1 Post-Fixation

1. Prepare the PFA and glutaraldehyde fixative solutions (*see* **Notes 10–15**).
2. Place a free-floating brain tissue section (i.e., a section contained within cryoprotectant and not fixed to a slide) into a Petri dish filled with cryoprotectant, and excise the regions of interest, using a razor blade (*see* **Notes 16** and **17**).
3. Transfer each section to one well of a 24-well plate (*see* **Note 18**) filled with PB, and place on an orbital shaker.
4. Wash 3 times for 10 min each in PB, to remove excess cryoprotectant.
5. Add 2 mL of the prepared fixative solution to clean, 50 mL glass vials.
6. Transfer the sections from the 24-well plate into the glass vials, using a fine paintbrush dedicated to PFA/glutaraldehyde use.
7. Incubate the sections in the fixative for 2 h at 4 °C.
8. Wash the tissues with PBS, 3 times for 10 min each, in the glass vials (*see* **Note 19**).
9. Transfer the sections to individual wells of a 24-well plate containing PBS, and place on an orbital shaker.

3.2 Tissue Preparation for Immunohistochemistry Staining

1. Remove the PBS and incubate the sections in 0.3% H_2O_2 for 5 min on an orbital shaker (*see* **Notes 20** and **21**).
2. Quickly flush out the H_2O_2 with PBS and wash the sections in PBS 3 times for 10 min each.
3. Remove the PBS from the wells, and incubate the sections in 0.1% $NaBH_4$ for 30 min on an orbital shaker.
4. Quickly flush out the $NaBH_4$ with PBS and wash the sections in PBS 3 times for 10 min each.
5. Remove the PBS and add the designated blocking buffer for CD68 or AIF1 to the wells and place on an orbital shaker for 1 h.
6. Remove the blocking buffer and incubate the sections overnight at 4 °C in the primary antibody solution of choice.

7. Allow the sections to warm to RT for 15 min.
8. Remove the primary antibody solution from the wells and wash the sections 3 times for 10 min with TBS (*see* **Note 22**).
9. Remove the TBS solution from the wells, and add the biotinylated secondary antibody of choice and incubate the sections for 90 min on an orbital shaker.
10. Quickly flush out the secondary antibody solution with TBS and wash the tissues 3 times for 10 min with TBS.
11. Remove the TBS and add the avidin-biotin complex solution to the sections and incubate for 1 h on an orbital shaker.
12. Quickly flush out the avidin-biotin solution with TBS and wash the sections 3 times for 10 min with TBS.
13. Remove the TBS and add the DAB solution in the wells.
14. Develop the staining using the DAB solution, under a fume hood (*see* **Notes 23** and **24**).
15. After optimal development of the DAB signal, immediately transfer the sections into new wells filled with PB, to stop the reaction.
16. Wash the sections with PB 3 times for 10 min.

3.3 Post-Fixation in OTO and Resin Embedding

1. Under a fume hood, mix an equal amount of 3% potassium ferrocyanide in PB with 4% osmium tetroxide (*see* **Note 25**).
2. Remove the PB from the wells, then incubate the sections for 1 h in the above osmium-potassium-ferrocyanide solution.
3. Discard solution, then wash the sections once with PB for 5 min.
4. Wash the sections once with a 50:50 mix of PB and MilliQ H_2O, for 5 min.
5. Wash the sections once in MilliQ H_2O, for 5 min.
6. Remove water, then incubate the sections in filtered 0.1% TCH for 20 min, under a fume hood.
7. Wash the sections with MilliQ H_2O, 3 times for 5 min each.
8. Remove water, then incubate the sections in 2% osmium tetroxide for 30 min, under a fume hood.
9. Repeat **Step** 7.
10. Carefully transfer the sections from the multi-well plate to individual glass vials (*see* **Note 26**).
11. Dehydrate the sections using increasing EtOH concentrations, for 10 min each: 35% twice, 50% once, 70% once, 80% once, 90% once, and then 100% 3 times.
12. Remove EtOH and incubate the sections with propylene oxide, 3 times for 10 min each.

13. Transfer the sections into an aluminum weighing dish containing resin and cover sections with fresh resin (*see* **Note 27**).
14. Keep the sections immersed in the resin overnight at RT, in a fume hood.
15. Cut two identical pieces of fluoropolymer film and label them using a permanent marker, on the outer edges.
16. On the side of the film with no writing, use a fine paintbrush to paint a thin layer of resin, ensuring there are no small bubbles (*see* **Note 28**).
17. Delicately place the sections over the resin layer.
18. Carefully place the second fluoropolymer film (that is *not* painted with resin) on top of the montage, starting from the middle and working outward.
19. Gently push any bubbles away from the tissues, using the blunt edge of a paintbrush or pipette tip (*see* **Note 29**).
20. Place the montage on an aluminum-covered plate, then rest small weights directly on top of the fluoropolymer films to ensure the resin is distributed evenly and polymerizes everywhere.
21. Incubate the final assembly at 55–60 °C for 5 days.

3.4 Ultramicrotomy and SEM Imaging

1. Separate the 2 fluoropolymer films.
2. Under a stereomicroscope, excise the resin-embedded area (s) of interest, using a razor blade (*see* **Note 30**).
3. Place a small drop of superglue on top of a resin block, and gently place the section on top of the glue.
4. Before the glue hardens, use the flat side of the razor blade to press the section onto the glue, to ensure it is properly placed (*see* **Note 31**).
5. Incubate the block overnight at 55–60 °C.
6. Use a razor blade to trim, in small increments, the end of the resin block, such that it takes on a pyramidal shape. Begin trimming from the outside edge of the block and progress toward the center.
7. Using a new razor blade, cut the region of interest into the shape of an isosceles trapezoid, using a new area of the razor blade for each cut (*see* **Note 32**).
8. Fill the boat of the diamond knife with filtered MilliQ H_2O until the reflection of the water is silver.
9. To remove excess material from the region of interest, cut off 300 μm sections with the semi-thin side of the diamond knife, until half of the cut section contains tissue.

10. Add a few drops of MilliQ H_2O to the diamond knife boat, and remove the excess semi-thin sections by passing a clean piece of bibulous paper from one edge of the knife to the other, without touching the blade.
11. Rinse out the remaining MilliQ H_2O and carefully dry the knife with compressed air.
12. Fill the boat again with filtered MilliQ H_2O until the reflection of the water is silver.
13. Cut ribbons of 70–75 μm sections using the ultrathin side of the diamond knife (*see* **Note 33**).
14. Flatten the sections by carefully and gently moving a bibulous paper, pre-dampened with xylene, over the sections, without ever touching the MilliQ H_2O (*see* **Note 34**).
15. Remove one silicon wafer from the packaging, using a new razor blade.
16. Carefully clean the surface of the silicon wafer using a delicate-task wiper, to remove the dust.
17. Place the silicon wafer underneath the water in the diamond knife boat using a fine pair of tweezers that is dedicated to the preparation of SEM samples.
18. Collect the sections by dragging the ribbons onto the top of the wafer while, at the same time, moving across the wafer.
19. Store the wafer in a dust-free environment until the sections are dry.
20. Apply a conductive carbon adhesive disk on top of a specimen mount and place the silicon wafer on top.
21. Image the area of interest at 25 nm, using a field-emission SEM equipped with a detector for backscattered electrons.
22. Scan the 25 nm mosaic to identify parenchymal microglial cells to use for ultrastructural analysis (*see* **Note 35**); disregard cells within the perivascular space if focusing on microglia (*see* Fig. 1).
23. Identify microglia by the patchy heterochromatin and euchromatin patterns of their nuclei, their long and narrow endoplasmic reticula, and the presence of inclusions (lysosomes, lipofuscin granules, and lipid bodies) (*see* **Note 36**).
24. If AIF1 staining was performed, image the microglial cells with an electron-dense peroxidase precipitate in their cytoplasm (*see* **Note 37**) and a patchy heterochromatin pattern (*see* Fig. 2).
25. If CD68 staining was performed, to investigate the phagolysosomal activity of microglia, image the microglial cells that have electron-dense staining, generally found at the plasma membrane, in secondary and tertiary lysosomes, or in phagosomes (*see* Fig. 3).

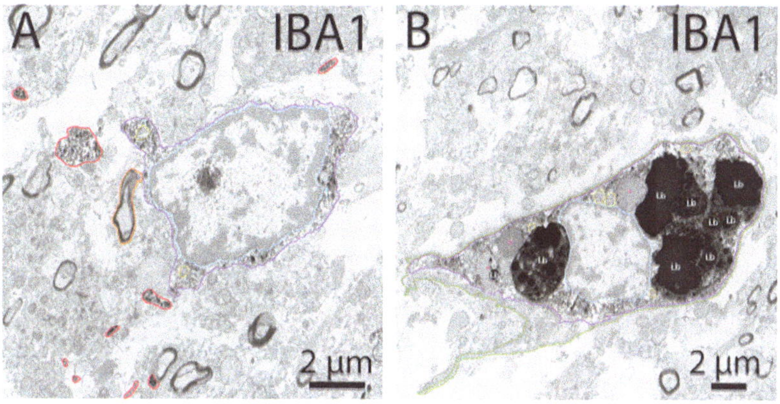

Fig. 1 Representative SEM images, of 5-nm resolution, of (**a**) a parenchymal AIF1 (IBA1)-immunopositive (+) microglial cell, and (**b**) a perivascular AIF1+ myeloid cell; the tissues originated in the head of the hippocampus from an 81-year-old female donor (PMI = 18 h). *Orange pseudo-coloring*, myelinated axon; *yellow outline*, mitochondrion; *blue outline*, nucleus; *purple outline*, cytoplasm; *green outline*, basement membrane; *red outline*, AIF1+ microglial processes; *blue arrow*, phagosome; *pink asterisk*, primary lysosomes; *Lb* lipid body

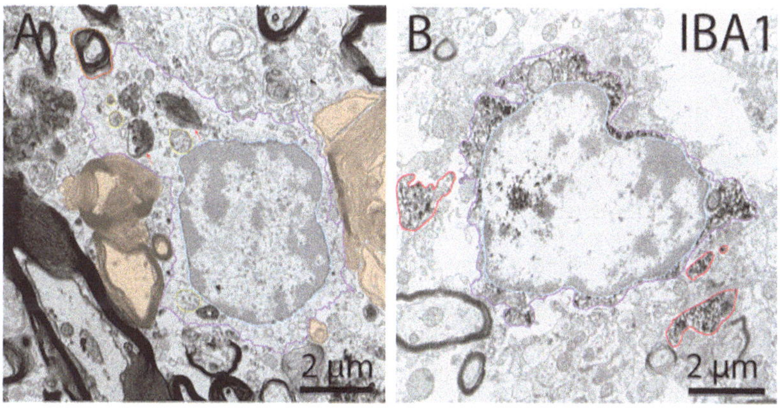

Fig. 2 Representative SEM images, of 5-nm resolution, of (**a**) an unstained microglial cell in the white matter of the forebrain, and (**b**) a stained AIF1 (IBA1)-immunopositive (+) microglial cell in the head of the hippocampus of an 81-year-old female donor (PMI = 18 h). *Orange pseudo-coloring*, myelinated axon; *yellow outline*, mitochondrion; *blue outline*, nucleus; *purple outline*, cytoplasm; *red outline*, AIF1+ microglial process; *red arrows*, digested myelinated axon; *white asterisk*, secondary lysosome

3.5 Ultrastructural Analysis of Microglia

Overview Quantitative ultrastructural analysis of microglia can take into account the prevalence of various subcellular elements (mitochondria, endoplasmic reticula, lipofuscin granules, lysosomes, lipid bodies, phagosomes, and Golgi apparati), as well as parenchymal interactions with neurons, synapses, glial cells, and the vasculature (*see* **Note 38**). Each time a microglial cell body contacts a parenchymal element (e.g., blood vessel, synapse, etc.) or contains a subcellular element of interest, it should be counted and totaled in a separate Excel spreadsheet. The relative frequency (i.e., the number of cells positive for the element analyzed) can be calculated

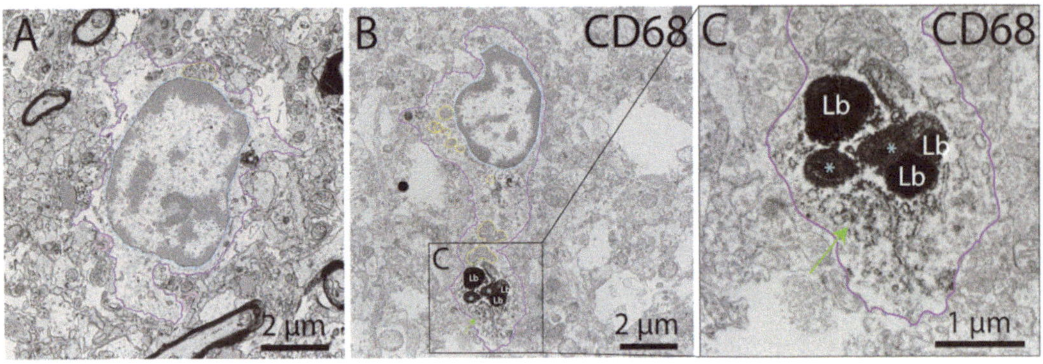

Fig. 3 Representative SEM images (5-nm resolution) from the head of the hippocampus: (**a**) an unstained microglial cell; (**b**) a microglial cell stained with CD68, a marker of phagolysosomal activity; and (**c**) a higher-magnification image of the CD68 staining in the *inset* of (B). *Yellow outline*, mitochondrion; *blue outline*, nucleus; *purple outline*, cytoplasm; *red outline*, axon terminal; *blue asterisk*, tertiary lysosome; *white asterisk*, secondary lysosome; *green arrow*, CD68 staining near tertiary lysosomes; *Lb* lipid body

afterward using Excel. Use software such as *ImageJ* to measure the elements analyzed (e.g., the dilation of the endoplasmic reticulum or the length of the mitochondrion or phagosome) (*see* **Note 39**). Making the images "blind" to the user is a crucial step prior to the ultrastructural analysis, to avoid bias.

1. Copy the SEM images that will be used for the analysis into a folder (*see* **Note 40**).
2. Copy the names of all the images into a separate Excel spreadsheet that will be used as a key for blinding.
3. Generate random, non-repeating numbers (e.g., using the option, *integer-set generation*, on software such as *random.org*). The resulting numbers should be displayed in a single column.
4. Copy the random numbers into the key sheet.
5. Change the names of the images in the folder to the newly generated numbers, and use only these new filenames in the analysis, to prevent introducing bias (*see* **Note 41**).

Microglial subcellular elements

6. Identify mitochondria by their oval shapes, electron-dense double membranes and cristae structures, which are small indentations of the inner membrane. Mitochondria that display a swollen appearance (an enlarged size with small cristae), dilated or degraded cristae (electron-lucent spaces), or a degraded outer membrane, can be characterized as "altered" and placed in a separate category.
7. Identify endoplasmic reticula as an electron-dense, long and narrow tubes with defined membranes, that are often found near the nucleus. Typically, endoplasmic reticulum is quite

narrow; dilation of the organelle (having a length over 100 nm), and the presence of electron-lucent spaces, are signs of cellular stress.

8. Identify Golgi apparati by their narrow tubes, with defined membrane structures that are stacked on top of each other.
9. Identify primary (immature) lysosomes by their circular shapes, and their having defined membranes and homogenous, granular, electron-dense (gray) interiors.
10. Identify secondary and tertiary (mature) lysosomes by their heterogenous appearance, being electron-dense (black areas with gray) with circular shapes that are associated with lipid bodies and vacuoles (for tertiary lysosomes only) (*see* **Note 42**).
11. Identify lipofuscin granules by their circular shapes, electron-dense interiors, and very distinct fingerprint-like patterns decorating the insides of the organelles.
12. Identify lipid bodies by their defined, electron-dense outlines surrounding homogeneous, electron-dense (appearing dark or in varying shades of gray) interiors.
13. Identify fully digested phagosomes by their defined, electron-dense membranes and electron-lucent, circular interiors (*see* **Note 43**).
14. Identify partially or undigested phagosomes by their defined electron-dense membranes containing parenchymal elements, such as axon terminals, myelinated axons, or dendritic spines.
15. Identify autophagosomes by their double membranes, with electron-lucent spaces in-between, and interiors with the same texture and color as the cellular cytoplasm. Autophagosomes can also contain mitochondria or other organelles (*see* **Note 44**).

Interactions with neuronal elements

16. Identify neuronal cell bodies by the granular appearance of their electron-lucent (light gray) nucleoplasms and cytoplasms. These cells are much larger than any glial cells (e.g., astrocytes, oligodendrocytes, microglia) and are circular (*see* **Note 45**).
17. Identify presynaptic elements (axon terminals) by their electron-lucent (light gray) cytoplasms, with the presence of several synaptic vesicles containing neurotransmitters (small circular elements).
18. Identify postsynaptic elements (dendritic spines) by their electron-lucent cytoplasms and the presence of postsynaptic densities, and electron-dense areas (dark gray) where neurotransmitter receptors are located (*see* **Notes 46** and **47**).

19. Identify myelinated axons by their electron-lucent cytoplasms, surrounded by electron-dense sheaths with either circular (cut coronally) or elongated (cut longitudinally) structures (*see* **Note 48**).

Interactions with other glial cells

20. Identify astrocytic cell bodies by their electron-lucent cytoplasms and nucleoplasms, the presence of intermediary filaments (long electron-dense lines clustered together), angular protrusions from their plasma membranes, and the presence of several inclusions (e.g., lysosomes and lipid bodies).
21. Identify oligodendrocytic cell bodies by their electron-dense cytoplasms, patchy heterochromatin patterns, rectangular shapes, and general distribution of the organelles toward one side of the cytoplasm.

Interactions with the vasculature

22. Identify microglial contacts with brain capillaries when you observe a cell body directly touching the basement membrane (an electron-dense outline surrounding the endothelial cells and pericytes).

Interactions with extracellular elements

23. Identify pockets of extracellular space by the long, narrow, electron-lucent regions located between the plasma membranes of the microglia and the adjacent parenchymal elements (*see* **Note 49**).
24. Identify extracellular digestion pockets by the presence of circular or oval electron-lucent areas where partially digested parenchymal elements can be identified (*see* **Note 50**).

4 Notes

1. Sodium borohydride should be prepared 5–20 min before use. As NaBH4 granules are highly reactive with water, the solution should be prepared under a fume hood.
2. Triton X-100 is a highly viscous liquid. It is more practical to prepare a larger quantity, diluted to 10% in MilliQ H_2O, rather than use small quantities of the pure solution.
3. The avidin-biotin solution must be prepared before just before use, according to the manufacturer's guidelines.
4. This protocol uses the tablet form of DAB, but liquid kits can also be purchased and used for this protocol. DAB is a suspected carcinogen and should be used under a fume hood. Large quantities of DAB (e.g., 20 mL) can be prepared and

frozen for future experiments (before the addition of H_2O_2). The H_2O_2 is added to activate the DAB reaction just before use, and should be added again, if needed, after 10 min.

5. Osmium tetroxide and potassium ferrocyanide are toxic compounds that should be used only under a fume hood. Once the osmium ampule is opened, transfer the solution to a glass vial whose surfaces are covered entirely by aluminum foil. Because of the highly photosensitive nature of osmium, the well plate should also be covered by aluminum foil. To neutralize osmium tetroxide, add an equal amount of corn oil and dispose of the mixture according to manufacturer and/or institutional guidelines. If you have leftover osmium, the glass vial can be kept at 4 °C for future experiments. Make sure that the color of the osmium (transparent liquid) has not changed. If so, add the corn oil and dispose of the solution.
6. Thiocarbohydrazide is a highly toxic compound and should be weighed and prepared under a fume hood. The TCH granules require heat to dissolve in MilliQ H_2O, with stirring for a minimum of 1 h, but do not increase the temperature above 60 °C, as this will generate unwanted secondary compounds. Because some granules will not be dissolved, the solution must be filtered before use. TCH is photosensitive and should be kept away from light by wrapping the solution container and the multi-well plate in aluminum foil.
7. Propylene oxide is highly corrosive and should be used under a fume hood. The user should wear appropriate protective gear (thick pairs of nitrile gloves). The consumables used for this step should all be made of glass, as propylene oxide dissolves plastics.
8. Resin is highly toxic and should be prepared and used only under a fume hood. It is best to prepare the resin 30 min before its use. It is important to add the components in alphabetical order (A, B, C, D), to ensure proper polymerization of the resin. Once weighed and placed in a plastic holder, make sure to properly mix the different components with a plastic transfer pipette. To neutralize the resin (by polymerization), keep it in an incubator at 55 °C for a minimum of 3 days. Leftover resin, or resin that was used to embed the tissues, can be poured into molds to create the resin blocs used for the ultramicrotomy, and incubated for 3 days at 55 °C.
9. One section of the diamond knife edge, or one diamond knife, should be dedicated to semi-thin sectioning, as this section is most likely to be damaged in the process of cutting (e.g., from the glue, wrong adjustments of the knife, and thicker sections). Another knife, or the other section of the knife edge, should be dedicated to the ultra-sectioning itself.

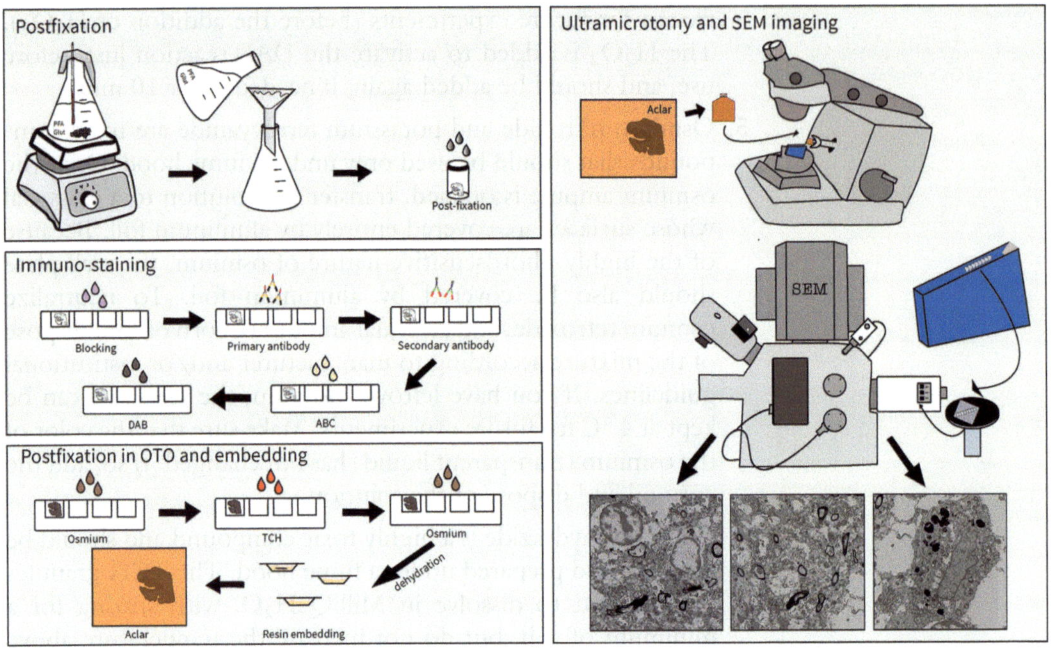

Fig. 4 Workflow for immunostaining and processing human postmortem brain samples for SEM. The process can be separated into four parts. *Upper left* panel shows how to prepare and apply fixative to the postmortem samples, which should be done entirely within a fume hood. *Middle left* panel summarizes the steps required to immunostain the tissue. *ABC* avidin-biotin-peroxidase complex, *DAB* diaminobenzidine. *Lower left* panel, the resulting samples are then processed with the osmium-thiocarbohydrazide-osmium (*OTO*) method and embedded in Durcupan resin. Finally (*right panel*), the region of interest is excised from the resin, and ultramicrotomy is used to generate ultrathin sections. These sections are then placed on a silicon chip and imaged by chip-mapping inside a field-emission SEM equipped with a backscattered-electron detector

10. The required amount of PFA and glutaraldehyde will vary based on the experimental design (i.e., the number of samples used for the experiments). However, as the samples will be transferred to 50 mL glass vials for the post-fixation, they will require approximately 2 mL of post-fixative per vial. An additional 10 mL should be prepared, in case of spillage. To prepare PFA, warm up fresh PB to 55 °C in a glass beaker covered with aluminum foil, using a hotplate stirrer and a magnetic stir bar under a fume hood (*see* Fig. 4 for a schematized workflow of the protocol). Weigh the required amount of granular PFA for a 2% solution, under the fume hood. Add the PFA to the 55 °C PB and stir until it is dissolved. Remove from heat and allow the solution to come to RT. Filter the solution, using a funnel and 25 μm filter paper, into a 500 mL bottle with a tight-fitting lid. Open a new glutaraldehyde ampule under a fume hood. Transfer the glutaraldehyde to a glass vial whose outer surfaces are covered entirely with aluminum foil. Slowly add the necessary volume of glutaraldehyde to the 2% PFA solution to obtain a 2% glutaraldehyde solution, while gently tilting the bottle after each mL added.

11. PFA, and the fumes resulting from heating it, are highly toxic. Always use PFA-dedicated tools (heating plate, thermometer, magnetic stir-bar, and scale). Additionally, the user should have a formalin neutralizer on hand to clean the PFA waste from the glassware.
12. The PFA solution should never go above 60–65 °C, as high temperatures can generate secondary aldehyde compounds that compromise the quality of the solution.
13. Small amounts of indissoluble PFA granules will remain regardless of the method used to dissolve the PFA.
14. It is important that the glutaraldehyde used is certified EM grade, so it will be free of polymers that could affect the quality and strength of the solution (polymerization reduces the cross-linking ability of glutaraldehyde).
15. A high concentration of glutaraldehyde can sometimes prevent or reduce immunogenicity. An alternative is to use 3.5% acrolein to postfix the samples. To prepare 1 L of 3.5% acrolein, slowly add 38.8 mL of EM-grade acrolein (from a 90% stock solution) to freshly made PB. Slowly tilt the bottle to mix the acrolein with the PB and filter the solution prior to its use. Acrolein is highly toxic and should be prepared under a fume hood.
16. The OTO post-fixation method darkens the samples, preventing anatomical differentiation of neuronal cell bodies and other landmarks, except for blood vessels. In the case of free-floating sections containing several regions, it is important to excise the area of interest before beginning. If there are multiple regions of interest in the same sample, they should be cut apart, then placed in distinct wells in a multi-well plate, and properly labeled.
17. The AIF1, CD68, OTO and resin-embedding protocols are optimized for tissues that are approximately 50 μm thick. Thicker samples will require periods of incubation that are proportionately increased (e.g., double the time if the samples are 100 μm thick).
18. Depending on the size or number of the regions of interest, the protocol can be performed in plates with a different number of wells (12 or 24). The volume of reagents required for a 24-well plate varies between 350–500 μL, and between 600–1000 μL for a 12-well plate. An online brain atlas can help the user find the correct region for the anatomical plane in which the tissues were cut (e.g., the *Allen Brain Atlas*).
19. After use, the aldehyde solutions, the PBS used for the washes after fixation, and the tools used for solution preparation, should be neutralized and disposed of according to the manufacturer's and/or institutional guidelines.

20. Bubbles will appear in the wells after adding the H_2O_2 solution. Make sure that they are completely removed from the tissues once they are washed with PBS.
21. Each condition or sample require their own well. It is preferable to separate each region of the same sample prior to the OTO post-fixation as the anatomical hallmarks will not be visible after these steps. If using a 24-well plate, each well should have a minimum of 350 μL of the prepared solutions.
22. The primary antibody solution can be stored in the −20 °C freezer to be reused for future experiments.
23. A dedicated paintbrush, stored in a fume hood, should be used for the DAB step.
24. As soon as the sections begin to turn beige, check the staining under a light microscope by immersing a tissue section in a drop of PB, on a slide. If the DAB reaction requires more time, the sections can be placed again in the same DAB well. The longer the DAB development, the higher the likelihood of having nonspecific staining. The DAB development for AIF1 and CD68 should be between 1 and 4 min. It is highly recommended to have a test section to optimize the timing.
25. The sections must be completely flat, prior to the addition of the osmium-potassium-ferrocyanide solution, as the sections will harden and retain their shape.
26. The sections become more fragile as they become more dehydrated. Therefore, placing the sections in the glass vial prior to dehydration will reduce the likelihood of breaking them when manipulating the tissues.
27. Make sure the tissues are completely covered by the resin, so they do not dry out overnight.
28. Resin removes the ink of the permanent marker. Make sure that the side with the marking does not come in contact with the resin.
29. While removing the bubbles, do not to press on the tissue sections, as they are fragile and will break with pressure.
30. The final trapezoid-shaped tissue used for semi- and ultrathin sectioning must be less than half of the length of the knife edge. This is also true for the area excised from the fluoropolymer film. If the tissue is too small (e.g., one-fourth of the knife edge), it will be more difficult to cut on the ultramicrotome.
31. It is possible that, while trimming or sectioning, the tissue will become detached from the resin block. It is crucial to ensure the tissue is stable and secured on the block prior to sectioning. If the tissue detaches during trimming, one can reglue it to a new block, then keep it in the incubator overnight.

32. Make certain that the top and bottom of the trapezoid are parallel, to obtain ribbons of ultrathin sections.
33. Be sure to adjust the thickness of the ultrathin cuts such that you obtain silver-colored samples. If the ultrathin sections are too thick, they will have a golden appearance. If they are too thin, they will have a dull gray color.
34. If the water in the diamond-knife boat comes into contact with xylene, the ultrathin sections will be contaminated. These contaminants can vary in size but are recognizable under SEM by their pitch-black appearances. It is best not to collect samples that touched the contaminated water, and to soak the boat thoroughly with MilliQ H_2O, overnight, before cutting again. Xylene is a corrosive, flammable and highly toxic compound. All consumables used to handle xylene should be made of glass and the user should wear proper protective gear in a ventilated area.
35. While imaging, be careful not to image and analyze cells that are ultrastructurally similar to parenchymal microglia but found within the perivascular space. Perivascular cells can be differentiated from parenchymal microglia by the presence of a basement membrane (thin and gray narrow outline) surrounding the perivascular cell. Additionally, although the heterochromatin and euchromatin patterns in the nucleus of the perivascular cell look similar to parenchymal microglia, they generally do not look as ruffled (parenchymal microglia will have small protrusions coming out of their cell bodies).
36. Parenchymal microglial cells possess heterochromatin and euchromatin patterns similar to those of oligodendrocytes. The user should make certain not to image or analyze oligodendrocytes. The latter can be differentiated from microglia by their larger sizes, the locations of their organelles (which are often clustered in one area of the cytoplasm), and their overall shapes (rectangular, as compared to the more elongated or rounded shapes of microglia).
37. Depending on the goal of the experiment, it is possible to stain the samples with the myeloid cell marker, AIF1, to distinguish parenchymal microglia from other brain cell types (e.g., neurons, cells of the oligodendrocytic lineage, and astrocytes) or perivascular cells. AIF1 is also useful for imaging microglial processes. In addition, this extra step enables faster localization of the general microglial cell population, since the contrast between the parenchymal elements and the electron-dense cytoplasm is easily identifiable in the 25 nm images.
38. Exclude all cells from the analysis that are partially cut from the frame, that are unfocused, or that have contaminants interfering with identification of their subcellular contents and

intercellular relationships with neighboring elements. The statistical power required for each experiment must be determined prior to imaging.

39. Set the scale based on the magnification and electron microscope used. This option can be found under the "analyze" tab. The element can then be measured using the "straight line" tool found in the main menu, followed by "analyze- > measure" (or type the letter "m" as a shortcut). The results appear in units of length.
40. A copy of all the images used for the analysis should be kept in a separate folder.
41. Once the analysis is completed, the images can be unblinded (i.e., adding the original name to the analysis spreadsheet, allowing the user to perform statistical analysis). Software such as *Prism* can be used for statistics.
42. Proper contrast and focus are crucial for the positive identification of lysosomes, because lipofuscin granules often have similar electron-dense interiors. If the contrast does not allow the user to confirm (or not) the presence of the fingerprint pattern that is typical for lipofuscin granules, the organelles might be misidentified.
43. Fully digested phagosomes should be completely round and have diameters of at least 100 nm to be included in this category.
44. Autophagosomes are often mistaken for altered mitochondria whose content is completely degraded. Make sure that the space between the membranes of the autophagosome is electron-lucent and not dense (mitochondria have an electron-dense space between their membranes).
45. Neuronal cell bodies are often misidentified as astrocytes, due to their similar electron-lucent appearances. However, astrocytes are not as circular as neurons, they are generally smaller, they contain intermediate filaments, and they possess angular protuberances.
46. Dendritic spines are ultrastructurally similar to electron-lucent astrocytic processes. Because astrocytic processes possess angular protuberances, they can be easily misidentified as dendritic spines. For this reason, only count the dendritic spines in which you are able to see the postsynaptic densities, along with direct contacts with axon terminals. Of note, it is possible to stain dendritic spines to positively identify them, which prevents under-counting these elements.
47. If microglia touch both the axon terminal and the dendritic spine at the same time, in conjunction with touching the synaptic cleft, it is possible to count them in a separate category. Alternatively, one can measure the proportion of the plasma membrane that comes into contact with each synaptic element.

48. A myelinated axon can have one or more sheaths tightly packed together. Degenerated myelin can sometimes appear similar to digested elements (where sheaths are separated and electron-lucent spaces are found between layers). However, degenerated myelin should be categorized separately from extracellular digested elements.
49. Extracellular space can often be mistaken for astrocytic processes. Astrocytic processes are also electron-lucent, but unlike the extracellular space, they have plasma membranes clearly outlining the processes. Moreover, they possess angular protrusions.
50. These elements are often difficult to identify positively. If the goal of the experiment is to determine the natures of these elements, immunostaining for the element of interest (e.g., presynaptic markers) could be performed.

Acknowledgments

We acknowledge that the University of Victoria is located on the territory of the Lekwungen peoples and that the Songhees, Esquimalt, and WSÁNEĆ peoples have relationships with this land. We would like to thank the brain bank at the CERVO Brain Research Center, Dr. Martin Parent and Ms. Marie-Josée Wallman for providing the human postmortem brain samples, and Fernando González Ibáñez for imaging the postmortem sections on the SEM. The brain bank at the CERVO Brain Research Center and handling of human postmortem tissues were approved by the Ethics Committee of the Institut Universitaire en Santé Mentale de Québec and Université Laval. In addition, written consent was obtained for the use of the human postmortem tissues and all experiments were performed in line with the Code of Ethics of the World Medical Association. M.K.S.P. is supported by a doctoral training award from the Canadian Institutes of Health Research (CIHR) and a Fond de Recherche du Québec-Santé (FRQS) doctoral award. E.S. is a recipient of a Faculty of Graduate Studies (FGS) award from the University of Victoria. M.C. is supported by a doctoral training award from FRQS. M.E.T. holds a Canada Research Chair (Tier 2) in *Neurobiology of Aging and Cognition*.

References

1. Carrier M, Robert M-È, González Ibáñez F et al (2020) Imaging the neuroimmune dynamics across space and time. Front Neurosci 14:903. https://doi.org/10.3389/fnins.2020.00903
2. Gosselin D, Skola D, Coufal NG et al (2017) An environment-dependent transcriptional network specifies human microglia identity. Science 356. https://doi.org/10.1126/science.aal3222

3. Savage JC, Picard K, González-Ibáñez F, Tremblay M-È (2018) A brief history of microglial ultrastructure: distinctive features, phenotypes, and functions discovered over the past 60 years by electron microscopy. Front Immunol 9:803. https://doi.org/10.3389/fimmu.2018.00803
4. Kay KR, Smith C, Wright AK et al (2013) Studying synapses in human brain with array tomography and electron microscopy. Nat Protoc 8:1366–1380. https://doi.org/10.1038/nprot.2013.078
5. Sele M, Wernitznig S, Lipovšek S et al (2019) Optimization of ultrastructural preservation of human brain for transmission electron microscopy after long post-mortem intervals. Acta Neuropathol Commun 7:144. https://doi.org/10.1186/s40478-019-0794-3
6. Krause M, Brüne M, Theiss C (2016) Preparation of human formalin-fixed brain slices for electron microscopic investigations. Ann Anat 206:27–33. https://doi.org/10.1016/j.aanat.2016.04.030
7. Glausier JR, Konanur A, Lewis DA (2019) Factors affecting ultrastructural quality in the prefrontal cortex of the postmortem human brain. J Histochem Cytochem 67:185–202. https://doi.org/10.1369/0022155418819481
8. Gage GJ, Kipke DR, Shain W (2012) Whole animal perfusion fixation for rodents. J Vis Exp. https://doi.org/10.3791/3564
9. McFadden WC, Walsh H, Richter F et al (2019) Perfusion fixation in brain banking: a systematic review. Acta Neuropathol Commun 7:146. https://doi.org/10.1186/s40478-019-0799-y
10. Fox CH, Johnson FB, Whiting J, Roller PP (1985) Formaldehyde fixation. J Histochem Cytochem 33:845–853. https://doi.org/10.1177/33.8.3894502
11. Kashi AM, Tahemanesh K, Chaichian S et al (2014) How to prepare biological samples and live tissues for scanning electron microscopy (SEM). Galen Med J 3:63–80
12. Tayri-Wilk T, Slavin M, Zamel J et al (2020) Mass spectrometry reveals the chemistry of formaldehyde cross-linking in structured proteins. Nat Commun 11:3128. https://doi.org/10.1038/s41467-020-16935-w
13. Korogod N, Petersen CCH, Knott GW (2015) Ultrastructural analysis of adult mouse neocortex comparing aldehyde perfusion with cryo fixation. elife 4. https://doi.org/10.7554/eLife.05793
14. St-Pierre M-K, Bordeleau M, Tremblay M-È (2019) Visualizing Dark Microglia. Methods Mol Biol 2034:97–110. https://doi.org/10.1007/978-1-4939-9658-2_8
15. Savage JC, St-Pierre M-K, Hui CW, Tremblay M-E (2019) Microglial ultrastructure in the hippocampus of a lipopolysaccharide-induced sickness mouse model. Front Neurosci 13:1340. https://doi.org/10.3389/fnins.2019.01340
16. El Hajj H, Savage JC, Bisht K et al (2019) Ultrastructural evidence of microglial heterogeneity in Alzheimer's disease amyloid pathology. J Neuroinflammation 16:87. https://doi.org/10.1186/s12974-019-1473-9
17. Hsu SM, Raine L, Fanger H (1981) Use of avidin-biotin-peroxidase complex (ABC) in immunoperoxidase techniques: a comparison between ABC and unlabeled antibody (PAP) procedures. J Histochem Cytochem 29:577–580. https://doi.org/10.1177/29.4.6166661
18. Goldberg MW, Fišerová J (2016) Immunogold labeling for scanning electron microscopy. Methods Mol Biol 1474:309–325. https://doi.org/10.1007/978-1-4939-6352-2_20
19. Ohsawa K, Imai Y, Kanazawa H et al (2000) Involvement of Iba1 in membrane ruffling and phagocytosis of macrophages/microglia. J Cell Sci 113(Pt 17):3073–3084
20. Ohsawa K, Imai Y, Sasaki Y, Kohsaka S (2004) Microglia/macrophage-specific protein Iba1 binds to fimbrin and enhances its actin-bundling activity. J Neurochem 88:844–856. https://doi.org/10.1046/j.1471-4159.2003.02213.x
21. Herz J, Filiano AJ, Smith A et al (2017) Myeloid cells in the central nervous system. Immunity 46:943–956. https://doi.org/10.1016/j.immuni.2017.06.007
22. Wohleb ES, Fenn AM, Pacenta AM et al (2012) Peripheral innate immune challenge exaggerated microglia activation, increased the number of inflammatory CNS macrophages, and prolonged social withdrawal in socially defeated mice. Psychoneuroendocrinology 37:1491–1505. https://doi.org/10.1016/j.psyneuen.2012.02.003
23. Tremblay M-È, Lowery RL, Majewska AK (2010) Microglial interactions with synapses are modulated by visual experience. PLoS Biol 8:e1000527. https://doi.org/10.1371/journal.pbio.1000527
24. González Ibanez F, Picard K, Bordeleau M et al (2019) Immunofluorescence staining using IBA1 and TMEM119 for microglial density, morphology and peripheral myeloid cell infiltration analysis in mouse brain. J Vis Exp. https://doi.org/10.3791/60510

25. Shahidehpour RK, Higdon RE, Crawford NG et al (2021) Dystrophic microglia are associated with neurodegenerative disease and not healthy aging in the human brain. Neurobiol Aging 99:19–27. https://doi.org/10.1016/j.neurobiolaging.2020.12.003
26. Tischer J, Krueger M, Mueller W et al (2016) Inhomogeneous distribution of Iba-1 characterizes microglial pathology in Alzheimer's disease. Glia 64:1562–1572. https://doi.org/10.1002/glia.23024
27. Keren-Shaul H, Spinrad A, Weiner A et al (2017) A unique microglia type associated with restricting development of Alzheimer's disease. Cell 169:1276–1290.e17. https://doi.org/10.1016/j.cell.2017.05.018
28. Bisht K, Sharma KP, Lecours C et al (2016) Dark microglia: a new phenotype predominantly associated with pathological states. Glia 64:826–839. https://doi.org/10.1002/glia.22966
29. St-Pierre M-K, Šimončičová E, Bögi E, Tremblay M-È (2020) Shedding light on the dark side of the microglia. ASN Neuro 12. https://doi.org/10.1177/1759091420925335
30. Lecours C, Bordeleau M, Cantin L et al (2018) Microglial implication in Parkinson's disease: loss of beneficial physiological roles or gain of inflammatory functions? Front Cell Neurosci 12. https://doi.org/10.3389/fncel.2018.00282
31. Savage JC, St-Pierre M-K, Carrier M et al (2020) Microglial physiological properties and interactions with synapses are altered at presymptomatic stages in a mouse model of Huntington's disease pathology. J Neuroinflammation 17. https://doi.org/10.1186/s12974-020-01782-9
32. Nelson LH, Warden S, Lenz KM (2017) Sex differences in microglial phagocytosis in the neonatal hippocampus. Brain Behav Immun 64:11–22. https://doi.org/10.1016/j.bbi.2017.03.010
33. Schafer DP, Lehrman EK, Kautzman AG et al (2012) Microglia sculpt postnatal neural circuits in an activity and complement-dependent manner. Neuron 74:691–705. https://doi.org/10.1016/j.neuron.2012.03.026
34. Acharjee S, Verbeek M, Gomez CD et al (2018) Reduced microglial activity and enhanced glutamate transmission in the basolateral amygdala in early CNS autoimmunity. J Neurosci 38:9019–9033. https://doi.org/10.1523/JNEUROSCI.0398-18.2018
35. Seligman AM, Wasserkrug HL, Hanker JS (1966) A new staining method (OTO) for enhancing contrast of lipid--containing membranes and droplets in osmium tetroxide--fixed tissue with osmiophilic thiocarbohydrazide (TCH). J Cell Biol 30:424–432. https://doi.org/10.1083/jcb.30.2.424
36. Palade GE (1952) A study of fixation for electron microscopy. J Exp Med 95:285–298. https://doi.org/10.1084/jem.95.3.285
37. Bahr GF (1954) Osmium tetroxide and ruthenium tetroxide and their reactions with biologically important substances. Electron stains. III. Exp Cell Res 7:457–479. https://doi.org/10.1016/s0014-4827(54)80091-7
38. Weibull C, Christiansson A, Carlemalm E (1983) Extraction of membrane lipids during fixation, dehydration and embedding of Acholeplasma laidlawii-cells for electron microscopy. J Microsc 129:201–207. https://doi.org/10.1111/j.1365-2818.1983.tb04174.x
39. Bahr GF, Bloom G, Friberg U (1957) Volume changes of tissues in physiological fluids during fixation in osmium tetroxide or formaldehyde and during subsequent treatment. Exp Cell Res 12:342–355. https://doi.org/10.1016/0014-4827(57)90148-9
40. Hui CW, St-Pierre M-K, Detuncq J et al (2018) Nonfunctional mutant Wrn protein leads to neurological deficits, neuronal stress, microglial alteration, and immune imbalance in a mouse model of Werner syndrome. Brain Behav Immun 73:450–469. https://doi.org/10.1016/j.bbi.2018.06.007
41. Nahirney PC, Tremblay M-E (2021) Brain ultrastructure: putting the pieces together. Front Cell Dev Biol 9. https://doi.org/10.3389/fcell.2021.629503
42. Tremblay M-È, Majewska AK (2019) Ultrastructural analyses of microglial interactions with synapses. Methods Mol Biol 2034:83–95. https://doi.org/10.1007/978-1-4939-9658-2_7

Chapter 5

Three-Dimensional Imaging of Fibrinogen and Neurovascular Alterations in Alzheimer's Disease

Mario Merlini, Elif G. Sozmen, Keshav S. Subramanian, Alissa L. Nana, William W. Seeley, and Katerina Akassoglou

Abstract

Cerebrovascular dysfunction is a hallmark of Alzheimer's disease (AD) that is linked to cognitive decline. However, blood-brain barrier (BBB) disruption in AD is focal and requires sensitive methods to detect extravasated blood proteins and vasculature in large brain volumes. Fibrinogen, a blood coagulation factor, is deposited in AD brains at sites of BBB disruption and cerebrovascular damage. This chapter presents the methodology of fibrinogen immunolabeling-enabled three-dimensional (3D) imaging of solvent-cleared organs (iDISCO) which, when combined with immunolabeling of amyloid β (Aβ) and vasculature, enables sensitive detection of focal BBB vascular abnormalities, and reveals the spatial distribution of Aβ plaques and fibrin deposits, in large tissue volumes from cleared human brains. Overall, fibrinogen iDISCO enables the investigation of neurovascular and neuroimmune mechanisms driving neurodegeneration in disease.

Key words iDISCO, Fibrinogen, Blood-brain Barrier, Biomarker, Alzheimer's disease (AD), Tissue Clearing, Microscopy, Neurodegeneration, Neurovascular, Neuroinflammation, 3D imaging

1 Introduction

Cerebrovascular damage, microbleeds, blood-brain barrier (BBB) disruption, and fibrinogen deposition are features of Alzheimer's disease (AD) pathology [1–6]. Given the contribution of cerebrovascular dysfunction to cognitive decline in AD [4–8], sensitive methods to detect cerebrovascular damage in large tissue volumes are critical the understanding the onset and progression of neurodegeneration, as well as the effects of therapies on the BBB. Recent advances in tissue-clearing methods, combined with immunolabeling, have enabled three-dimensional (3D) imaging of vasculature by fluorescence microscopy using iDISCO [9–11]. However,

Supplementary Information The online version contains supplementary material available at [https://doi.org/10.1007/978-1-0716-2655-9_5].

Jerold Chun (ed.), *Alzheimer's Disease: Methods and Protocols*,
Methods in Molecular Biology, vol. 2561, https://doi.org/10.1007/978-1-0716-2655-9_5,

iDISCO imaging of brain vasculature alone does not capture plasma protein extravasation into the CNS. Here, we describe fibrinogen iDISCO as a highly sensitive imaging method for BBB disruption in human brains.

Fibrinogen is deposited in the AD brain at sites of neurovascular abnormalities and neuroinflammation [7, 12, 13]. It increases in the brain and the cerebrospinal fluid (CSF) of AD patients who are carriers for *ApoE4*, the major genetic risk factor for AD [5, 14]. Fibrinogen activates microglia and is deposited at sites of microglial activation in human AD brains [7, 8, 12, 13, 15]; this activation promotes oxidative stress and spine elimination, which cause cognitive impairment in mouse models of AD [7, 16]. Fibrinogen is unique among other blood proteins in that it clots and forms insoluble fibrin deposits in the CNS that activate glia and immune cells [3, 8, 15]. It binds amyloid β (Aβ), leading to inhibition of fibrin degradation and, therefore, sustained fibrin deposits [17].

For fibrinogen iDISCO, the areas of interest in the human brain must be dissected and fixed in formalin. The tissue is then sectioned, dehydrated, and immunolabeled for fibrinogen, Aβ, and a marker of the vasculature, then cleared and imaged. Fibrinogen iDISCO reveals distinct patterns of BBB abnormalities in AD brains, including focal fibrin deposits around tortuous vascular structures and deposits in the brain parenchyma, together with or independent of Aβ deposition [3, 7]. Thus, fibrinogen iDISCO may be used for the detection of BBB abnormalities during aging, or in AD and other neurological diseases, including multiple sclerosis, traumatic brain injury, stroke, epilepsy, and psychiatric disorders.

We describe a protocol for triple-immunolabeling of fibrinogen, Aβ, and vasculature in human brain tissue, using iDISCO. The iDISCO protocol for human brains described herein was developed based on the iDISCO protocol for mouse tissue [10] described in http://idisco.info.

2 Materials

2.1 Human Tissue Fixation and Processing

1. Neutral-buffered formalin (NBF), 10%, 50 mL per tissue block.
2. *10× PBS stock* (~ approximately 400 mL of 1× PBS required per assay): 1.37 M NaCl, 27 mM KCl, 80 mM Na_2HPO_4, and 20 mM KH_2PO_4, in double-distilled water (ddH_2O), pH 7.4.
3. Sodium azide, 0.02% in 1× PBS (200 mL per tissue block).
4. Forceps, flat-tipped, non-serrated.
5. Specimen jars, 50–100 mL, flat-bottomed.

2.2 Tissue Sectioning and Dehydration

1. Eppendorf tubes, 5 mL.
2. Surgical scalpel, #11, disposable.
3. Nutator mixer.
4. Methanol (MeOH), 100% (20 mL per sample).
5. MeOH dilution series in ddH_2O (5 mL each): 20%, 40%, 60%, and 80%, kept at room temperature (RT).
6. Syringes, disposable, 10–20 cc.
7. Disposable needles, 18–20 gauge.
8. Dichloromethane (DCM)/methanol (MeOH) solution, 66%/33% (v/v) (5 mL per sample); prepare just before use, keep at RT.
9. Hydrogen peroxide (H_2O_2), 5% in methanol (5 mL per sample); prepare just before use, keep at 4 °C.

2.3 Immunolabeling

1. Incubator, 37 °C, into which the nutator mixer can be placed.
2. Eppendorf tubes, 5 mL.
3. MeOH dilution series in ddH_2O (5 mL each): 20%, 40%, 60%, and 80%, kept at room temperature (RT).
4. 1× PBS, pH 7.4.
5. Triton X-100.
6. Tween-20.
7. Glycine.
8. Dimethyl sulfoxide (DMSO).
9. Heparin, 10 mg/mL in ddH_2O (10 mL of stock solution); can be stored for 6 mos. at 4 °C.
10. Normal donkey serum.
11. *PTx.2 (PBS, Triton X-100, 2 mL) Washing Solution* (1 L): 10× PBS, 100 mL, Triton X-100, 2 mL, to 1 L with ddH_2O. Can be stored for 6 mos. at RT.
12. *Permeabilization solution* (500 mL): PTx.2, 400 mL, glycine, 11.5 g, DMSO, 100 mL, prepared fresh for each experiment and kept at RT.
13. *Blocking solution* (50 mL for 5 mL per sample): PTx.2, 42 mL, normal donkey serum, 3 mL, DMSO, 5 mL. Can be prepared 7 d ahead of each experiment; keep at 4 °C.
14. *PTwH (PBS, Tween with heparin)*: 10× PBS, 100 mL, Tween-20, 2 mL, heparin, 1 mL of a 10 mg/mL solution, to 1 L with ddH_2O. Can be stored for 6 mos. at RT.
15. *Primary antibody solution* (5 mL per sample): DMSO, 5%, donkey serum, 3% in PTwH. Can be prepared 7 d ahead of each experiment and kept at 4 °C.

16. *Secondary Antibody Solution* (5 mL per sample): donkey serum, 3% in PTwH. Can be prepared 7 d ahead of each experiment and kept at 4 °C.
17. Guinea pig anti-human CD31/PECAM-1 antibody (Synaptic Systems), diluted 1:100 to add 20 μL of antibody per sample. The antibody must be from the indicated supplier (*see* **Note 5**).
18. Rabbit anti-human β amyloid antibody (IBL-America), (1 μg/ mL final concentration) (20 μL per sample). The antibody must be from the indicated supplier (*see* **Note 5**).
19. Sheep anti-human fibrinogen (US Biological), 30 μg/mL final concentration (20 μL per sample). The antibody must be from the indicated supplier (*see* **Note 5**).
20. Alexa 488-conjugated, donkey anti-guinea pig secondary antibody (Jackson ImmunoResearch), 10 μg/mL final concentration (40 μL per sample). The antibody must be from the indicated supplier (*see* **Note 6**).
21. Alexa 647-conjugated, donkey anti-rabbit secondary antibody (Jackson ImmunoResearch), 10 μg/mL final concentration (40 μL per sample). The antibody must be from the indicated supplier (*see* **Note 6**).
22. Cy3-conjugated, donkey anti-sheep secondary antibody (Jackson ImmunoResearch), (10 μg/mL final concentration (40 μL per sample). The antibody must be from the indicated supplier (*see* **Note 6**).

2.4 Tissue Clearing

1. Methanol (MeOH), 100% (20 mL).
2. MeOH dilution series in ddH_2O (5 mL each): 20%, 40%, 60%, and 80%, kept at RT.
3. Disposable syringe, 10–20 cc.
4. Disposable needles, 18–20 gauge.
5. Dichloromethane (DCM)/methanol (MeOH) solution, 66%/ 33% (v/v) (5 mL per sample); prepare just before use, keep at RT.
6. DCM, 100% (10 mL per sample).
7. Dibenzyl ether (DBE) (5 mL per sample).

2.5 Image Acquisition and Processing

1. Caps, flat-topped, for 15 mL centrifuge tubes.
2. Kwik-Sil® silicone adhesive, or equivalent.
3. Coverslip, glass, square or circular.
4. Kimwipes®.
5. Pipets, disposable, 2 mL.
6. Olympus FluoView 1000 MPE platform, or equivalent (*see* **Note 11**).
7. *ImageJ* software.

3 Methods

Obtain 8–10 mm thick coronal slabs of brains from AD patients and controls [18] (*see* **Note 1**)

3.1 Fixation of Human Brain Tissue

1. Immerse coronal slabs in 10% NBF for 72 h at RT. Ensure slabs are placed flat and arranged in anatomical position during fixation, using flat-tipped forceps to minimize mechanical damage.
2. After 72 h, rinse the slabs 3 times in 0.02% sodium azide, and then store them in 0.02% sodium azide at 4 °C.
3. Use a scalpel to cut 5–10-mm-thick blocks (approximately 1 × 2 cm) of the region of interest from the fixed coronal slabs, including dementia-related regions [19].
4. Place tissue blocks into 50–100 mL flat-bottomed specimen jars and store in 0.02% sodium azide at 4 °C until needed.

3.2 Tissue Sectioning and Dehydration

1. Perform all steps described below in 5 mL Eppendorf tubes. Cut the formalin-fixed tissue blocks into 1.5–2-mm-thick sections with a scalpel (*see* **Note 2**).
2. Dehydrate the tissue in the following freshly prepared, consecutive MeOH/ddH_2O solutions, for 1 h each at RT: 20%, 40%, 60%, 80%, 100%.
3. Cool the tissues at 4 °C for 1 h (*see* **Note 3**).
4. Incubate the tissues in DCM/MeOH, 66/33% on a nutator mixer at RT, overnight. Use large-bore needle attached to the disposable syringe to retrieve DCM from the secure-cap container.
5. Wash the samples twice in 100% MeOH for 30 min at RT on the nutator mixer, then cool the samples at 4 °C for 1 h.
6. Bleach the brain tissue samples in freshly made, pre-cooled (4 °C) 5% H_2O_2 in MeOH for 36 h at 4 °C.

3.3 Immunolabeling

Use 2 mL Eppendorf tubes for the primary and secondary antibody incubations (*see* **Steps 3** and **5**, below); use 5 mL Eppendorf tubes for all other steps to ensure sufficient coverage. For larger tissue samples, use 5 mL tubes (*see* **Notes 4–6**).

1. Rehydrate the samples in the following consecutive MeOH solutions for 1 h each: 80%, 60%, 40%, 20%, ddH_2O.
2. Wash samples in *PTx.2* twice for 1 h at RT under continuous shaking.

3. Incubate the samples in permeabilization solution at 37 °C for 36 h on the nutator mixer.
4. Incubate the samples in blocking solution at 37 °C for 48 h on nutator mixer.
5. Incubate the samples in primary antibody solution containing sheep anti-human fibrinogen, rabbit anti-human β amyloid, and guinea pig anti-human CD31/PECAM-1 (all antibodies at a 1:100 dilution), at 37 °C for 72 h on the nutator mixer (*see* **Note 5**).
6. Wash the samples 5 times in PTwH on the nutator mixer, changing the solution every 2 h. Leave in PTwH overnight at 37 °C on the nutator mixer.
7. Incubate the samples in Secondary antibody solution containing Alexa 488-conjugated donkey anti-guinea pig, Alexa 647-conjugated donkey anti-rabbit, and Cy3-conjugated donkey anti-sheep antibodies (all at 1:200 dilution), at 37 °C, for 72 h on the nutator mixer (*see* **Note 6**).
8. Wash the samples 5 times in PTwH, at RT, on the nutator mixer, changing the solution every 2 h. Leave in PTwH overnight at RT on the nutator mixer.

3.4 Tissue Clearing

Perform all steps described below in 5 mL Eppendorf tubes (*see* **Note 4**).

1. Dehydrate the tissue in the following freshly prepared, consecutive MeOH/ddH_2O solutions, for 1 h each at RT: 20%, 40%, 60%, 80%, 100% (*see* **Note 7**).
2. Using forceps, transfer the samples into freshly prepared DCM/MeOH, 66/33%. Incubate on the nutator mixer at RT for 3 h.
3. Wash twice in 100% DCM for 30 min each, on the nutator mixer at RT. Decant DCM between washes.
4. Using forceps, carefully place the samples in 100% DBE-containing (new) Eppendorf tubes. Gently invert the tubes to ensure adequate mixing of the DBE. The samples will become translucent within 30–60 s (*see* **Note 8**).
5. Store the cleared tissue at RT in an upright position and shield from light. Ensure that the Eppendorf tubes are fully filled and closed to prevent oxidation of the samples. Samples can be stored for several months without an appreciable decrease in quality.

3.5 Preparation of Imaging Chambers for Image Acquisition

The tissue samples must be immersed in DBE at all times, including during imaging. Therefore, an imaging chamber containing DBE and the sample must be prepared before acquiring images.

1. Flat-topped screw caps of 15 mL centrifuge tubes will serve as imaging chambers for the tissue samples. Prepare separate imaging chambers for each sample. Alternatively, the lids of 5 mL Eppendorf tubes can be used for smaller samples.
2. Fill each lid with 100% DBE using a disposable pipet.
3. With a pair of forceps, gently remove the sample from its storage tube and place in the DBE-filled lid prepared in **Step 2**.
4. Add extra DBE to the lid until the tissue is completely submerged.
5. Place a square or circular glass coverslip on top of the filled lid. The size of the coverslip should be larger than the lid. Make sure no air bubbles become trapped between the coverslip and the samples (*see* **Note 9**).
6. Seal the coverslip onto the lid using Kwik-Sil® silicone adhesive, verifying that there are no DBE leaks from the seal. Carefully clean the coverslip glass and the lid, removing all traces of DBE with a Kimwipe (*see* **Notes 10** and **11**).

3.6 Image Acquisition and Processing

1. Use the 543-nm HeNe gas laser to excite Cy3-labeled secondary antibodies and the 635-nm diode laser to excite Alexa 647-labeled secondary antibodies. Tune the two-photon laser to 940 nm to excite Alexa 488-labeled secondary antibodies (*see* **Notes 12–14**).
2. Separate the Cy3 (543 nm) and Alexa 647 (635 nm) fluorescence emission spectra using a 640-nm dichroic mirror, with a 560–620-nm (Cy3) and a 655–755-nm (Alexa 647) emission filter mounted before the detector. For the Alexa 488 secondary antibodies, pass the fluorescence emission from the two-photon laser through a 495–540-nm emission filter mounted before the non-de-scanned detector.
3. Acquire Z-stacks in the sequential imaging mode at a 1 μm step resolution, 4.0 μs pixel dwell-time, and 512 × 512 or 1024 × 1024 pixel resolution.
4. Process images with *ImageJ* software. Load the combined Alexa 488, Cy3, and Alexa 647 image Z-stack files as separate Z-stacks into ImageJ. Convert images to 8-bit grey-scale, and de-noise the images using the "remove outliers" plugin with the radius set to 2.0 pixels and the threshold set to 50 internal units. Subtract the Alexa 488 image stack (CD31 signal) from the Cy3 image stack (fibrinogen signal) to ensure removal of possible nonspecific, false-positive fluorescence signals arising from lipofuscin and other non-fibrinogen deposits.

5. Combine the processed-image Z-stacks as pseudo-colored RGB stacks, then re-process using the "Unsharp mask" plugin with the radius/sigma set between 3.0 and 5.0 and the mask weight set at 1/10th of the radius/sigma, e.g., the radius/sigma = 4.0 and the mask weight = 0.4.
6. Process the RGB stacks into 3D volume projections using the "3D viewer" plugin, followed by a 360° rotation recording, to yield a .tiff 3D volume projection image file (*see* **Note 15**).

3.7 Qualitative Analysis of iDISCO-Processed Tissue

1. Following image acquisition and processing, analyze the images qualitatively for the presence of fibrinogen in intravascular, vessel-associated, and parenchymal regions, with and without Aβ deposits, as well as tortuous vascular structures [20] with vessel-associated fibrinogen deposits in AD brains, using the representative images shown here serving as a guide (Figs. 1a–d and Videos 1, 2, and 3) (*see* **Notes 1, 16–18**).
2. Define vessel-associated fibrinogen as vascular or perivascular fibrin deposits present at the abluminal side of CD31-stained vessels in non-demented control (NDCTRL) brains, using the representative images shown here serving as a guide (Fig. 2 and Video 1) (*see* **Notes 1, 16–18**).

4 Notes

1. The current protocol has been used for iDISCO processing of blocks of the lateral temporo-occipital cortex dissected from 5 patients with AD and 3 non-demented control (NDCTRL) subjects. The brains were procured and processed according to the UCSF Neurodegenerative Disease Brain Bank protocol [18]. The age range was 70–94 years; female/male percentage was 38%; Braak stage and Thal Phase NDCTRLs: II and 0–2, respectively, for NDCTRLs and IV–VI and 4–5, respectively, for AD. The brains were obtained within a postmortem interval of 6–15 h. Neuropathological diagnoses were made following consensus histological and diagnostic criteria for AD [19, 21, 22].
2. Because of the lack of a species barrier, handle all human tissue using the appropriate personal protective equipment and by performing all steps in a biosafety hood, as required by your institutional guidelines.
3. Cooling down the brain tissue samples is required to counterbalance the heat generation by the DCM/MeOH.
4. Perform all steps with tubes fully filled and tightly closed to prevent oxidation of the samples.

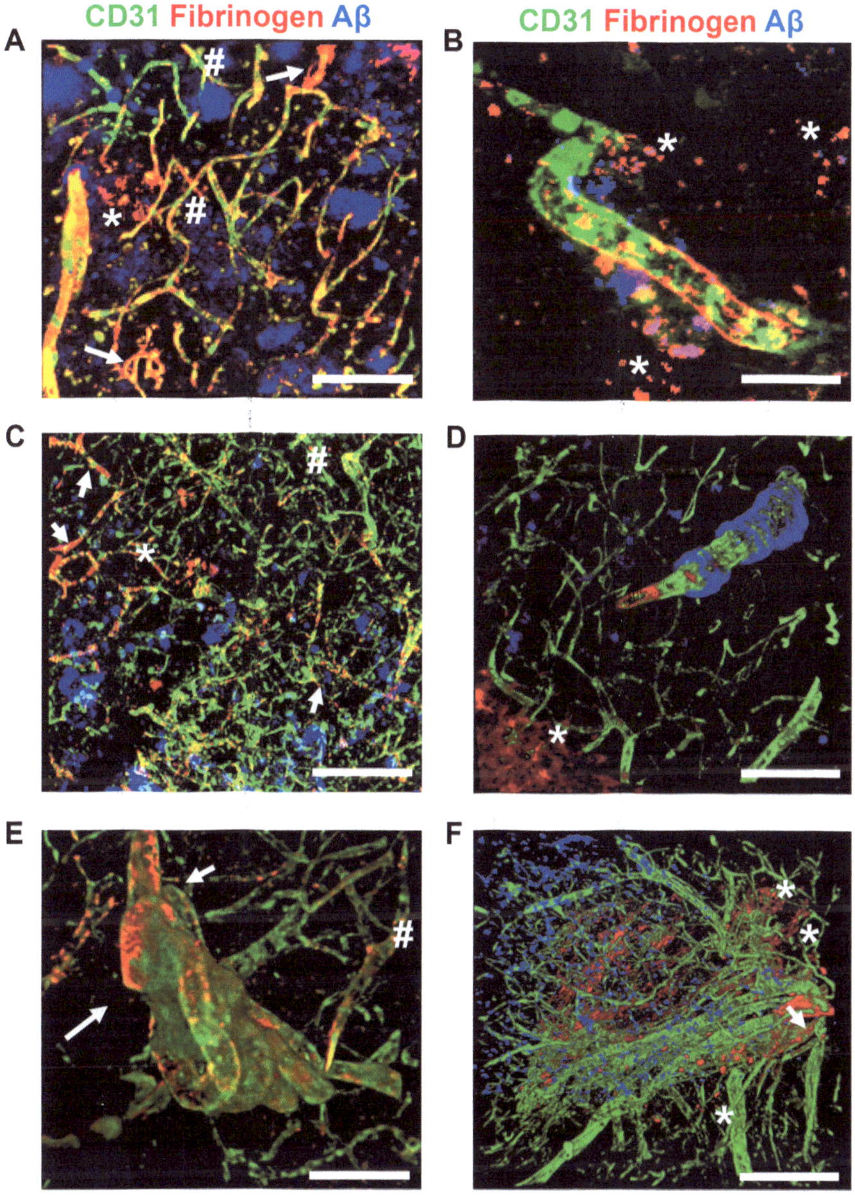

Fig. 1 Fibrinogen iDISCO in AD Brain. (**a–d**) 3D immunolabeling of cleared temporo-occipital brain tissue from three AD brains, stained for the vascular marker, CD31 (*green*), fibrinogen (*red*), and Aβ (*blue*). *Arrows* and *asterisks* indicate vessel-associated and parenchymal fibrinogen, respectively; *# sign* indicates intravascular fibrinogen. *Magenta* in (**b**) indicates Aβ-associated fibrinogen. (**e**, **f**) Representative 3D volume projections of iDISCO-cleared temporo-occipital AD brain tissue, stained for CD31, fibrinogen, and Aβ, showing examples of vascular tortuosity observed in AD brain. Vascular tortuosity was observed in 3 out of the 5 AD brains. *Scale bars* = 75 μm (**a**), 25 μm (**b**), 50 μm (**c**), and 25 μm (**d–f**). (Images reprinted from Merlini et al. [7])

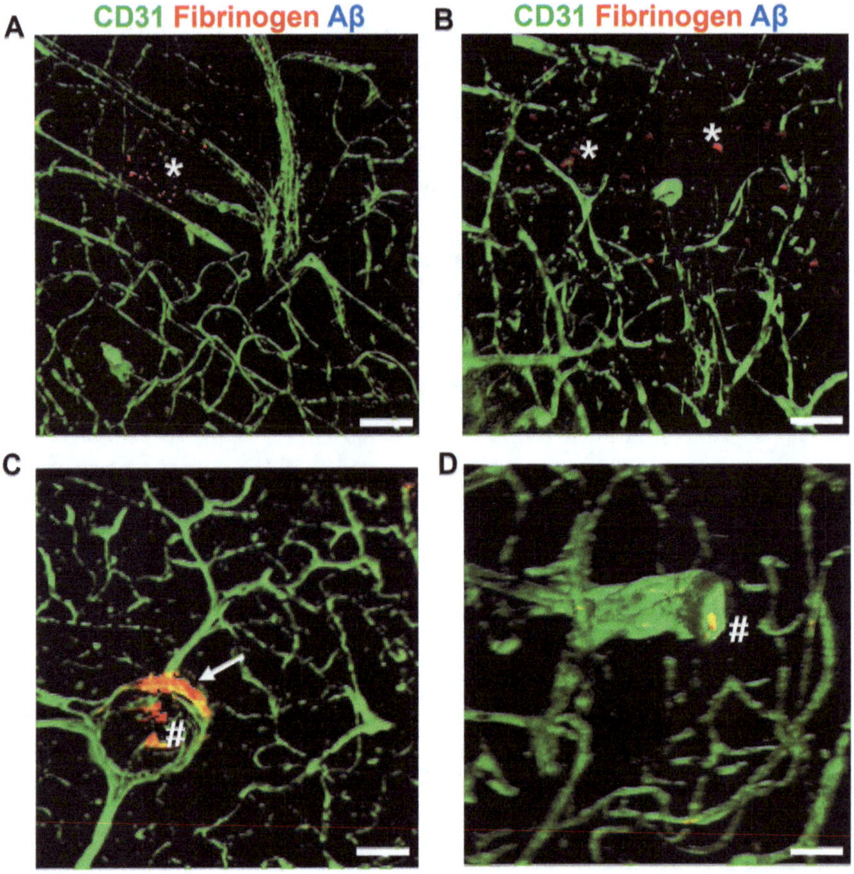

Fig. 2 Fibrinogen iDISCO in non-demented control (NDCTRL) brain tissue. (**a–d**) Representative 3D volume projections of iDISCO-cleared temporo-occipital brain tissue from age-matched NDCTRL subjects stained for the vascular marker, CD31, fibrinogen, and Aβ. Images are representative from 3 NDCTRL brains. Vessel-associated fibrinogen deposits are present on the abluminal side of CD31-stained vessels. *Arrows* and *asterisks* indicate vessel-associated and parenchymal fibrinogen deposits, respectively; *# sign* indicates intravascular fibrinogen. *Scale bars* = 50 μm (**a**, **b**) and 25 μm (**c**, **d**). (Images reprinted from Merlini et al. [7])

5. If using primary antibodies different from those described in this protocol, studies will be needed to test for compatibility with MeOH and optimal antibody concentration. Although a so-called "alternative method" based on MeOH-free solutions is available (https://idisco.info), in our hands, iDISCO processing of human brain tissue requires the MeOH-based solutions to yield high signal-to-noise ratios during imaging. Use the following method to confirm MeOH compatibility of primary antibodies:
   - Prepare 20 μm thick frozen sections of the formalin-fixed brain tissue and mount on SuperFrost® slides.

- Incubate the tissue slides in 100% MeOH for 3 h. Rehydrate the brain tissue directly in PBS and proceed with the regular immunostaining protocol used for staining frozen sections. Use non-MeOH-treated slides as positive controls.

6. If using secondary antibodies different from those described in this protocol, optimum antibody concentration needs to be determined by the researcher. Always include secondary antibody control tissue samples for each brain in the staining series, i.e., brain tissue samples that are processed along with all other samples from the same brain but for which the primary antibodies are omitted. These controls are essential for evaluating successful staining, determining staining quality in relation to background noise/autofluorescence, and detecting potential false-positive immunofluorescence signals.
7. As an optional pause in the protocol, the tissues may be left, at this point, in 100% MeOH at RT overnight.
8. DBE is highly corrosive; take all safety precautionary measures, including adequate personal protective equipment, when handling.
9. If air bubbles become trapped inside the imaging chamber, they may be removed by slightly lifting the coverslip at a shallow angle. Subsequently, gently add DBE through a 23–27-gauge needle attached to a syringe, until a convex meniscus has formed and the air bubbles are expelled. Gently lower the coverslip and absorb any spilled DBE with a tissue. Seal with Kwik-Sil® as described.
10. Careful but thorough cleaning of the assembled imaging chamber is required to prevent damage, by DBE, to the confocal and two-photon microscope lenses. Use 70% isopropyl alcohol and soft tissues, e.g., Kimwipes®.
11. Do not use any type of glue or silicon sealants, as they will chemically react with DBE, resulting in loss of the seal and subsequent leakage of DBE from the imaging chamber.
12. We have used the Olympus FluoView 1000 MPE platform, consisting of an Olympus BX61WI microscope with integrated one-photon and two-photon laser lines and an Olympus 20×, 0.5-NA water-immersion lens for fluorescence image acquisition. Any similar type of confocal with a two-photon microscope platform combined with similar lasers and appropriate multi-alkali and/or GaAsP photomultiplier tubes may be used.
13. One-photon lasers: a 543-nm HeNe gas laser and a 635-nm diode laser. Two-photon laser: Spectra-Physics MaiTai DeepSee *e*HP, 690–1040 nm Ti:Sapphire femtosecond laser. Dichroic mirror and emission filters: 640-nm dichroic mirror and 495–540-nm, 560–620-nm, and 655–755-nm emission

filters. Because of the relatively high photon scattering and autofluorescence in the human brain, especially in white matter, it is highly recommended to excite Alexa 488-labeled proteins/structures of interest with a two-photon laser as described. The Cy3- and Alexa 647-labeled proteins/structures can be imaged with conventional one-photon lasers, as their excitation and emission wavelengths do not significantly overlap with those of brain tissue and, thus, yield less autofluorescence.

14. Although this protocol describes the use of a combined confocal and two-photon imaging platform for image acquisition, images may be acquired readily with an adequate light-sheet microscope setup that is suitable for volume imaging of DBE-immersed brain tissue.
15. This Note is also related to **Notes 5** and **6**. Aged (human) brain has increased intracellular lipofuscin. The pervasive autofluorescence of lipofuscin may also be detected at the longer wavelengths and with the corresponding emission filters described here, and may thus be apparent in the stained brain tissue samples. It is therefore necessary to compare the immunofluorescence signals in each stained sample to those in the corresponding secondary-antibody controls, for proper validation of true- and false-positive signals. The Subtract Background and/or Remove Outliers plugins of ImageJ should be used and set according to the immunofluorescence signals in the respective secondary antibody control samples, for each specific brain.
16. Neuropathological studies have identified fibrinogen deposition in AD patients, with increased deposition correlating with the *ApoE4* mutation [13, 14, 23] and with increased BBB permeability [5]. Fibrinogen iDISCO is ideally suited to decipher the patterns of fibrinogen deposition and cerebrovascular abnormalities in different AD patient populations and to determine their correlation with the *ApoE4* mutation, cerebral amyloid angiopathy, white-matter hyperintensities, and microbleeds. As loss of BBB integrity is focal in AD, 3D imaging of plasma protein extravasation in relatively thick tissue sections is advantageous, compared with methods that rely only on thin tissue sections, since the decreased number, length, and complexity/branching of vascular structures inherent to the latter reduce BBB leak detection, possibly leading to false-negative assessments of BBB permeability. As BBB disruption also occurs focally in other neurological diseases, fibrinogen iDISCO could be applied to neurovascular studies of cortical and white-matter lesions from patients with multiple sclerosis, Parkinson's disease, amyotrophic lateral sclerosis, and some psychiatric disorders [3].

17. Fibrinogen in the CSF is a biomarker in AD that positively correlates with p-tau and soluble PDGFRβ, a pericyte marker indicative of loss of BBB integrity [5, 24]. Thus, fibrinogen iDISCO in AD brains can be used to co-register neuropathological findings with fluid biomarkers in patients with AD. Furthermore, fibrinogen iDISCO could guide the development of fibrin–PET probes for the direct detection of vascular damage and BBB disruption in human AD brains.
18. Fibrinogen binds Aβ, and this interaction inhibits fibrin degradation, leading to sustained fibrin deposits [17]. Fibrinogen iDISCO reveals different patterns of fibrin deposits, either perivascular, together with Aβ, or distant from Aβ deposits [7]. Fibrinogen induces microglial activation and correlates with activated microglia in human AD brains [7, 8, 12, 13, 15]. Furthermore, fibrinogen activation of microglia induces the elimination of dendritic spines and causes cognitive impairment in mouse models of AD [7]. Therefore, it could be useful to adapt the fibrinogen iDISCO protocol for triple immunolabeling of fibrinogen, microglia, and vasculature in brains from human AD patients. Overall, 3D imaging of fibrinogen in human brains can be used to investigate neurovascular and neuroimmune mechanisms driving neurodegeneration and cognitive decline in AD and other neurological diseases.

Acknowledgments

We thank Pamela E. Rios Coronado for expert technical assistance and Kathryn Claiborn for expert editorial assistance. Human tissue samples were provided by the Neurodegenerative Disease Brain Bank at the University of California, San Francisco, which receives funding support from NIH grants P01AG019724 and P50AG023501, the Consortium for Frontotemporal Dementia Research, and the Tau Consortium. The work described herein was supported by a Swiss National Science Foundation Early Postdoc Mobility Fellowship to M.M., the National MS Society Award FAN-2008-37045 to E.G.S. and grants from the Simon Family Trust, the Dagmar Dolby Family Fund, Edward and Pearl Fein, the Conrad N. Hilton Foundation 17348, NIA RF1 AG064926, and NIH/NINDS R35 NS097976 to K.A.

Competing Interests K.A. is a co-founder, scientific advisor, and board director of Therini Bio. Her interests are managed in accordance with the Gladstone Institutes' conflict-of-interest policy. All other authors declare no conflict of interest.

References

1. Sweeney MD, Sagare AP, Zlokovic BV (2018) Blood-brain barrier breakdown in Alzheimer disease and other neurodegenerative disorders. Nat Rev Neurol. https://doi.org/10.1038/nrneurol.2017.188
2. Montagne A, Zhao Z, Zlokovic BV (2017) Alzheimer's disease: A matter of blood-brain barrier dysfunction? J Exp Med 214(11): 3151–3169. https://doi.org/10.1084/jem.20171406
3. Petersen MA, Ryu JK, Akassoglou K (2018) Fibrinogen in neurological diseases: mechanisms, imaging and therapeutics. Nat Rev Neurosci 19(5):283–301. https://doi.org/10.1038/nrn.2018.13
4. Strickland S (2018) Blood will out: vascular contributions to Alzheimer's disease. J Clin Invest 128(2):556–563. https://doi.org/10.1172/JCI97509
5. Montagne A, Nation DA, Sagare AP, Barisano G, Sweeney MD, Chakhoyan A et al (2020) APOE4 leads to blood-brain barrier dysfunction predicting cognitive decline. Nature 581(7806):71–76. https://doi.org/10.1038/s41586-020-2247-3
6. Iadecola C (2013) The pathobiology of vascular dementia. Neuron 80(4):844–866. https://doi.org/10.1016/j.neuron.2013.10.008
7. Merlini M, Rafalski VA, Rios Coronado PE, Gill MT, Ellisman M, Muthukumar G et al (2019) Fibrinogen induces microglia-mediated spine elimination and cognitive impairment in Alzheimer's disease. Neuron 101(6): 1099–1108
8. Akassoglou K (2020) The immunology of blood: connecting the dots at the neurovascular interface. Nat Immunol 21(7):710–712. https://doi.org/10.1038/s41590-020-0671-z
9. Liebmann T, Renier N, Bettayeb K, Greengard P, Tessier-Lavigne M, Flajolet M (2016) Three-dimensional study of Alzheimer's disease hallmarks using the iDISCO clearing method. Cell Rep 16(4):1138–1152. https://doi.org/10.1016/j.celrep.2016.06.060
10. Renier N, Wu Z, Simon DJ, Yang J, Ariel P, Tessier-Lavigne M (2014) iDISCO: a simple, rapid method to immunolabel large tissue samples for volume imaging. Cell 159(4): 896–910. https://doi.org/10.1016/j.cell.2014.10.010
11. Molbay M, Kolabas ZI, Todorov MI, Ohn TL, Erturk A (2021) A guidebook for DISCO tissue clearing. Mol Syst Biol 17(3):e9807. https://doi.org/10.15252/msb.20209807
12. Paul J, Strickland S, Melchor JP (2007) Fibrin deposition accelerates neurovascular damage and neuroinflammation in mouse models of Alzheimer's disease. J Exp Med 204(8): 1999–2008
13. Ryu JK, McLarnon JG (2008) A leaky blood-brain barrier, fibrinogen infiltration and microglial reactivity in inflamed Alzheimer's disease brain. J Cell Mol Med 13(9A):2911–2925
14. Hultman K, Strickland S, Norris EH (2013) The APOE varepsilon4/varepsilon4 genotype potentiates vascular fibrin(ogen) deposition in amyloid-laden vessels in the brains of Alzheimer's disease patients. J Cereb Blood Flow Metab 33(8):1251–1258. https://doi.org/10.1038/jcbfm.2013.76
15. Adams RA, Bauer J, Flick MJ, Sikorski SL, Nuriel T, Lassmann H et al (2007) The fibrin-derived gamma377-395 peptide inhibits microglia activation and suppresses relapsing paralysis in central nervous system autoimmune disease. J Exp Med 204(3):571–582
16. Ryu JK, Rafalski VA, Meyer-Franke A, Adams RA, Poda SB, Rios Coronado PE et al (2018) Fibrin-targeting immunotherapy protects against neuroinflammation and neurodegeneration. Nat Immunol 19(11):1212–1223. https://doi.org/10.1038/s41590-018-0232-x
17. Cortes-Canteli M, Paul J, Norris EH, Bronstein R, Ahn HJ, Zamolodchikov D et al (2010) Fibrinogen and beta-amyloid association alters thrombosis and fibrinolysis: a possible contributing factor to Alzheimer's disease. Neuron 66(5):695–709. https://doi.org/10.1016/j.neuron.2010.05.014
18. Spina S, La Joie R, Petersen C, Nolan AL, Cuevas D, Cosme C et al (2021) Comorbid neuropathological diagnoses in early versus late-onset Alzheimer's disease. Brain 144(7): 2186–2198. https://doi.org/10.1093/brain/awab099
19. Tartaglia MC, Sidhu M, Laluz V, Racine C, Rabinovici GD, Creighton K et al (2010) Sporadic corticobasal syndrome due to FTLD-TDP. Acta Neuropathol 119(3): 365–374. https://doi.org/10.1007/s00401-009-0605-1
20. Brown WR, Thore CR (2011) Review: cerebral microvascular pathology in ageing and neurodegeneration. Neuropathol Appl Neurobiol 37(1):56–74. https://doi.org/10.1111/j.1365-2990.2010.01139.x

21. Kim EJ, Sidhu M, Gaus SE, Huang EJ, Hof PR, Miller BL et al (2016) Selective frontoinsular von economo neuron and fork cell loss in early behavioral variant frontotemporal dementia. Cereb Cortex 26(4):1843. https://doi.org/10.1093/cercor/bhw012
22. Montine TJ, Phelps CH, Beach TG, Bigio EH, Cairns NJ, Dickson DW et al (2012) National Institute on Aging-Alzheimer's Association guidelines for the neuropathologic assessment of Alzheimer's disease: a practical approach. Acta Neuropathol 123(1):1–11. https://doi.org/10.1007/s00401-011-0910-3
23. Cortes-Canteli M, Mattei L, Richards AT, Norris EH, Strickland S (2015) Fibrin deposited in the Alzheimer's disease brain promotes neuronal degeneration. Neurobiol Aging 36(2):608–617. https://doi.org/10.1016/j.neurobiolaging.2014.10.030
24. Fan DY, Sun HL, Sun PY, Jian JM, Li WW, Shen YY et al (2020) The correlations between plasma fibrinogen with amyloid-beta and tau levels in patients with Alzheimer's disease. Front Neurosci 14:625844. https://doi.org/10.3389/fnins.2020.625844

Part II

Human Neural Cellular Models

Chapter 6

Human-Induced Pluripotent Stem Cell (hiPSC)-Derived Neurons and Glia for the Elucidation of Pathogenic Mechanisms in Alzheimer's Disease

Jessica E. Young and Lawrence S. B. Goldstein

Abstract

Alzheimer's disease (AD) is a common neurodegenerative disorder and a mechanistically complex disease. For the last decade, human models of AD using induced pluripotent stem cells (iPSCs) have emerged as a powerful way to understand disease pathogenesis in relevant human cell types. In this review, we summarize the state of the field and how this technology can apply to studies of both familial and sporadic studies of AD. We discuss patient-derived iPSCs, genome editing, differentiation of neural cell types, and three-dimensional organoids, and speculate on the future of this type of work for increasing our understanding of, and improving therapeutic development for, this devastating disease.

Key words Human-induced pluripotent stem cells, Neuronal differentiation, Informed consent, Alzheimer's disease, Organoids

1 Introduction

Alzheimer's disease and related dementias (AD/ADRD) are devastating neurodegenerative conditions with no current treatments that slow or halt disease progression. Neuropathologic characterizations of AD reveal extracellular senile plaques, composed of small amyloid beta (Aβ) peptides and intraneuronal accumulations of hyperphosphorylated tau protein (pTau), called neurofibrillary tangles. Other features of the AD brain include enlarged early endosomes, neuroinflammation, and dystrophic neurites [1, 2]. However, many of these pathological characteristics are primarily present at the end-stage of a decades-long disease process. Therefore, understanding early biological mechanisms and biochemical pathways that contribute to cellular dysfunction, and ultimately pathology, is crucial.

Human-induced pluripotent stem cell (hiPSC) technology has been employed for the past decade to address this challenge in AD

Jerold Chun (ed.), *Alzheimer's Disease: Methods and Protocols*,
Methods in Molecular Biology, vol. 2561, https://doi.org/10.1007/978-1-0716-2655-9_6,

research. These cellular models provide an experimental platform to interrogate cell type-specific mechanisms, including early biochemical and cellular pathways, in patient-specific genetic backgrounds. These approaches can be combined with recent rapid advances in genome editing technology, which allow precise molecular dissection of disease-associated genetic mutations and population variants in isogenic settings. Protocols are also now available to differentiate most major cell types of the central nervous system (CNS), and methods to generate three-dimensional, multicellular brain organoids are being developed [3, 4]. Such approaches may provide clues to disease-specific interactions among different types of brain cells. However, challenges remain in the reliability and reproducibility of protocols and in generating three-dimensional models that include cells from all brain lineages.

Here, we discuss the current advantages and challenges of using hiPSC models to study AD and highlight some new insights regarding AD disease biology that have emerged from the use of these technologies.

2 hiPSC-Generation Methods: Reprogramming

In 2007, Yamanaka and colleagues identified four transcription factors, Oct4, Klf4, Sox2, and c-myc, that successfully converted human fibroblasts to human-induced pluripotent stem cells (hiPSCs) with characteristics of human embryonic stem cells [5, 6]. Early reprogramming strategies used retroviral vectors to introduce the reprogramming factors into fibroblasts. However, issues with random insertion of reprogramming factors and incomplete silencing of those factors raised concern about using cells with integrated factors for clinical and other applications [7]. We note, however, that many studies published before non-integrating methods were developed used hiPSC lines with integrated retroviruses carrying reprogramming vectors. Although hiPSC lines were generally developed that used fluorescent protein markers to find or select lines that had silenced the integrated factors, reactivation of these factors (indicated by activation of a fluorescent protein marker) is a well-known, though relatively uncommon, phenomenon [8]. It is thus important to exert some caution when using these older hiPSC lines.

Non-integrating methods for generating hiPSC lines are now standard for reprogramming and are in use at many core facilities. These methods include using non-integrating viruses such as Adenovirus or Sendai virus, episomal vectors, synthetic mRNA, protein transduction [7], chemical agents [9] or CRISPR/Cas9 methods to activate endogenous genes [10]. Non-integrating methods are now in use by most research groups and stem-cell core facilities and are generally considered to yield lines that are more stable than integrating methods. Several protocols now exclude the reprogramming factor c-myc for additional safety reasons [11]; many plasmids for reprogramming somatic cells to hiPSCs are available from Addgene. A potential complication with Sendai vectors is that they can remain present in hiPSC lines for many passages. Estimates are that it may take as many as 9 passages to remove these vectors, so some care and caution is recommended [12].

In general, reprogramming is expensive and time-consuming. For large numbers of unique patient lines, it may not be feasible for an academic laboratory to reprogram cells of interest on their own. However, many academic centers now support stem-cell "Core" facilities where this process is run as a fee-for-service endeavor. This is a strong advantage, as Core facilities are often staffed by professionals who use the most up-to-date methodology combined with relevant experience.

An important approach using existing hiPSC lines is to generate lines that are isogenic except for mutations or variants of interest. Gene-editing technology, such as TALEN or CRISPR methods, can be used to induce AD mutations or variants in control cell lines or to correct AD mutations in patient cells. These methods generate control reference lines with introduced mutations, or patient reference lines with corrected mutations. In both cases, the comparisons are made in an "isogenic" background.

3 hiPSC Analysis Methods

An important consideration in hiPSC generation is the integrity of the genomes after reprogramming. Our groups published early work reporting that hiPSCs can accumulate oncogenic mutations at a rate tenfold higher than normal somatic mutations [13]. Other groups have documented aneuploidies, translocations, and copy-number variants following reprogramming [7]. Some of these abnormalities can be detected by traditional G-band karyotyping, although this technique can be expensive and will miss many sub-chromosomal abnormalities. Digital karyotyping using PCR-based methods provides a higher-resolution analysis of genomic integrity and often is more cost-effective, although it cannot distinguish balanced chromosomal translocations or inversions that do not delete or duplicate genomic materials. Recently, Assou and colleagues developed a digital droplet PCR test that can screen for genetic abnormalities in culture supernatant, thereby providing a scalable platform for screening hiPSC lines [14].

An important standard practice in hiPSC characterization is the assessment of pluripotency or the ability to differentiate into the three germ layers. Previously, pluripotency was assessed by transplanting hiPSCs into various anatomical sites in a mouse and assaying for an experimental tumor called a teratoma that is usually comprised of endoderm, mesoderm, and ectoderm [15]. However, standardization of this technique among labs was not consistent, and it was considered an undue burden and stress on animals used. Currently, molecular techniques are used to assess pluripotency and involve generating embryoid bodies from hiPSCs and measuring the expression of germ-layer genes or proteins by quantitative RT-PCR or immunostaining [16, 17]. In general, for use of hiPSCs in a laboratory setting, it is good practice to confirm both genomic integrity and pluripotency of the cells prior to establishing a disease model.

4 Patient-Derived Cells: Ethics and Informed Consent

We strongly recommend that any investigator planning use of hiPSC lines consult their local IRB (or equivalent) to obtain guidance on local ethics rules and informed consent requirements. We recommend one of the two following general approaches, depending on the nature of the patients who are involved. The first method starts with a non-demented control: a not as-yet demented patient carrying an autosomal dominant, familial AD (FAD) mutation from which blood or skin cells will be harvested for reprogramming; both blood and skin can be used as source material for reprogramming to the hiPSC state. When starting with a non-demented patient, standard methods of IRB approval and IRB-approved informed consent documents can be used. These documents should generally spell out, in clear language, that serving as a donor of cellular materials is voluntary and free of coercion. The risks of biopsy or blood collection and cellular donation should also be clearly specified. It is generally regarded as appropriate to include consent for all genetic/genomic analyses to be carried out on donated materials, as well as permission for original and reprogrammed materials to be deposited in generally accessible cell banks for use by the broad research community. It is common and accepted to state that commercial rights are waived by the donating individual. We have included a sample informed consent document in **Note**. Although this version of a consent was suitable up through approximately 2021, recent technical developments in making neuronal organoids, neuronal chimeras in different species of animal, and pseudo embryos or blastoids are prompting a reconsideration of general consent for all forms of research. It is possible that, moving forward, specific consent for generating any of these novel constructs will be required. If a donor is suffering from dementia, then the approach outlined below should be followed.

A second method for obtaining useful hiPSC lines relies on biopsies or blood samples from human patients that have developed dementia caused by FAD mutations, or who suffer from sporadic

AD (SAD). Obtaining these samples for reprogramming requires IRB approval and the use of IRB-approved informed-consent documents. Procedures and acquisition of consent, comparable to those described above, should be used in obtaining and reprogramming cellular materials from these patients. Once an individual is severely demented, and possibly not communicative, it is generally possible to secure consent from an individual assigned as healthcare proxy or equivalent. For individuals with mild to moderate dementia, and who are also responsive and communicative, we have used an instrument referred to as a *cued consent*. With this method, individuals who retain normal intelligence, but who suffer from dementia, are "walked" step by step through a consent document while being asked at each step whether they understand and whether they freely give consent. We strongly recommend that colleagues wishing to obtain cellular materials from potential donors be very careful and work with an IRB or its equivalent on developing robust protocols and consent requirements as ethics rules and practices may vary internationally. Finally, obtaining cells from de-identified sources where consent forms have been obtained for other procedures, including cells obtained via autopsies, generally does not require informed consent or IRB approval.

4.1 Obtaining hiPSC from Cell Banks

Cell banks for storage and distribution are developing and are much needed. The costs of growing, testing, and distributing a few published cell lines are not terribly onerous. But in the long term, the needs of the field are best served if published hiPSC lines are available, via cell banks, to all investigators. Fortunately, there are two banks that serve these needs, at least in part. One, the National Centralized Repository for Alzheimer's Disease and Related Dementias (NCRAD), is funded by the NIH, and another, located in California, is funded by the California Institute for Regenerative Medicine (CIRM). We urge investigators to deposit lines in these banks, or others, and to support them when their funding is up for competitive renewal. Banks such as these need long-term community support so that they don't disappear every 5 years, only to be replaced by a new effort.

5 Neuronal Differentiation from hiPSCs

While AD affects all cell types of the CNS, neurons derived from hiPSCs have been the most widely studied. Many early studies documented AD cellular phenotypes from neurons generated from patients with autosomal-dominant FAD mutations, and these studies will be discussed in more detail below. Here, we highlight two widely used methods to generate neurons from hiPSCs. In our experience, published differentiation methods vary considerably in their reproducibility or the degree to which they require modification in order to replicate published results. The field is still developing standards and testing published methods. The current situation is evolving rapidly as there are many new methods being tested in various labs. We give below a description of methods that seem to work well and reproducibly.

5.1 Directed Neuronal Differentiation by Dual SMAD Inhibition

Directed differentiation of hiPSC into neurons begins with induction of neuroectoderm by growth factors. One reliable method requires inhibition of the activities of the transforming growth factor beta (TGFβ) and bone morphogenetic protein (BMP) superfamily. In 2009, Chambers and colleagues published work showing that inhibition of TGFβ signaling by recombinant Noggin, a BMP antagonist, and the small molecule SB431542, which inhibits activin and nodal (two members of the TGFβ pathway), led to highly efficient induction of neuroectoderm in both human embryonic stem cells and hiPSCs [18]. They further showed that, whereas the default led to mainly cortical excitatory neurons, the addition of other patterning factors such as sonic hedgehog (SHH), retinoic acid and FGF8, led to generation of midbrain and hindbrain cell types. This protocol, called dual SMAD inhibition (dSMADi) because of its effect on stopping the actions of the SMAD signal transducers of the TGFβ pathway, has been refined to include small molecules such as dorsomorphin or LDN 193189 rather than recombinant noggin protein [19, 20]. In 2012, Shi et al. published a comprehensive protocol to differentiate hiPSCs into cortical neurons [21] and demonstrated that cortical neurons made from Down syndrome patients developed AD pathologies in culture, such as extracellular amyloid aggregates and intracellular accumulations of hyperphosphorylated tau protein [22]. While these differentiations primarily yield excitatory neurons, this protocol can also be modified to generate populations of inhibitory interneurons [23]. In general, these differentiations from hiPSCs yield a mixed population of cells and the efficiency of differentiation can vary between cell lines such that the final relative numbers of neurons or glia can be variable among lines and between different rounds of differentiation.

To address the problem of variability in the efficiency of neuronal differentiation, our group developed a comprehensive panel of cell-surface markers to purify neural progenitor cells and neurons [24], thus allowing the study of cellular phenotypes in a more homogenous population of cells. In particular, this study identified a population of $CD184^{+}/CD24^{+}$ $/CD271^{-}/CD44^{-}$ neural stem cells that would go on to yield $CD24^{+}/CD44^{-}/CD184^{-}$ neurons. This sorting protocol was used by our group to generate pure neurons from FAD and SAD patients and from which we detected increases in both Aβ peptides and pTau levels [25]. Current protocols from our labs combine induction of dSMADi differentiation and cell-surface marker selection to isolate neurons from hiPSCs for a range of AD phenotypic studies [17, 26–30]. These methods also generate a proliferative neural stem/progenitor population prior to terminal differentiation. These NSCs can be expanded and cryopreserved so that researchers do not need to start from the pluripotent stem cells for each differentiation experiment.

5.2 Direct hiPSC-Neuron Transdifferentiation by NGN2 Expression

An intriguing alternative differentiation method derives from transdifferentiation methods used to generate induced neurons (iNs) by forced expression of transcription factors in fibroblasts. This method was initially described in 2010 by Vierbuchen and colleagues, who found that expression of three transcription factors, Brn2, ASCL1, and Mytl1, could induce transdifferentiation of mouse embryonic fibroblasts to functional neurons, bypassing a pluripotent intermediate [31]. Further work in human cells demonstrated that directly transdifferentiated iNs may retain a more faithful age-dependent transcriptomic signature than that found in hiPSC-derived neurons [32] (Mertens, others). A comparison of fibroblast-to-neuron iNs and hiPSC-Ns has been recently published [33].

Similar methods have been employed to transdifferentiate pluripotent stem cells into neurons, also called iNs. In 2013, Zhang and colleagues published a protocol showing that forced expression of the transcription factor NGN2 in either hiPSCs or hESCs can convert the pluripotent stem cells to neurons with functional properties in approximately 2 weeks [34]. Current protocols using NGN2 involve an inducible vector system with the NGN2 transgene driven by the reverse tetracycline-controlled transactivator (rtTA). In this system, inducible expression is turned on with tetracycline or a derivative such as doxycycline [34]. These constructs are introduced into the hiPSC/hESC population using lentivirus vectors and can then be subcloned to generate a stable line. This protocol generates cortical excitatory neurons that are electrophysiologically active and bypasses some of the more time-consuming parts of dSMADi-based differentiations. Elegant work using this system generated an hiPSC line where the NGN2 gene is integrated into the adeno-associated virus integration site 1 (AAVS1), a safe-harbor locus in human pluripotent stem cells [35]. This cell line was used successfully in a high-content screen for Tau-lowering compounds where levels of endogenous Tau were measured with an anti-human Tau antibody. Adrenergic receptor agonists were identified as lowering Tau in these neurons [35]. It is important to note that this system represents only one human genetic background; thus, the normal phenotypic heterogeneity present in human cells, and how they respond to a given treatment, is not recapitulated.

Although the NGN2 protocol defines a powerful system, some additional limitations must be considered. This protocol bypasses a progenitor stage and therefore bypasses developmental programs that may be important, even for cells involved in an adult-onset neurodegenerative disorder. For example, impaired neurogenesis has been implicated in AD [36, 37], and investigations into this mechanism in humans would benefit from a model that generates neural stem/progenitor cells. While AD affects both excitatory glutamatergic neurons and GABAergic inhibitory interneurons [38], the NGN2 protocol primarily generates cortical excitatory neurons. Similar methods are described to generate GABAergic neurons using the transcription factors Ascl1 and Dlx2 [39]; however, this model has not been used in any AD studies to date. Many considerations are needed to develop an in vitro model for AD using hiPSCs. The differentiation scheme should be chosen to address most accurately the scientific question that is being asked.

5.3 Generating Non-neuronal Cell Types from hiPSC

A key challenge to probing AD cellular mechanisms using hiPSC technology is that the human brain is made up of many cell types, some of which are not derived from common neuronal progenitors. Thus, while astrocytes and oligodendrocytes can be generated from lineages shared with neurons, i.e., neuroectoderm, other cell types, such as microglia and vasculature, derive from endodermal and mesodermal lineages, requiring different differentiation methods. For example, in 2017, Abud and colleagues reported a differentiation strategy to generate microglia from pluripotent stem cells by first generating hematopoietic progenitor cells (HPCs) and then further differentiating to microglial-like cells. Cells derived by this method are phagocytic and secrete inflammatory cytokines. Furthermore, they morphologically resemble human microglia when transplanted into mice or brain organoids [40]. Modifications have been published to the original protocol to make it more streamlined [41] (McQuade), and STEMCELL Technologies provides a commercialized kit based on these methods.

Astrocyte differentiation protocols from hiPSCs are now robust and can also be initiated from neural ectoderm-derived neural progenitors [42, 43]. Following these protocols, our group demonstrated that astrocytes could be differentiated from $CD184^{+}/CD24^{+}/CD271^{-}/CD44^{-}$ neural stem cells by scraping NPCs into suspension culture on an orbital shaker, to form neurospheres, and culturing in Astrocyte Growth Medium, available from Lonza [44]. This study utilized gene-edited *APP* knockout (KO) and FAD mutant (APP Swe/Swe) hiPSCs and isogenic controls to demonstrate a role for full-length *APP* in LDL endocytosis and cholesterol homeostasis in human astrocytes [44]. Similar to microglia, astrocyte differentiation media and growth factors have been commercialized (STEMCELL Technologies).

Oligodendrocyte differentiation has been described by using a chemically defined protocol as well as by transdifferentiation induced by transcription factors. In the first method, neural progenitors are induced from hiPSCs by dual SMAD inhibition, and then oligodendrocyte progenitors are differentiated by activating sonic hedgehog (SHH) and retinoic acid signaling pathways [45]. Cells are matured by adding key factors promoting oligodendrocyte survival, i.e., PDGF, NT3, T3, IGF, and HGF, and expression of myelin basic protein [45]. In the transdifferentiation method, forced expression of the transcription factors SOX10, OLIG2, and NKX6.2 in neural progenitor cells yielded nearly 70% of oligodendrocytes in approximately 1 month [46]. These cells expressed myelin basic protein and extended multiple processes around both nanofibers and neuronal axons in co-culture [46].

As more groups routinely turn to hiPSC-based models, commercially defined kits and factors will likely become the default methods for the differentiation of CNS cell types. We refer to the discussion in the previous section regarding whether to choose developmentally programmed methods or transdifferentiation methods to generate desired cell types. Additional complexity comes from generating cultures containing more than one cell type under conditions conducive to viability and function of the mixed cells. A potential problem is that culture conditions supporting differentiation of each one of multiple different cell types can be quite different and not necessarily conducive to normal function or viability of multiple cell types under a single set of 2D culture conditions. Overcoming these problems is an important goal for future research, but, in spite of these problems, good-quality differentiation methods for multiple brain cell types have been reported and are leading to important first steps in the in vitro analyses of the contributions of each cell type to the initiation of AD.

5.4 Organoids and 3D Cultures

Two-dimensional models, generally consisting of one cell type, have provided insight into mechanisms of AD and other neurodegenerative disorders. However, cells grown in monolayer culture lack the maturity, three-dimensional architecture, and cell-cell dynamics of the adult brain. An important recent development is the establishment of methods for generating brain organoids consisting of multiple cell types in a 3D cluster. In these methods, pluripotent stem cells generate neuronal progenitors from which clusters of differentiating progenitors generate 3D structures. The original methods relied on self-organization of cells into neuroepithelial structures [47, 48], whereas later methods included dSMADi to increase neurogenic potential and reduce heterogeneity [49]. Recent advances include fusion of spheroids representing different brain regions to monitor neuronal migration [50] and elegant methods to monitor oscillatory networks that indicate formation of spatial and temporal neuronal activity [51]. These organoid models generate behaviors reminiscent of fetal brain and may ultimately be superior to the 2D models that are most commonly used in cell-biological analyses. Thus far, however, the organoid models consist primarily of cells derived from the neural ectoderm lineage. Initial efforts to "seed" neuronal organoids with cells from different lineages, in particular microglia, are underway [40, 52]. The major advances in three-dimensional cultures have largely been in the development and functionality of organoid technology, although they are also used to model neurologic disease (reviewed in [53]). This area will surely grow quickly, and we look forward to the insights to be gleaned from these methods.

Three-dimensional hiPSC-based models for AD are present but are not as common as the 2D models of the disorder. Choi and colleagues described the first 3D culture of AD human neural progenitor cells, where cells were cultured while suspended in a support matrix consisting of Matrigel [54]. These hNPCs were engineered to overexpress human *APP* and *PSEN1* harboring FAD mutations and demonstrated Aβ and phospho-tau aggregates [54]. More recently, hiPSCs with FAD mutations or from Down syndrome were differentiated into cortical organoids that also showed accumulation of Aβ and pTau aggregates and a higher incidence of cell death than control organoids [55]. These studies indicate that high levels of APP expression and increased amyloidogenic cleavage of APP can mimic some of the neuropathologic features of AD in a 3D system. How these models will relate to SAD is presently unclear; however, organoids made from APOE e4 hiPSC also showed increased Aβ and pTau levels in 3D [56].

6 Insights from FAD and SAD Models Using hiPSC Models

A critical issue in understanding AD and developing therapeutics is whether the amyloid plaques (composed of amyloid beta, Aβ) and neurofibrillary tangles (composed of aggregated phosphorylated Tau, pTau) found postmortem in AD brains are the key causes of the abnormal neuronal behavior and cognitive decline that are characteristic of AD. The failure, thus far, of therapies targeting Aβ and Tau has led to multiple avenues of research into different cellular targets. It is then important to determine whether cellular changes such as endocytic or transcytotic abnormalities, which may be caused by autosomal dominant AD mutations, are early pathogenic events or off-pathway changes unrelated to later pathologies, are not yet understood. Thus, a current controversy derives from observations that hiPSC models may exhibit early phenotypes but do not develop established neuropathology. However, alterations in cellular signaling pathways, endocytosis, and cholesterol modulation have the potential to reveal novel therapeutic targets, although it is still too early to tell whether these findings are predictive of AD mechanisms.

6.1 FAD Models

Early-onset familial AD (FAD) is an autosomal dominant disease and is primarily due to mutations in three genes: the amyloid precursor protein (APP) gene and the presenilin 1 and 2 genes (PSEN1/2). Clinically, FAD shows early-onset dementia, usually occurring at less than 65 years of age. For the most part, these mutations all affect the cleavage of APP by β or γ secretases, generating more Aβ42 peptides and C-terminal fragments of APP and often changing the Aβ 42:40 ratio, such that the more amyloidogenic form of the peptide (Aβ1–42) is produced. In addition to mutations changing the enzymatic processing of APP, mutations that give rise to extra copies of *APP*, such as duplications (APP^{Dp}) or trisomy 21, as seen in Down syndrome, are also sufficient to cause early-onset FAD.

The first work to generate an FAD model using hiPSCs from patients was by Yagi and colleagues in 2011 [58] where they demonstrated that neurons derived from patients with either *PSEN1* or *PSEN 2* mutations showed increased Aβ1–42 that responded to inhibitors of γ-secretase. Several other papers followed showing that FAD mutations, either derived from patient fibroblasts or generated through genome engineering of *PSEN* mutations, showed similar increases in Aβ [29, 59–61]. Ortiz-Virumbales and colleagues generated human basal forebrain cholinergic neurons from *PSEN2 N141I* subjects and demonstrated an increase in the Aβ 42:40 ratio as well as electrophysiological defects in these neurons. CRISPR/Cas9-mediated gene correction of this mutation corrected these phenotypes in an isogenic system, linking both the amyloidogenic phenotype and the electrophysiologic defects to this *PSEN2* variant [62].

Mutations in *APP* that promote β-cleavage or even a gene-dose increase in *APP* also increase Aβ secretion and phospho-Tau. For example, Muratore and colleagues documented increased Aβ peptides and increases in total and phosphorylated Tau from hiPSC-neurons harboring the *APP* London mutation, V717I [63]. Israel and colleagues derived neurons from patients with a duplication of *APP* (APP^{Dp}) and also showed increased Aβ and phospho-Tau. Interestingly, in these experiments, the increased levels of pTau were rescued by treatment with an inhibitor of the β secretase, which reduces the first cleavage of APP, but not the γ secretase, which generates the Aβ peptide. These data suggest that the first amyloidogenic cleavage of APP that yields β-c-terminal fragments may be more toxic to neurons [25]. Additional work has shown that hiPSC-derived neurons from Down syndrome patients, which contain an extra copy of *APP* on Chr. 21, produced aggregated forms of Aβ and pTau after long-term culture (>90 days) [22]. Indeed, longer culture time may be necessary to observe aggregations of Aβ and pTau that more closely mimic AD neuropathology. For example, Raja and colleagues documented putative extracellular accumulations of Aβ and increased pTau from APP^{Dp} hiPSCs grown in 3D culture for 90 days [64].

There are multiple cellular pathways impacted in AD pathogenesis that may be upstream or even independent of Aβ and pTau aggregation. Early endosome abnormalities are an early cytopathogical hallmark of AD, reported in both FAD and SAD, prior to the appearance of amyloid plaques [1, 65]. Israel and colleagues documented increased size of Rab5+ early endosomes from APP^{Dp} grown in 2D cultures [25], and Raja and colleagues also observed enlarged endosomes in neurons from an *APP*^{Dp} patient grown in 3D [64]. A large study generating an allelic series of isogenic lines harboring one or two copies of FAD mutations in *APP* and *PSEN1* demonstrated enlarged early endosomes in all genotypes studied [57]. This phenotype was gene-dose dependent with a larger increase in size in lines with homozygous mutations. In this study, the endosome size phenotype was rescued by treating with β-secretase inhibitors, suggesting that enlargement of these structures is dependent upon APP processing. Further examination of the functionality of the endosomal network by our group, Woodruff and colleagues, showed that neurons with FAD mutations have defective endocytosis of APP and LDL and that this impacts transcytosis of these proteins, a process by which cargo are endocytosed and then trafficked to the neuronal axon [28]. Taken together, this work suggests that abnormalities in vesicle trafficking may represent one of the earlier stages of dysfunction in FAD. Recent drug screening on *APP*^{Dp} neurons found that several inhibitors of cholesterol synthesis or storage led to reduced levels of pTau made by these cells. Interestingly, these compounds also reduced pTau independently of *APP* copy number or Aβ levels, suggesting that an independent pathway may control pTau levels [27]. Further work with these inhibitors established that storage forms of cholesterol lead indirectly to control of APP processing via a cholesterol binding pocket in APP [66].

6.2 SAD Models

Late-onset SAD is the most common form of AD and generally develops after 65 years of age. Although the established neuropathology of AD, amyloid plaques and neurofibrillary tangles, is similar to FAD, the age of the patients upon death generally presents more comorbidities in the brains, such as vascular pathology. Genetically, SAD germline genomes do not harbor causative mutations in *APP* or *PSEN1/2* but likely contain a polygenic mix of variants in multiple loci that confer a small degree or risk or protection against the disease. This genetic heterogeneity and unknown contributions of environmental and lifestyle factors make it more challenging to model SAD in vitro as the phenotypic results can be more variable [17, 25, 59].

SAD risk is effectively a complex genetic trait. Thus, there are several dozen loci now consistently associated with increased SAD risk. Broadly, these loci fall into pathways that involve immune regulation, endocytosis and vesicular trafficking, and cholesterol

Lawrence Goldstein, PhD, a professor in the Department of Cellular and Molecular Medicine, director of the UCSD Stem Cell Program, and faculty member with the Shiley-Marcos Alzheimer's Disease Research Center (ADRC), and his collaborators are conducting a research study to find out more about the genetic factors that may cause Alzheimer's disease or contribute to its progression, as well as the factors underlying dementia with Lewy bodies. You have been asked to take part because of the following reasons:

- You have Alzheimer's disease.
- You have dementia with Lewy bodies.
- You are known to carry a genetic change that causes Alzheimer's disease.
- You are related to a person who is a carrier of a genetic change that may cause Alzheimer's disease.
- You are related to a person who has dementia with Lewy bodies.
- You are a healthy adult with no memory problems and are suitable to be part of a control or comparison group.
- You are a research participant in the longitudinal study of the Shiley-Marcos ADRC or the UCLA Mary S. Easton Center for Alzheimer's Disease Research.

9 Project Information

You are being asked to provide skin cells from your forearm for the purpose of generating *human pluripotent stem cells.* Human pluripotent stem cells are unique cells that have the potential to become many kinds of cells in the body. For example, they can become muscle cells, blood cells, or even brain cells, called "neurons." The specific type of cell that a stem cell becomes is dependent on the way in which they are grown in the laboratory. These stem cells will be used to generate human brain cells (neurons) in the laboratory that have the same genetic makeup as you do. None of the skin cells you provide will be used to produce a human pregnancy or a cloned human person. We hope to create a "laboratory model" using all of the genes from someone who has Alzheimer's disease, who is at risk for Alzheimer's disease, who has dementia with Lewy bodies, or healthy elderly controls. Since we do not know what causes Alzheimer's disease or dementia with Lewy bodies, laboratory models may help us study the causes of the disease and test new medications and therapies. As part of this study, we will also be determining your genetic makeup by analyzing your DNA (DNA is the genetic material inside of your cells).

We will be enrolling approximately 50 subjects with Alzheimer's disease, 10 subjects at risk for Alzheimer's disease,10 subjects with dementia with Lewy bodies, and 50 normal controls in this study. Persons affected by Alzheimer's disease, at risk for Alzheimer's disease, those with dementia with Lewy bodies, and those serving as healthy controls will follow the same procedures in this study. This study will be done in one single visit of no more than 1 h in length. Most of the time will be spent in a discussion of the study as described in this informed consent document.

There is no guarantee that stem cells will be successfully created from your skin cells. If you are curious and wish to know later whether any of your skin cells produced stem cells you may contact *Dr. Lawrence Goldstein at 858-534-9700.*

10 Voluntary Choice

Providing your skin cells for this research project is completely voluntary. You have the right to agree or to refuse to provide your skin cells for this project. The quality of your current or future medical care and your relationship with UCSD and the Shiley-Marcos Alzheimer's Disease Research Center or UCLA and the UCLA Mary S. Easton Center for Alzheimer's Disease Research will NOT change in any way whether you agree or refuse to provide any cells for this research project.

11 What Will Happen to the Collected Stem Cells?

It is likely that the retrieved stem cells, which would be genetically matched to you, will be stored for many years. These stored stem cells are also likely to be used by researchers at other institutions for either nonprofit or profit ventures and for other research purposes. However, Dr. Goldstein will be responsible for deciding how your cells will be used. The specimens collected from you and the DNA that they contain may also be used in additional research to be conducted by the University of California and J. Craig Venter Institute personnel or other researchers outside of academia collaborating in this research. These specimens, DNA, and their derivatives may have significant therapeutic or commercial value. You consent to such uses.

One possible research use of these stored stem cells might involve changing some of their genes, i.e., genetically manipulating them. Another possible research use might be to study some of the stem cells by analyzing their DNA or by placing them into laboratory animals. In addition, the stored stem cells might be used in research related to human transplantation. These are just three common examples of what might happen to the stored stem cells. But there are many other future possible research uses that are simply unknown or impossible to predict at this time. You will have no say as to which institutions, nonprofit or commercial, or researchers may share the stem cells made from your skin cells or will have access to your coded genetic information. You will also have no say in who may be the recipient of transplanted cells if that ever becomes possible. If medically appropriate, and if cells are available, they will be available to you for transplantation.

Local ethical and scientific review committees must approve and future uses of stored stem cells, insuring their use in scientifically, ethically, and legally appropriate ways.

Dr. Lawrence Goldstein will use the cells or the genetic material derived from your skin cells for the procedures described above. It is possible that there might be additional research procedures in the future that could use this material. If Dr. Goldstein or another researcher wishes to use your sample for a research purpose not described in this consent form, they will send a request to the UCSD Institutional Review Board. This Board protects the rights and welfare of research subjects like you. The Board will determine if the research proposed is permissible.

The genetic material and products derived from it, i.e., genetically engineered stem cell lines, might be of research value to other investigators. If you agree to the use of your genetic material for future research conducted by someone other than Dr. Lawrence Goldstein who is located at the University of California, please sign the statement below authorizing such use.

__________________________ You agree that the genetic material and cells derived from your skin cell biopsy may be used for future research conducted by the University of California and regulated by the California Institute of Regenerative Medicine and the appropriate institutional review boards.

Even though this genetic material and future products derived from it may have some commercial value, you will not receive any compensation. The stem cells that are generated may have significant commercial potential in the future. However, by signing this form, you understand that there are no plans for you to receive any direct financial benefits from any future commercial development and scientific patents of discoveries made through the use of these stem cells. Dr. Goldstein will be responsible for deciding how these cells and their DNA will be used. The specimens collected from you and the DNA that they contain may also be used in additional research to be conducted by the University of California personnel collaborating in this research. These specimens, DNA, and their derivatives may have significant therapeutic or commercial value. You consent to such uses.

By signing this consent, you authorize the use and storage of your skin samples and the genetic material derived from them by Dr. Lawrence Goldstein. Dr. Goldstein, his associates, or his successors in these studies will keep your DNA specimen and/or the information derived from it indefinitely. You may revoke this authorization by submitting a statement in writing to Dr. Lawrence Goldstein, PhD (UCSD, 9500 Gilman Drive, La Jolla, CA 92093-0683).

You may also revoke your authorization to share your genetic material with other researchers to be used in future research not described here.

You may withdraw your consent for whatever reason at any time before your skin cells are used to create new stem cell lines or your genetic material is analyzed. If you decide after providing your cells that you do not want the specimens collected from you to be used for future research, you must inform Dr. Goldstein in writing, who will use his/her best efforts to stop any additional studies. However, in some cases, such as if your cells are grown up and are found to be generally useful or if their DNA has already been analyzed, it may be impossible to locate and stop such future research once the materials have been widely shared with other researchers.

12 How Will Your Skin Cells Be Collected?

If you agree to be in this study, the following will happen to you:

1. Skin biopsy procedure (done at the UCSD Shiley-Marcos ADRC, the UCSD Altman Clinical and Translational Research Institute (ACTRI), the UCLA Mary S. Easton Center for Alzheimer's Disease Research or at the subject's home depending on what is most convenient for the patient).
   (a) The area where the skin will be removed is first cleaned with disinfectant.
   (b) A numbing solution of 1% xylocaine with epinephrine will be injected just under the skin near the site of skin removal. This is much like a dentist injecting novacaine for dental work.
   (c) When the area of skin becomes numb after several minutes, a small punch is used to cut the skin into a small circle about the size of the head of a pencil eraser. This punch works similar to a cookie cutter. The loose piece of skin can then be picked up with a pair of tweezers and placed in a specially prepared tube.
   (d) You and the staff member doing the biopsy will be given a choice of hemostasis methods (described below) depending on your convenience and amount of bleeding (we expect little to no bleeding following a skin biopsy since the epinephrine causes vasoconstriction).

(i) If bleeding is minimal and you prefer, Polysporin ointment and a Band-Aid will be applied. A follow-up appointment will be made 10–14 days later at the UCSD ADRC or ACTRI to evaluate the skin biopsy site for healing. Alternatively, depending on what is most convenient for you, you will be advised to see your primary care physician to inspect the biopsy site. At this stage, the skin will have healed and only antibiotic ointment and a Band-Aid will be necessary for about 1 week.

(ii) If bleeding is significant, or if you prefer, hemostasis will be maintained by closing the small biopsy wound with 1–2 nonabsorbable nylon sutures. A follow-up appointment will be made 10–14 days later at the UCSD ADRC or ACTRI to remove the sutures and to evaluate the skin biopsy site for healing. Alternatively, depending on what is most convenient for you, you will be advised to see your primary care physician to remove the sutures and to inspect the biopsy site. At this stage, the skin will have healed and only antibiotic ointment and a Band-Aid will be necessary for about 1 week.

(iii) If bleeding is significant, or if you prefer, hemostasis will be maintained by closing the small biopsy wound with 1–2 gut chromic sutures, which are absorbed quickly ~2–3 weeks. A follow-up appointment will be made 10–14 days later at the UCSD ADRC or ACTRI to evaluate the skin biopsy site for healing. Alternatively, depending on what is most convenient for you, you will be advised to see your primary care physician to inspect the biopsy site. At this stage, the skin will have healed and only antibiotic ointment and a Band-Aid will be necessary for about 1 week.

(e) In the unlikely event that you feel pain, we recommend an over-the-counter pain killer such as acetaminophen or ibuprofen. You should not exceed 2 g of acetaminophen or 2400 mg of ibuprofen per day.

(f) You will be told to watch the site for any signs of redness or swelling. If any of these occur, please contact the ADRC nurse practitioner at 858-822-4800 or the ACTRI nurse at 858-657-5115 by phone immediately.

(g) Evidence of infection will be referred for antibiotic treatment.

(h) This procedure should take no more than 15 min.

The skin sample and its genetic material will be coded by number and no identifying information will be stored with it. It will be accompanied by research data already collected at the ADRC as part of your annual visits, such as your ability to carry out simple and complex household tasks, your ability to take care of personal hygiene, scores on different tests measuring your memory, language function, ability to reason, and your attention. No additional testing of this kind will be given as part of this study. This data is coded as well, and no identifying information is shared.

If you are a person carrying or at risk for carrying a genetic alteration that causes Alzheimer's Disease, such genetic information as well as clinical information (e.g., scores on tests of memory, etc. as above) will be communicated to Dr. Lawrence Goldstein by the referring doctor or researcher. This information will be associated with your cells according to the code number described above. It will therefore not be possible for study personnel to associate your identity with your genetic status.

2. The skin cell sample will be transferred to the laboratory of Dr. Lawrence Goldstein (or Dr. Fred Gage in the case of samples from patients with dementia with Lewy bodies). Your cells will then either be fused with existing human embryonic stem cells that were previously derived from excess embryos donated with permission of the donors or they may have special genes inserted into the skin cells to reprogram them to stem cells or brain cells. Your skin sample will be used for research purposes only.

Risks or discomforts from skin biopsy procedure Participation in this study may involve some physical risks or discomforts. These include:

1. *There is a slight risk that an individual might be allergic to the local anesthetic. In order to minimize this risk, each participant will be asked if they know of any reaction to anesthetic used at a dentist's office which is similar to what is being used in this study. If a person describes a reaction they will not be enrolled in the study.*
2. *A slight pinprick sensation will be felt when the anesthetic is first injected under the skin at the site of the skin biopsy. There might be some further pain as the anesthetic wears off. It will be necessary to keep the area clean and dry for several days. A small Band-Aid will be placed over the area. If the area is not kept clean, there is a possibility of a local infection.*
3. *A small scar may develop at the site of the skin removal. The size of the scar will be no larger than the size of the skin removed, approximately 1/8–1/4 inch. This is about the size of the head of a pencil eraser.*

4. *There may be psychological risks. Some people who provide skin cells for stem cell research might experience feelings of anxiety or regret, especially when considering the possibility that their skin cells may result in the creation of genetically matched stem cells. Some may also feel vulnerable and anxious during the consent process.*

Providing skin cells for this project involves some risk to your privacy. Efforts to protect you against this risk are discussed below. Because this is an investigational study there may be some unknown risks that are currently unforeseeable [NOTE: The biopsy procedure is a standard clinical procedure and not experimental nor investigational]. You will be informed of any significant new findings.

If you are injured as a direct result of participation in this research, the University of California will provide any medical care you need to treat those injuries. The University will not provide any other form of compensation to you if you are injured. *You may call the UCSD Human Research Protections Program office at (858) 246-4777* for more information about this, or to inquire about your rights as a research subject, or to report research-related problems.

13 What Are the Alternatives to Providing Your Skin Cells for This Project?

One of your alternatives is to not participate at all in this research project. You may decide to do nothing, or you may decide to provide your skin cells to another research project.

In the case that your skin cells are included in this research but you decide later not to participate any longer in this research project, you may (1) have your skin cells discarded, (2) provide your skin cells to another research project, or (3) return your skin sample to storage. Please note that these alternatives are possible only if your skin cells have not already been used by researchers for experiments.

14 What Are the Potential Benefits of Providing Your Skin Cells for This Project?

You would be providing your skin cells solely for the advancement of this research project and stem cell research in general. There will be no direct benefit to you from these procedures. There is no cost to participate in this study nor is there any compensation. The investigator, however, may learn more about the cause of Alzheimer's disease and dementia with Lewy bodies, and the effectiveness of possible new medications designed to slow the progression of these diseases. This knowledge may help other people with Alzheimer's disease or dementia with Lewy bodies in the future.

_______________ has explained this study to you and answered your questions. If you have other questions about the research, your rights as a research subject, or questions about any research-related problems, you may reach *Dr. Lawrence Goldstein at 858-534-9700*.

Participation in research is entirely voluntary. You may refuse to participate or withdraw at any time without jeopardy to the medical care you will receive at this institution or loss of benefits to which you are entitled.

15 How Will Your Privacy Be Protected?

The records of your involvement with this research project will be kept confidential to the extent allowed by law. All samples will be stored in the laboratory of Dr. Lawrence Goldstein and maintained in a coded manner without any identifying information. No information will be entered into your medical record. In addition to Dr. Goldstein, and members of the Goldstein laboratory, your DNA will also be studied by the nonprofit J. Craig Venter Institute. There may be future collaborators that are not known at this time who will have access to your coded genetic information.

If your skin cells are used to create new stem cell lines, then the resulting stem cells and any new stem cells that they produce will be a complete genetic match to you. To protect your genetic privacy, only your identification code, not your name, will be discoverable to the researchers who generate stem cells and the researchers who may later work with the resulting stored stem cells or your genetic information. The results of your DNA analyses will also be confidentially handled through the use of identification codes instead of names.

If as a result of participation in this study we obtain information that could significantly affect your health or well-being, we will attempt to inform you of the existence of this information. You may then decide if you wish to know what we have learned.

If you carry or are at risk for carrying genetic alterations that cause Alzheimer's disease, you will not learn whether or not you are actually a carrier through this study. Testing for genetic alterations that cause this form of Alzheimer's disease is available through a genetic counselor outside of this study. Personnel from the Shiley-Marcos ADRC or other referring doctors or researchers will inform you about this option if you request it.

Local and other regulatory agencies and project sponsors and funding agencies may review the research project records to ensure that your rights as a skin cell donor are being adequately protected.

Any report that the researchers publish will not include any information that will make it possible for readers to identify you as a skin cell donor.

Some people involved in genetic studies have felt anxious about the possibility of carrying an altered gene that they could possibly pass on to their children. Even though we will do our best to keep your information confidential, there is a possibility that your genetic risk for certain diseases is accidentally divulged to the wrong source, and if that happens, you might be discriminated against in obtaining life or health insurance, employment, or the ability to adopt children. You should also be aware that we may detect instances of nonpaternity. For example, if a person you believe is one of your parents is not actually your biological parent, the testing may inadvertently detect this. Ordinarily, you will not be informed of this, if it occurs. Instances are known in which a subject in research has been required to furnish genetic information as a precondition for in applying for health insurance and/or a job. Participation in this study does not mean that you have had genetic testing. Genetic testing means having a test performed and the results provided to you and your doctor. If you are interested in having genetic testing performed you should consult your doctor, as some commercial tests are available. Your doctor can provide you with the necessary information to determine if such a test would be appropriate for you.

16 Will You Receive Compensation?

You will not receive any cash, payment with goods or services, or receive any other valuable consideration for the cells you provide to this research project.

You have received a copy of this consent document and the Experimental Subject's Bill of Rights to keep.

You agree to participate in this study.

Subject's signature	Printed Name of Subject	Date

For Adults Incapable of Giving Consent:

Signature	*Printed Name*	*Date*
Legally Authorized Representative	*Legally Authorized Representative*	

Signature of Witness Presenting Consent	Printed Name of Witness Presenting Consent	Date

References

1. Cataldo AM et al (2000) Endocytic pathway abnormalities precede amyloid beta deposition in sporadic Alzheimer's disease and Down syndrome: differential effects of APOE genotype and presenilin mutations. Am J Pathol 157(1): 277–286
2. Serrano-Pozo A et al (2011) Neuropathological alterations in Alzheimer disease. Cold Spring Harb Perspect Med 1(1):a006189
3. Giandomenico SL, Sutcliffe M, Lancaster MA (2021) Generation and long-term culture of advanced cerebral organoids for studying later stages of neural development. Nat Protoc 16(2):579–602
4. Shou Y et al (2020) The application of brain organoids: from neuronal development to neurological diseases. Front Cell Dev Biol 8: 579659
5. Takahashi K et al (2007) Induction of pluripotent stem cells from adult human fibroblasts by defined factors. Cell 131(5):861–872
6. Takahashi K, Yamanaka S (2006) Induction of pluripotent stem cells from mouse embryonic and adult fibroblast cultures by defined factors. Cell 126(4):663–676
7. Kang X et al (2015) Effects of integrating and non-integrating reprogramming methods on copy number variation and genomic stability of human induced pluripotent stem cells. PLoS One 10(7):e0131128
8. Miura K et al (2009) Variation in the safety of induced pluripotent stem cell lines. Nat Biotechnol 27(8):743–745
9. Ma X, Kong L, Zhu S (2017) Reprogramming cell fates by small molecules. Protein Cell 8(5): 328–348
10. Weltner J, Trokovic R (2021) Reprogramming of fibroblasts to human iPSCs by CRISPR activators. Methods Mol Biol 2239:175–198
11. Nakagawa M et al (2008) Generation of induced pluripotent stem cells without Myc from mouse and human fibroblasts. Nat Biotechnol 26(1):101–106
12. Ye L et al (2013) Blood cell-derived induced pluripotent stem cells free of reprogramming factors generated by Sendai viral vectors. Stem Cells Transl Med 2(8):558–566
13. Gore A et al (2011) Somatic coding mutations in human induced pluripotent stem cells. Nature 471(7336):63–67
14. Assou S et al (2020) Recurrent genetic abnormalities in human pluripotent stem cells: definition and routine detection in culture supernatant by targeted droplet digital PCR. Stem Cell Rep 14(1):1–8
15. Buta C et al (2013) Reconsidering pluripotency tests: do we still need teratoma assays? Stem Cell Res 11(1):552–562
16. Rose SE et al (2018) Leptomeninges-derived induced pluripotent stem cells and directly converted neurons from autopsy cases with varying neuropathologic backgrounds. J Neuropathol Exp Neurol 77:353–360
17. Young JE et al (2015) Elucidating molecular phenotypes caused by the SORL1 Alzheimer's disease genetic risk factor using human induced pluripotent stem cells. Cell Stem Cell 16(4): 373–385
18. Chambers SM et al (2009) Highly efficient neural conversion of human ES and iPS cells by dual inhibition of SMAD signaling. Nat Biotechnol 27(3):275–280
19. Kim DS et al (2010) Robust enhancement of neural differentiation from human ES and iPS cells regardless of their innate difference in differentiation propensity. Stem Cell Rev Rep 6(2):270–281
20. Shi Y et al (2012) Human cerebral cortex development from pluripotent stem cells to functional excitatory synapses. Nat Neurosci 15(3):477–486. S1
21. Shi Y, Kirwan P, Livesey FJ (2012) Directed differentiation of human pluripotent stem cells to cerebral cortex neurons and neural networks. Nat Protoc 7(10):1836–1846
22. Shi Y et al (2012) A human stem cell model of early Alzheimer's disease pathology in Down syndrome. Sci Transl Med 4(124):124ra29
23. Maroof AM et al (2013) Directed differentiation and functional maturation of cortical interneurons from human embryonic stem cells. Cell Stem Cell 12(5):559–572
24. Yuan SH et al (2011) Cell-surface marker signatures for the isolation of neural stem cells, glia and neurons derived from human pluripotent stem cells. PLoS One 6(3):e17540
25. Israel MA et al (2012) Probing sporadic and familial Alzheimer's disease using induced pluripotent stem cells. Nature 482(7384): 216–220
26. Knupp A et al (2020) Depletion of the AD risk gene SORL1 selectively impairs neuronal endosomal traffic independent of amyloidogenic APP processing. Cell Rep 31(9):107719

27. van der Kant R et al (2019) Cholesterol metabolism is a druggable axis that independently regulates tau and amyloid-beta in iPSC-derived Alzheimer's disease neurons. Cell Stem Cell 24(3):363–375. e9
28. Woodruff G et al (2016) Defective transcytosis of APP and lipoproteins in human iPSC-derived neurons with familial Alzheimer's disease mutations. Cell Rep 17(3):759–773
29. Woodruff G et al (2013) The presenilin-1 DeltaE9 mutation results in reduced gamma-secretase activity, but not total loss of PS1 function, in isogenic human stem cells. Cell Rep 5(4):974–985
30. Young JE et al (2018) Stabilizing the Retromer complex in a human stem cell model of Alzheimer's disease reduces TAU phosphorylation independently of amyloid precursor protein. Stem Cell Rep 10(3):1046–1058
31. Vierbuchen T et al (2010) Direct conversion of fibroblasts to functional neurons by defined factors. Nature 463(7284):1035–1041
32. Mertens J et al (2015) Directly reprogrammed human neurons retain aging-associated transcriptomic signatures and reveal age-related nucleocytoplasmic defects. Cell Stem Cell 17(6):705–718
33. D'Souza GX et al (2020) The application of in vitro-derived human neurons in neurodegenerative disease modeling. J Neurosci Res 99:124–140
34. Zhang Y et al (2013) Rapid single-step induction of functional neurons from human pluripotent stem cells. Neuron 78(5):785–798
35. Wang C et al (2017) Scalable production of iPSC-derived human neurons to identify tau-lowering compounds by high-content screening. Stem Cell Rep 9(4):1221–1233
36. Babcock KR et al (2021) Adult hippocampal neurogenesis in aging and Alzheimer's disease. Stem Cell Reports 16:681–693
37. Li Puma DD, Piacentini R, Grassi C (2020) Does impairment of adult neurogenesis contribute to pathophysiology of Alzheimer's disease? A still open question. Front Mol Neurosci 13:578211
38. Xu Y et al (2020) GABAergic inhibitory interneuron deficits in Alzheimer's disease: implications for treatment. Front Neurosci 14:660
39. Yang N et al (2017) Generation of pure GABAergic neurons by transcription factor programming. Nat Methods 14(6):621–628
40. Abud EM et al (2017) iPSC-derived human microglia-like cells to study neurological diseases. Neuron 94(2):278–293. e9
41. McQuade A et al (2018) Development and validation of a simplified method to generate human microglia from pluripotent stem cells. Mol Neurodegener 13(1):67
42. Russo FB et al (2018) Modeling the interplay between neurons and astrocytes in autism using human induced pluripotent stem cells. Biol Psychiatry 83(7):569–578
43. Yao H et al (2016) The Na(+)/HCO3(−) co-transporter is protective during ischemia in astrocytes. Neuroscience 339:329–337
44. Fong LK et al (2018) Full-length amyloid precursor protein regulates lipoprotein metabolism and amyloid-beta clearance in human astrocytes. J Biol Chem 293(29): 11341–11357
45. Douvaras P, Fossati V (2015) Generation and isolation of oligodendrocyte progenitor cells from human pluripotent stem cells. Nat Protoc 10(8):1143–1154
46. Ehrlich M et al (2017) Rapid and efficient generation of oligodendrocytes from human induced pluripotent stem cells using transcription factors. Proc Natl Acad Sci U S A 114(11): E2243–E2252
47. Lancaster MA et al (2013) Cerebral organoids model human brain development and microcephaly. Nature 501(7467):373–379
48. Kadoshima T et al (2013) Self-organization of axial polarity, inside-out layer pattern, and species-specific progenitor dynamics in human ES cell-derived neocortex. Proc Natl Acad Sci U S A 110(50):20284–20289
49. Krefft O et al (2018) Generation of standardized and reproducible forebrain-type cerebral organoids from human induced pluripotent stem cells. J Vis Exp 131:e56768
50. Birey F et al (2017) Assembly of functionally integrated human forebrain spheroids. Nature 545(7652):54–59
51. Trujillo CA et al (2019) Complex oscillatory waves emerging from cortical organoids model early human brain network development. Cell Stem Cell 25(4):558–569. e7
52. Ormel PR et al (2018) Microglia innately develop within cerebral organoids. Nat Commun 9(1):4167
53. Zhang DY, Song H, Ming G (2021) Modeling neurological disorders using brain organoids. Semin Cell Dev Biol 111:4–14
54. Choi SH et al (2014) A three-dimensional human neural cell culture model of Alzheimer's disease. Nature 515(7526):274–278

55. Gonzalez C et al (2018) Modeling amyloid beta and tau pathology in human cerebral organoids. Mol Psychiatry 23(12):2363–2374
56. Lin YT et al (2018) APOE4 causes widespread molecular and cellular alterations associated with Alzheimer's disease phenotypes in human iPSC-derived brain cell types. Neuron 98:1141–1154
57. Kwart D et al (2019) A large panel of isogenic APP and PSEN1 mutant human iPSC neurons reveals shared endosomal abnormalities mediated by APP beta-CTFs, not Aβ. Neuron 104(2):256–270. e5
58. Yagi T et al (2011) Modeling familial Alzheimer's disease with induced pluripotent stem cells. Hum Mol Genet 20(23):4530–4539
59. Kondo T et al (2013) Modeling Alzheimer's disease with iPSCs reveals stress phenotypes associated with intracellular Abeta and differential drug responsiveness. Cell Stem Cell 12(4):487–496
60. Liu Q et al (2014) Effect of potent gamma-secretase modulator in human neurons derived from multiple presenilin 1-induced pluripotent stem cell mutant carriers. JAMA Neurol 71(12):1481–1489
61. Paquet D et al (2016) Efficient introduction of specific homozygous and heterozygous mutations using CRISPR/Cas9. Nature 533(7601):125–129
62. Ortiz-Virumbrales M et al (2017) CRISPR/Cas9-Correctable mutation-related molecular and physiological phenotypes in iPSC-derived Alzheimer's PSEN2 (N141I) neurons. Acta Neuropathol Commun 5(1):77
63. Muratore CR et al (2014) The familial Alzheimer's disease APPV717I mutation alters APP processing and Tau expression in iPSC-derived neurons. Hum Mol Genet 23(13):3523–3536
64. Raja WK et al (2016) Self-organizing 3D human neural tissue derived from induced pluripotent stem cells recapitulate Alzheimer's disease phenotypes. PLoS One 11(9):e0161969
65. Cataldo AM et al (2008) Down syndrome fibroblast model of Alzheimer-related endosome pathology: accelerated endocytosis promotes late endocytic defects. Am J Pathol 173(2):370–384
66. Langness VF et al (2021) Cholesterol-lowering drugs reduce APP processing to Abeta by inducing APP dimerization. Mol Biol Cell 32(3):247–259
67. Karch CM, Goate AM (2015) Alzheimer's disease risk genes and mechanisms of disease pathogenesis. Biol Psychiatry 77(1):43–51
68. Flamier A et al (2018) Modeling late-onset sporadic Alzheimer's disease through BMI1 deficiency. Cell Rep 23(9):2653–2666
69. Meyer K et al (2019) REST and neural gene network dysregulation in iPSC models of Alzheimer's disease. Cell Rep 26(5): 1112–1127. e9
70. Corder EH et al (1994) Protective effect of apolipoprotein E type 2 allele for late onset Alzheimer disease. Nat Genet 7(2):180–184
71. Farrer LA et al (1997) Effects of age, sex, and ethnicity on the association between apolipoprotein E genotype and Alzheimer disease. A meta-analysis. APOE and Alzheimer Disease Meta Analysis Consortium. JAMA 278(16): 1349–1356
72. Holtzman DM, Herz J, Bu G (2012) Apolipoprotein E and apolipoprotein E receptors: normal biology and roles in Alzheimer disease. Cold Spring Harb Perspect Med 2(3):a006312
73. Wang C et al (2018) Gain of toxic apolipoprotein E4 effects in human iPSC-derived neurons is ameliorated by a small-molecule structure corrector. Nat Med 24(5):647–657
74. Sullivan SE et al (2019) Candidate-based screening via gene modulation in human neurons and astrocytes implicates FERMT2 in Abeta and TAU proteostasis. Hum Mol Genet 28(5):718–735
75. Fjorback AW et al (2012) Retromer binds the FANSHY sorting motif in SorLA to regulate amyloid precursor protein sorting and processing. J Neurosci 32(4):1467–1480
76. Robbins JP et al (2018) Clusterin is required for beta-amyloid toxicity in human iPSC-derived neurons. Front Neurosci 12:504
77. Lidstrom AM et al (1998) Clusterin (apolipoprotein J) protein levels are increased in hippocampus and in frontal cortex in Alzheimer's disease. Exp Neurol 154(2):511–521
78. Pottier C et al (2013) TREM2 R47H variant as a risk factor for early-onset Alzheimer's disease. J Alzheimers Dis 35(1):45–49
79. Brownjohn PW et al (2018) Functional studies of missense TREM2 mutations in human stem cell-derived microglia. Stem Cell Rep 10(4): 1294–1307
80. Martins S et al (2020) IPSC-derived neuronal cultures carrying the Alzheimer's disease asso-

ciated TREM2 R47H variant enables the construction of an Aβ-induced gene regulatory network. Int J Mol Sci 21(12):4516
81. Miller JD et al (2013) Human iPSC-based modeling of late-onset disease via progerin-induced aging. Cell Stem Cell 13(6):691–705
82. Tagliafierro L, Zamora ME, Chiba-Falek O (2019) Multiplication of the SNCA locus exacerbates neuronal nuclear aging. Hum Mol Genet 28(3):407–421
83. Cohen-Carmon D et al (2020) Progerin-induced transcriptional changes in Huntington's disease human pluripotent stem cell-derived neurons. Mol Neurobiol 57(3): 1768–1777

Chapter 7

Modeling Alzheimer's Disease Using Human Brain Organoids

Karina Karmirian, Mariana Holubiec, Livia Goto-Silva, Ivan Fernandez Bessone, Gabriela Vitória, Beatriz Mello, Matias Alloatti, Bart Vanderborght, Tomás L. Falzone, and Stevens Rehen

Abstract

Alzheimer's disease (AD) is the primary cause of dementia, to date. The urgent need to understand the biological and biochemical processes related to this condition, as well as the demand for reliable in vitro models for drug screening, has led to the development of novel techniques, among which stem cell methods are of utmost relevance for AD research, particularly the development of human brain organoids. Brain organoids are three-dimensional cellular aggregates derived from induced pluripotent stem cells (iPSCs) that recreate different neural cell interactions and tissue characteristics in culture. Here, we describe the protocol for the generation of brain organoids derived from AD patients and for the analysis of AD-derived pathology. AD organoids can recapitulate beta-amyloid and tau pathological features, making them a promising model for studying the molecular mechanisms underlying disease and for in vitro drug testing.

Key words Brain organoids, Induced pluripotent stem cells, Amyloidosis, Tauopathies, neurodegenerative diseases, Alzheimer's disease

1 Introduction

Alzheimer's disease (AD) is pathologically distinguished by the presence of amyloid plaques (extracellular deposits of amyloid-β (Aβ) peptide aggregates), neuritic plaques, and neurofibrillary tangles (intracellular deposits of hyperphosphorylated tau). Other changes include synaptic loss, neuronal death, and astrogliosis [1, 2].

Aβ is produced from the sequential cleavage of the amyloid precursor protein (APP) by β- and γ-secretases [3]. From this processing, Aβ1–40 and Aβ1–42 peptides can be generated and

Karina Karmirian and Mariana Holubiec contributed equally to this work.

Jerold Chun (ed.), *Alzheimer's Disease: Methods and Protocols*,
Methods in Molecular Biology, vol. 2561, https://doi.org/10.1007/978-1-0716-2655-9_7,

lead to aggregation in β-sheet conformations to form oligomers and fibrils. Aβ1–42 has an extended C terminus that makes it more hydrophobic than Aβ1–40 and thus more likely to generate aggregates. APP can also be cleaved by α- and γ-secretases in the non-amyloidogenic pathway, producing the soluble APPα and preventing the generation of Aβ [3]. Increases in Aβ1–42 processing are detected by changes in the Aβ1–42/40 ratio observed in the cerebrospinal fluid of AD patients and result in its accumulation into amyloid plaques, a distinctive hallmark of AD pathology in the brain [4–6].

The microtubule-associated protein, tau, is mainly expressed in neurons and astrocytes and is the main component of neurofibrillary tangles. Tau is subject to different posttranslational modifications; of these, hyperphosphorylation is the most studied and relevant for disease [7, 8], because tau found in pathological lesions is highly phosphorylated. These modifications affect its conformation and, more importantly, its function [8].

Many widely used AD models are based on transgenic animals or synthetic Aβ peptide exposure in immortalized cells or animal-derived cultures [9–11]. Throughout the years, animal models have been instrumental in shedding light on the mechanisms underlying AD, but they have some limitations: (1) mice do not develop AD pathological features spontaneously during aging, thereby requiring the incorporation of several concomitant genetic mutations of human genes, which rarely occur in humans [12]; (2) although amyloidosis can be recapitulated with *APP* and presenilin 1 (*PSEN1*) mutations, tau pathology requires specific mutations or expression of human tau, since most APP/PS1 mice do not exhibit tauopathies, nor degeneration, during AD-like disease progression [13]; and (3) animal models do not include a complete human background, in that they lack known and unknown protein-protein interactions that may be key for the development of pathology. These factors may underlie the poor translational potential obtained from preclinical studies using rodents.

Since 2007, human-induced pluripotent stem cells (iPSC) have dramatically changed in vitro disease modeling. It became possible to reprogram somatic cells from patients and healthy donors (usually epithelial cells from urine samples or fibroblasts from skin biopsies) and generate iPSCs maintaining the same donor genetic background [14]. iPSC lines derived from familial AD (FAD) patients carrying *APP*, *PSEN1*, or *PSEN2* mutations have already been established [15–20]. iPSCs carrying *PSEN1/PSEN2* mutations have been differentiated into neurons that exhibited higher levels of Aβ 1–42 peptides than in control neurons [21]. In addition, neurons derived from iPSCs carrying the *APP* mutation (D678H) exhibited low levels of synaptophysin expression, deficits in neurite outgrowth, increased caspase-1 activity, tau hyperphosphorylation (at Thr181 and Ser396), as well as higher levels of Aβ

1–42 and Aβ 1–40, compared to control neurons [22]. These studies demonstrated that some pathological hallmarks of disease are reproduced in neuronal cultures derived from AD patient cells in vitro, effectively recapitulating Aβ and tau pathologies within a human genetic background. However, these in vitro models lack amyloid plaque deposition, neurofibrillary tangles, and complex cell-cell interactions within the extracellular matrix [23].

The formation of organized cellular clusters from iPSCs, which can be induced to form differentiated neural tissue, is a powerful strategy to overcome the aforementioned challenges. These structures, named brain organoids, can recapitulate some of the histology and distinct cell types present in the developing brain, such as ventricular proliferative zones and cortical layers. The protocol for brain organoid generation was first described in 2013 and was initially applied to model mutation-based diseases, such as microcephaly [24, 25]. Since then, brain organoids have been used to study several neurological disorders [26, 27].

Since brain organoids were conceived as a model for the study of the developing brain, it was debated whether they could model neurodegenerative diseases that appear late in life. Recent studies have demonstrated their potential for neurodegenerative disease modeling, including AD [23, 28, 29]. Furthermore, to overcome the limitations related to cell maturation and development in vitro, these organoids can be cultured for longer periods to allow cellular maturation and onset of disease phenotypes [30]. Today, a handful of publications have shown that FAD-derived iPSC brain organoids can reproduce APP and tau pathologies, along with synaptic loss [20, 31].

The dissemination of accessible protocols for culturing brain organoids derived from AD patients is of paramount importance. A detailed description of the techniques used to evaluate the disease phenotype, the sensitivity ranges of these techniques, and the amount of material and number of replicates required for robust analysis may reduce the time and resources required to progress from the bench to the bedside. Here, we describe a simplified protocol, with possible alternatives, for AD-brain organoid generation, utilized with an iPSC line carrying the *APP* Swedish mutation (APP^{SWE}) [32]. Validated reagents and methods used to evaluate AD pathology are described, including Western blot (WB), immunofluorescence, ELISA, and classical histochemistry. Using these tools, we demonstrated an increase in Aβ 1–42 levels, Aβ plaque deposition, and tau hyperphosphorylation in human brain organoids after 90 days in culture. The ability to effectively recapitulate the molecular hallmarks of AD in brain organoids, with the possibility of monitoring them in vitro, make these techniques valuable for AD pathology research and for drug screening.

2 Materials

2.1 Induced Pluripotent Stem Cell Culture

All reagents and cell culture media must be prepared under sterile conditions in biological safety hoods.

1. mTeSR® medium: Prepare according to manufacturer's instructions (using basal medium and supplements), then aliquot (in 50 mL volumes) into 50 mL Falcon tubes and store at 4 °C for up to 2 weeks. For long-term use, aliquot in 40 mL volumes and store at 20 °C for up to 6 months. Alternatively, StemFlex medium can be used as an equivalent stem cell culture medium.
2. Matrigel matrix: Thaw on ice, aliquot in 500 μL volumes, into 1.5 mL microtubes, and store at −20 °C. Alternatively, Geltrex™ LDEV-Free matrix can be used.
3. *Phosphate-buffered saline (PBS), pH 7.5 (10×)*: Add 80 g NaCl, 2 g KCl, 17.8 g Na_2HPO_4, and 2.4 g KH_2PO_4 to 800 mL of sterile dH_20, then adjust pH using either HCl or NaOH, and dilute to 1 L. Filter using a 0.2 μm acetate filter. Alternatively, purchase sterile-filtered 10× PBS that is suitable for cell culture.
4. 1× PBS: Dilute 10× PBS solution with sterile water.
5. Petri dish, 60 mm × 15 mm, tissue-culture-treated, sterile.
6. Syringe, 1 mL with insulin (31 gauge) needle, sterile.
7. Plastic cell scraper, sterile.
8. Polyethersulfone (PES) syringe filter, 0.22 μm, sterile.

2.2 Brain Organoid Formation

All reagents and cell culture media must be prepared under sterile conditions in biological safety hoods.

1. Accutase cell-detachment solution: Thawed at room temperature (RT), aliquoted in 10 mL volumes into 15 mL Falcon tubes, and stored at −20 °C. After the first use, store the aliquot at 4 °C.
2. Rho-kinase inhibitor solution (1 mM) (ROCKi: Y-27632): Reconstitute 5 mg in 14.78 mL sterile dH_20, aliquot 500 μL into 1.5 mL microtubes, and store at −20 °C for up to 1 year.
3. mTeSR® medium or StemFlex medium, prepared per the instructions in **item 1** (Subheading 2.1), and stored at 4 °C.
4. Dulbecco's modified eagle medium: Nutrient Mixture F-12 (DMEM/F-12) or Advanced DMEM/F-12, stored at 4 °C.
5. Neurobasal™ medium, stored at 4 °C.
6. KnockOut™ serum replacement (KOSR), stored at −20 °C.
7. Fetal bovine serum, embryonic stem cell quality, stored at −20 °C.

8. GlutaMAX™ supplement (100×).
9. *Penicillin-streptomycin solution* (*Pen-Strep*): 10,000 units penicillin and 10 mg streptomycin per mL, in 0.9% NaCl in sterile dH_20. Aliquot 10 mL in 15 mL Falcon tubes and store at −20 °C. After the first use, store the aliquot at 4 °C.
10. Minimum essential medium-nonessential amino acid solution (MEM-NEAA) (100×), stored at 4 °C.
11. Collagenase IV, 100 U/μM (1000× stock solution): 1 g of Collagenase IV in 1 mL DMEM, stored at −20 °C.
12. 2-Mercaptoethanol (≥99.0%), stored at 4 °C. When needed, dilute in DMEM/F12 immediately before use.
13. Heparin, grade I-A (≥180 UPS units/mg), 5 mg/mL in 1× PBS, stored at 4 °C.
14. N-2 supplement (100×), aliquoted in 1 mL volumes into 1.5 mL microtubes and stored at −20 °C.
15. B-27™ supplement (50×), serum-free, without vitamin A, aliquoted in 1 mL volumes into 1.5 mL microtubes, and stored at −20 °C.
16. B-27™ supplement (50×), serum-free, aliquoted in 1 mL volumes into 1.5 mL microtubes, and stored at −20 °C.
17. Human insulin solution, 9.5–11.5 mg/mL, sterile-filtered, suitable for cell culture, stored at 4 °C.
18. Matrigel® matrix, thawed on ice, aliquoted in 500 μL volumes into 1.5 mL microtubes, and stored at −20 °C.
19. Bovine serum albumin (BSA) in dH_20 (5 mL), sterile, 0.1%, stored at −20 °C.
20. Human fibroblast growth factor (FGF) basic recombinant protein (b-FGF) (10 μg/mL): Reconstitute 50 μg of b-FGF in 5 mL of sterile 0.1% BSA; store at −20 °C.
21. Ultra-low attachment multiple-well plates, 96 wells, with round bottoms, clear, sterile.
22. Ultra-low attachment 24-well plates, sterile.
23. Non-treated 60-mm culture dish, sterile.
24. Optional: Non-treated 100 mm culture dish, sterile (*see* **Step 1** in Subheading 3.1.3).
25. Tissue-culture plates, 6-well (two), sterile, for each batch of 96 brain organoids.
26. CO_2-resistant orbital shaker.
27. Cell culture flasks, 75 cm, sterile.
28. Cell culture incubator, 37 °C, humidified with 5% CO_2.
29. Falcon tubes, 15 mL, sterile.
30. Falcon tubes, 50 mL, sterile.

31. Pipettes, 5 mL, sterile.
32. Multichannel micropipette.
33. Centrifuge with rotors for 15 mL Falcon tubes and 96-well plates.
34. Inverted tissue-culture microscope.
35. Non-filtered pipette tips, 1000 μL, with cut edges, autoclaved, sterile.
36. P200 tip *box*, autoclaved, sterile.
37. Parafilm.
38. Hemocytometer.
39. *hESC medium*: DMEM/F12, 10% knockout serum replacement (KOSR), 3% ESC-quality FBS, 1% Pen-Strep, 1% GlutaMAX, 1% MEM-NEAA, 100 μM 2-mercaptoethanol, sterile-filtered (at 0.2 μm). Store at 4 °C for up to 2 weeks (*see* **Note 1**).
40. *Neuro-induction medium* (NIM): DMEM/F12, 1% Pen-Strep, 1% N-2, 1% GlutaMAX, 1% MEM-NEAA, 1 μg/mL heparin, sterile-filtered (at 0.2 μm). Store at 4 °C for up to 2 weeks (*see* **Note 2**).
41. *Neural differentiation medium with B27, without Vitamin* A (NDM(-VA)): DMEM/F12 and Neurobasal medium (1:1), 1% Pen-Strep, 0.5% N-2, 1% B-27 without vitamin A, 1% GlutaMAX, 1% MEM-NEAA, 50 μM 2-mercaptoethanol, 1: 4000 insulin. Store at 4 °C for up to 2 weeks (*see* **Note 2**).
42. *Neural differentiation medium with B27 and Vitamin A* (NDM(+VA)): DMEM/F12 and Neurobasal medium (1:1), 1% Pen-Strep, 0.5% N-2, 1% B-27, 1% GlutaMAX, 1% MEM-NEAA, 50 μM 2-mercaptoethanol, 1:4000 insulin. Store at 4 °C for up to 2 weeks (*see* **Notes 2** and **3**).

2.3 Western Blots

1. Microtubes, 1.5 mL, low protein-binding.
2. Liquid nitrogen.
3. *RIPA Buffer*: 1.5 mL of 1 M NaCl, 0.1 mL of 1% Nonidet P-40, 0.05 mL of 0.5% sodium deoxycholate (SDC), 0.01 mL of 0.1% SDS and 5 mL of 50 mM Tris–HCl pH 7.4; bring to 10 mL with dH_2O, store at 4 °C.
4. Polypropylene pellet pestles.
5. Na_3VO_4 (sodium orthovanadate) and NaF (sodium fluoride) (phosphatase inhibitors), stored at –20 °C.
6. *Protease inhibitor cocktail*: 104 mM 4-(2-aminoethyl)benzene-sulfonyl fluoride hydrochloride (AEBSF), 80 μM aprotinin, 4 mM bestatin, 1.4 mM E-64, 2 mM leupeptin, and 1.5 mM pepstatin A. Store at –20 °C.

7. Insulin syringes, 1 mL, with 31G needles (one for every 4 organoids).
8. Microtube centrifuge capable of at least 12,000 × *g*.
9. Pierce™ BCA protein assay kit. Alternatively, use a Bradford protein assay.
10. *Laemmli sample buffer* (*4×*): 0.8 g SDS, 4 mg bromophenol blue, 4 mL glycerol, 5 mL 0.5 M Tris–HCl, pH 6.8; bring to 10 mL with dH_2O. Add β-mercaptoethanol (to reach 2%) immediately before use.
11. Mini-Protean Tetra Cell glass plates (short and spacer), or equivalent, for 1–1.5 mm thick running gels.
12. Mini-Protean Tetra Cell tank, or equivalent.
13. Combs, 15-lane, for 1 or 1.5 mm gels.
14. Mini-Protean casting stands and frames (or equivalents) for 1 or 1.5 mm gels.
15. Acrylamide/bisacrylamide solution, 30%, 29:1, stored at 4 °C.
16. Tris–HCl, 1.5 M, pH 8.8.
17. Tris–HCl, 0.5 M, pH 6.8.
18. Ammonium persulfate solution (APS), 10% (g/100 mL) in dH_2O, stored in 50 μL aliquots at −20 °C.
19. Tetramethylethylenediamine (TEMED).
20. *Resolving gel* (12%): 4 mL of 30% acrylamide/bisacrylamide solution, 2.5 mL of pH 8.8 Tris–HCl buffer, 3.4 mL dH_20, 0.1 mL 10% sodium dodecyl sulfate (SDS), 50 μL 10% APS solution and 5 μL TEMED.
21. *Stacking gel* (4%): 0.52 mL of 30% acrylamide/bisacrylamide solution, 1 mL of Tris–HCl buffer, pH 6.8, 2.44 mL dH_20, 0.04 mL 10% SDS, 30 μL 10% APS solution and 3 μL TEMED.
22. See Blue Plus2 prestained protein standard, or equivalent molecular weight ladders.
23. *Running buffer for Tris-glycine SDS-PAGE* (*10×*): Mix 30.3 g Tris-base, 144.1 g glycine, 10 g SDS in 500 mL dH_20, bring to pH 7.4 and dilute to 1 L. Store at 4 °C.
24. Polyvinylidene difluoride (PVDF) membrane.
25. Wet/tank blotting system.
26. *Transfer buffer* (*10×*): Add 30.3 g Tris-base and 144.1 g glycine to 500 mL dH_20, then dilute to 1 L. Add methanol (to reach 20%) when preparing the 1× solution. Store at 4 °C.
27. *Ponceau Red staining solution*: 0.1 g Ponceau S (Anedra), 5 mL glacial acetic acid, diluted to 100 mL with dH_2O; store at RT.

Table 1
Primary antibodies used for Western blot analysis

Antibody	Dilution	Host	Supplier
APP	1:1000	Mouse	MAB348 Millipore
Aβ	1:500	Rabbit	PA5-32263 Thermo Fisher
Tau	1:1000	Mouse	ADx 201—ADx Neurosciences
p-Tau S396/404 (PHF1)	1:150	Mouse	ALZFORUM, Peter Davies
p-Tau T231	1:1000	Mouse	ADx 253—ADx Neurosciences

28. *Tris-buffer saline* (TBS), pH 7.5 (20×): Add 121 g Tris-base, 175.2 g NaCl to 800 mL dH_20, then adjust pH using either HCl or NaOH and dilute to 1 L.
29. *Tris-buffer saline/Tween 20* (TBS-T): 500 μL Tween 20 in 500 mL of 1× TBS.
30. BSA (5%) in TBS-T.
31. BSA (1% and 5%) in dH_20, stored at −20 °C.
32. SDS (10%) in dH_20.
33. Electrophoresis power supply (250 V).
34. Primary antibodies: *see* Table 1.
35. Secondary antibodies: polyclonal anti-rabbit and anti-mouse immunoglobulins, horseradish peroxidase-conjugated.
36. Pierce™ ECL Western blotting substrate, or equivalent chemiluminescence reagents.

2.4 Immunofluorescence

1. Cryostat.
2. Disposable, plastic 37 × 24 × 5 mm cast for tissue inclusion.
3. Tissue-Tek® O.C.T.™ compound.
4. Gelatin-coated 75 × 25 mm glass slides.
5. Mowiol mounting medium, stored at −20 °C.
6. TritonX-100.
7. Normal goat serum (NGS), stored at −20 °C.
8. BSA (1%).
9. *PBS, pH 7.5 (10×)*: Add 80 g NaCl, 2 g KCl, 17.8 g Na_2HPO_4, and 2.4 g KH_2PO_4 to 800 mL of dH20, then adjust pH using either HCl or NaOH and dilute to 1 L.
10. Sucrose solutions (5% and 30%) in PBS.
11. *Paraformaldehyde* (PFA; 4%): Dissolve 4 g of paraformaldehyde powder in 50 mL dH_20, heat to 80 °C while stirring. Add 1 M NaOH in drops until the liquid becomes clear, dilute to 100 mL with PBS and cool to 4 °C; can be stored at −20 °C.

Table 2
Primary antibodies used for immunofluorescence analysis

Antibody	Dilution	Host	Supplier
Nestin	1:750	Rabbit	RA22125 Neuromics
MAP2	1:300	Mouse	M1406 Sigma Aldrich
βIII-Tubulin (TUJ1)	1:500	Mouse	MRB-435P BioLegend
NeuN	1:500	Mouse	MAB377 Millipore
GFAP	1:500	Rabbit	Z0334 Dako
PAX6	1:100	Rabbit	42-6600 Thermo Fisher
APP	1:500	Mouse	MAB348 Millipore
Aβ	1:500	Rabbit	PA5-32263 Thermo Fisher
Tau	1:100	Mouse	ADx 201—ADx Neurosciences
p-Tau S396/404 (PHF1)	1:250	Mouse	ALZFORUM, Peter Davies

12. Microtubes, 1.5 mL.
13. Primary antibodies: *see* Table 2.
14. Secondary antibodies: (a) Alexa fluor 488 goat anti-rabbit IgG (H + L) conjugate, (b) Alexa fluor 568 goat anti-rabbit IgG (H + L) conjugate, (c) Alexa fluor 488 goat anti-mouse IgG (H + L) conjugate, and (d) Alexa fluor 568 goat anti-mouse IgG (H + L) conjugate.
15. Hoechst staining solution, stored at −20 °C.
16. Wet chamber: humidified blot paper in a closed plastic box with a grid that allows the positioning of slides.
17. Epifluorescence microscope, such as an inverted Olympus spinning-disk IX-83 (or equivalent), or a confocal fluorescence microscope.

2.5 ELISA Assay

1. ELISA kit (EuroImmun EQ. 6521-9601 L) for β-amyloid 1–42 detection, containing antibody-coated microplates, standard-curve calibrators and controls, biotin, enzyme conjugate, wash buffer, chromogen/substrate solution, and stop solution (0.5 M sulfuric acid). Equivalent ELISA kits suitable for Aβ 1–42 detection in conditioned medium can be used.
2. Pipette tips, low-binding, 200 μL.
3. Microtubes, low-binding, 1.5 mL.
4. Multichannel micropipette.
5. Microplate washer.
6. Microplate absorbance reader.

2.6 Staining

1. EtOH, 80%.
2. NaCl.
3. Cellulose acetate filters, 0.2 μm pore-size.
4. *Congo Red solution*: Add enough NaCl to 50 mL of 80% ethanol, with stirring, to reach saturation. Add 0.1 g of Congo Red and stir overnight. Filter the solution using a 0.2 μm cellulose acetate filter.
5. *Thioflavin solution*: 0.05 g Thioflavin-S in 10 mL dH_2O, filtered with a 0.2 μm cellulose acetate filter before use.
6. NaOH (1%) in dH_2O.

3 Methods

3.1 Generation of Human Brain Organoids

All cell culture procedures must be performed under sterile conditions in biological safety hoods.

3.1.1 Embryoid Body (EB) Formation

1. Use iPSC colonies cultured on Matrigel- or Geltrex-coated plates with defined borders and few differentiated cells. Use manual passage before beginning, to improve colony morphology (*see* **Note 4**). One 60 mm plate is enough to prepare organoids for a 96-well plate.
2. Observe the iPSC colonies under an inverted tissue-culture microscope (10× magnification) and manually remove differentiated cells using a sterile 20 μL tip attached to a 1000 μL tip to scratch the cells and detach them from the plate.
3. Wash once with PBS.
4. Add 1 mL Accutase solution to a 60 mm dish and place it in the incubator, at 37 °C, for 4–5 min.
5. After 4–5 min, transfer 1 mL from the plate to a 15 mL Falcon tube. Add 1 mL PBS to the plate, gently removing loosely attached cells by rinsing the plate.
6. Using a 1000 μL tip, transfer 1 mL from the plate to a 15 mL Falcon tube and dissociate the colonies until single cells or small clusters are formed.
7. Immediately add 20 μL of a 1 mM ROCKi aliquot to achieve a 10 μM final concentration and mix gently.
8. Count 10 μL of the cell solution using a hemocytometer: Add 10 μL of the cell suspension between the hemocytometer and the cover glass and count the number of cells in all four outer squares. Divide this number by four to obtain the mean of the number of cells/square. Multiply the number of cells per square by 10^4 and by 2 (dilution factor) to obtain the number of cells/mL of suspension. *Use this number in* **Step 11**, below.

If the cell density is too high to allow an accurate count, perform additional dilutions with PBS in **Step 6**, and multiply by the corresponding dilution factor.

9. Centrifuge at 300 × *g* for 4 min.
10. Remove the supernatant and, with a 5 mL pipette, add 1 mL of mTeSR® medium or StemFlex and gently resuspend the cell pellet. Add 10 μM ROCKi.
11. Add the appropriate volume of mTeSR® medium or StemFlex to achieve a final concentration of 60,000 cells/mL.
12. Based on the volume of medium added in **Steps 10** and **11**, add sufficient ROCKi solution to reach 50 μM final concentration.
13. Add b-FGF to a final concentration of 4 ng/mL. If using StemFlex, it is not necessary to add b-FGF.
14. Place 150 μL of the cell solution into each well of an ultra-low attachment 96-well plate using a multichannel pipette (this volume means that 9000 cells will be plated into each well, based on the 60,000 cells/mL solution prepared in **Step 11**). An embryoid body (EB) will form in each well.
15. Centrifuge the plate at 300 × *g* for 1 min.
16. Incubate at 37 °C, in 5% CO_2 (consider this to be Day 0).
17. On Day 1, remove and discard 75 μL of medium from each well and add 150 μL of hESC medium + 50 μM ROCKi solution + 4 ng/mL b-FGF (*see* **Note 5** for an alternative protocol).

    On Day 3, the EBs should have a diameter greater than 350–400 μm. Change the medium by removing 150 μL of the old medium and adding 150 μL of new hESC medium, but *do not add* more ROCKi and b-FGF (*see* **Note 6**).

3.1.2 Neuroepithelium Formation

1. On Day 6, the EBs should have a diameter greater than 500–600 μm. Transfer them, using a 1000 μL pipette tip with a cut edge, to a ultra-low attachment 24-well plate with 0.5 mL/well of *Neuro-induction medium* (NIM). Each well should have only one EB.
2. On Day 8, change the medium by removing 0.5 mL/well of the old medium and adding 1.0 mL/well of new NIM.
3. At this point, the EBs should have formed small protuberances on their surfaces. Healthy EBs should have smooth edges, and neuroepithelia should start to appear (as a translucent border).

3.1.3 Organoid Differentiation

1. After 4 days of neuroepithelium formation (Day 10), the organoids must be transferred to Matrigel. Prepare 3 mL Matrigel, diluted in ice-cold DMEM/F12 (following manufacturer's instructions), and add to a 60 mm non-adherent plate. For each 60 mm plate, transfer all the organoids from the 24-well

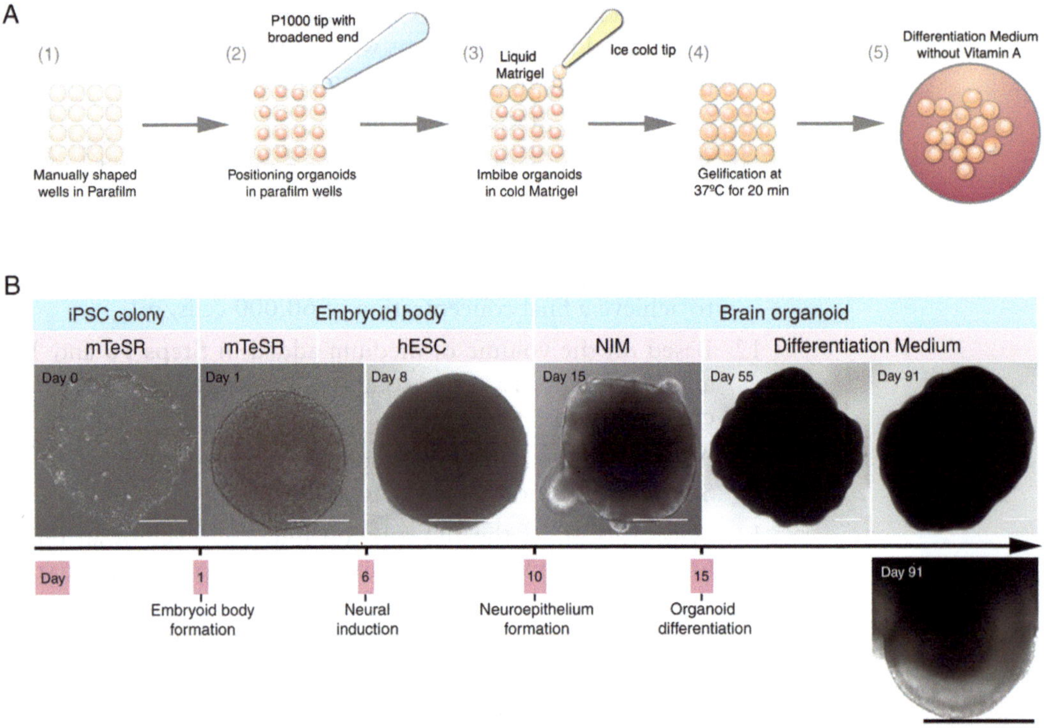

Fig. 1 Alternative inclusion protocol and timeline of brain organoid formation. (**a**) Procedure for an alternative manual Matrigel inclusion. All steps are carried out under strict sterile conditions and Matrigel is maintained on ice until the organoids are included. (1) Manually generate small "wells" in Parafilm, using an empty P200-tip sterile box as a mold. (2) Choose organoids with care, and use a P1000 tip with a broadened (cut) end for transfer, carefully positioning the organoids in each well. (3) Add 20 μL of cold Matrigel in each well using an ice-cold pipette tip; each organoid should be totally immersed in Matrigel. (4) Incubate at 37 °C to solidify the Matrigel. (5) Transfer organoids to a plate containing Differentiation medium. (**b**) Timeline depicting brain organoid formation. Representative micrographs illustrate different stages in organoid growth: iPSCs colonies (Day 0), EB formation from cellular aggregates (Days 1 and 8), EB with budding neuralization structures (Day 15), and differentiated brain organoids with neuroepithelial structures (Days 55 and 91) that are particularly visible at a higher magnification (*lower panel* of Day 91). *Scale bars*: 400 μm

plate; alternatively, organoids from four 24-well plates can be transferred to a 100 mm non-adherent plate (in this case, use 8 mL Matrigel).

For an alternative Matrigel inclusion method (*see* **Note 7** and Fig. 1a).

2. Incubate at 37 °C and 5% CO_2 for 1 h.
3. Prepare a 24-well plate by adding 0.5 mL of *Neural differentiation medium with B27, without Vitamin A* (NDM(-VA)) to each well.
4. After the 1-h incubation in Matrigel, transfer the organoids to the wells of the 24-well plate containing the NDM(-VA). Each well should have only one organoid.

5. After 3 days (Day 13), change the medium.
6. On Day 15, transfer the organoids to 6-well plates using cut 1000 μL pipette tips. 8–10 organoids should be placed per well in 3 mL of *Neural differentiation medium with B27 and Vitamin A* (NDM(+VA)). By that time, the organoids should start exhibiting a more complex neuroepithelium.
7. Place the organoids on an orbital shaker, set at 90 rpm, inside the 37 °C incubator (with 5% CO_2). From then on, brain organoids will be grown in suspension.
8. Change the medium twice a week (every 4 days) using NDM (+VA) and continue to culture the organoids until Day 90. On Day 45, consider placing some organoids in other wells to avoid attachment to each other as they become larger.

   See Fig. 1b for a schematic timeline of organoid development using this technique. *See* **Note 8**, regarding alternative protocols.

3.2 Analysis of APP and Tau Pathology

3.2.1 Immunofluorescence for Organoid Characterization and APP, Tau, and Amyloid Staining

1. Collect a subset of organoids into 1.5 mL microtubes (up to 4 organoids per microtube), decant the medium, and wash them twice with 37 °C PBS, mixing gently.
2. Carefully remove the remaining PBS using a pipette, then add 1 mL of 4% PFA and leave it at 4 °C overnight.
3. Using a pipette, carefully remove the PFA, then add 1 mL of 5% sucrose in PBS to each microtube. Repeat this procedure for a total of 3 times. Remove the 5% sucrose, using a pipette, then add 30% sucrose in PBS and incubate for 4 h.
4. Remove the sucrose solution and transfer the tissues into plastic casts (4 organoids per cast). Immediately add Tissue-Tek at RT and freeze at −20 °C.
5. Cut 20 μm sections, using the cryostat. Place each tissue section on gelatin-coated glass slides (6 Control (CTL) and 6 APP^{SWE} organoids per slide) and store them at −20 °C.
6. Before staining, allow the NGS and other staining solutions to come to RT.
7. Let the slides warm to RT, then wash the sections twice with PBS, for 10 min per wash (*see* **Note 9**).
8. Incubate the sections for 1 h at RT in a PBS solution containing 0.5% TritonX-100 and 5% NGS. During incubation, prepare the primary antibodies for **Step 9**.
9. Remove the PBS/TritonX/NGS solution from the slides, then incubate the sections with primary antibodies (anti-APP, anti-Tau, anti-Aβ, and anti-p-Tau S396/404) diluted in a PBS solution (*see* Table 2 for recommended concentrations) containing 0.1% Triton X-100 and 1% NGS. Leave them at 4 °C overnight in a wet chamber.

10. Recover the primary antibodies (if needed), then wash the slides, three times for 10 min each, with PBS.
11. Incubate for 2 h at RT with secondary antibodies diluted 1:500 in PBS. Keep the slides in a dark, wet chamber from this step forward.
12. Wash slides as in **Step 10**.
13. Counterstain nuclei with Hoechst staining solution diluted 1: 1000 in PBS, for 5 min at RT.
14. Wash three times with PBS, as above, then twice with dH_20.
15. Carefully place cover slips over the slides, using Mowiol as the mounting medium.
16. Let the slides air dry in darkness, then seal the edges using nail varnish.
17. The slides can be stored at −20 °C in a dark box until they are analyzed.
18. View slides using a confocal fluorescence microscope or a spinning-disk microscope, and use a digital camera to record images.
19. Perform reactive area and intensity analysis using ImageJ software.

    See Fig. 2 for the morphological characterization of brain organoids, including staining against: a neuronal progenitor marker (Pax6), an immature-neuron marker (Nestin), three different neuronal markers (NeuN, MAP-2, and βIII-Tubulin), and an astrocyte marker (GFAP). Figure 3a shows representative images of APP and Aβ staining, and Fig. 4a,b shows representative images of Tau and p-Tau S396/404 staining in Control and APP^{SWE} brain organoids.

3.2.2 Western Blot for APP and Tau Detection

1. Collect 4 organoids in a 1.5 mL microtube using a sterilized spatula. Carefully remove the medium and wash them twice with 1× PBS, mixing gently.
2. Quickly transfer the tissue to dry microtubes and fast-freeze them in liquid nitrogen.
3. Thaw the tissues on ice, then add 100 μL of *RIPA Buffer* to each tube.
4. Add protease inhibitor cocktail (2.5 μL) and phosphatase inhibitor solution (1 μL sodium orthovanadate), for every 4 organoids, and gently homogenize using a small plastic pestle against the microtube (*see* **Note 10**).
5. Further homogenize by passing the tissue through a 31G needle attached to a 1 mL insulin syringe (*see* **Note 11**).
6. Centrifuge at 12,100 × *g* for 30 min, at 4 °C.

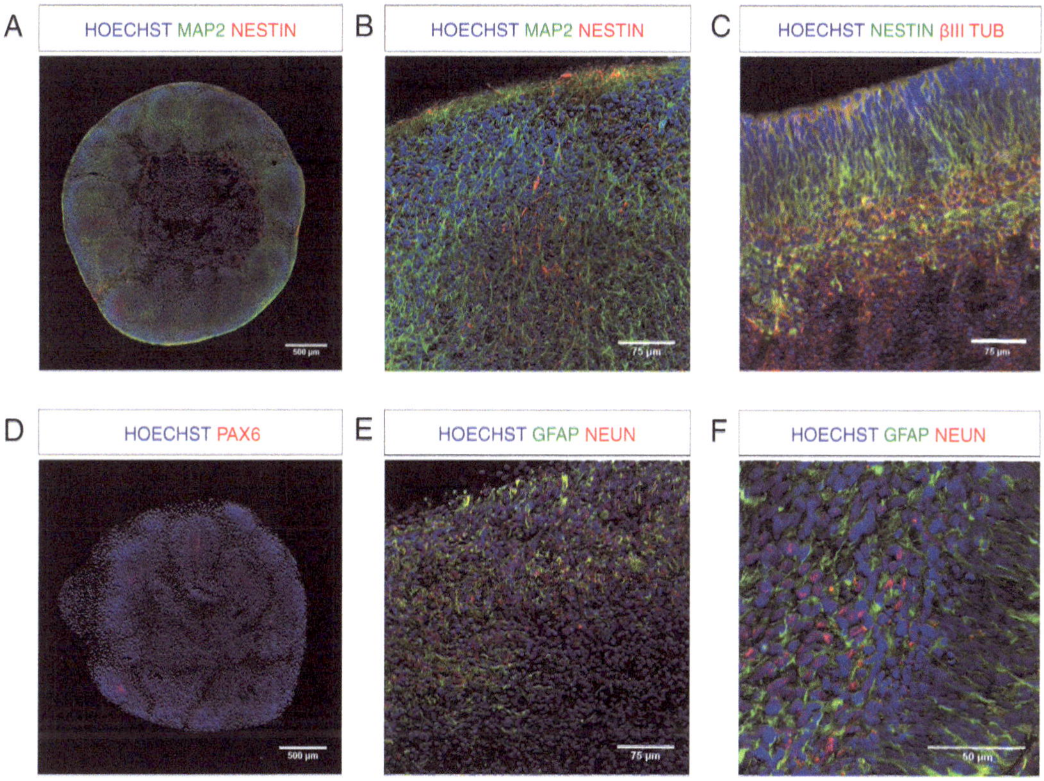

Fig. 2 Morphological characterization of brain organoids at Day 90. Representative micrographs depicting organoids immunostained for: (**a**, **b**) MAP2 (*green*) and Nestin (*red*), (**c**) Nestin (*green*) and βIII-tubulin (*red*), (**d**) Pax6 (*red*), (**e**, **f**) GFAP (*green*) and NeuN (*red*), and counterstained with Hoechst for nuclear staining (*blue*). MAP2 and Nestin staining are mainly localized in the periphery of the organoids. There are small, scattered areas with strong Pax6 immunostaining associated with neural precursors in neuroepithelia, while GFAP and NeuN are present in the periphery of organoids. NeuN, MAP2, and βIII-tubulin are representative markers of mature neurons, whereas GFAP identifies mainly astrocytes. Nestin and Pax6 indicate the presence of immature neurons and neural progenitor cells

7. Collect the supernatant and quantify its protein using the BCA protein assay kit or Bradford assay.
8. Dilute the supernatants in 4× Laemmli sample buffer, such that the final solution contains 40 μg of total protein for every 20 μL.
9. Prepare 12% electrophoresis resolving gel using the Mini-Protean Tetra Cell system or an equivalent.
10. Place the liquid resolving gel into a cast that is 1–1.5 mm thick, and wait for it to polymerize, about 15–30 min.
11. When the gel is polymerized, add the 4% stacking gel to the top, and quickly set the appropriate comb.
12. When the stacking gel is polymerized, the glass cast with the gel can be used immediately or be saved in a plastic bag with a small amount of dH_20 at 4 °C, for up to 30 days.

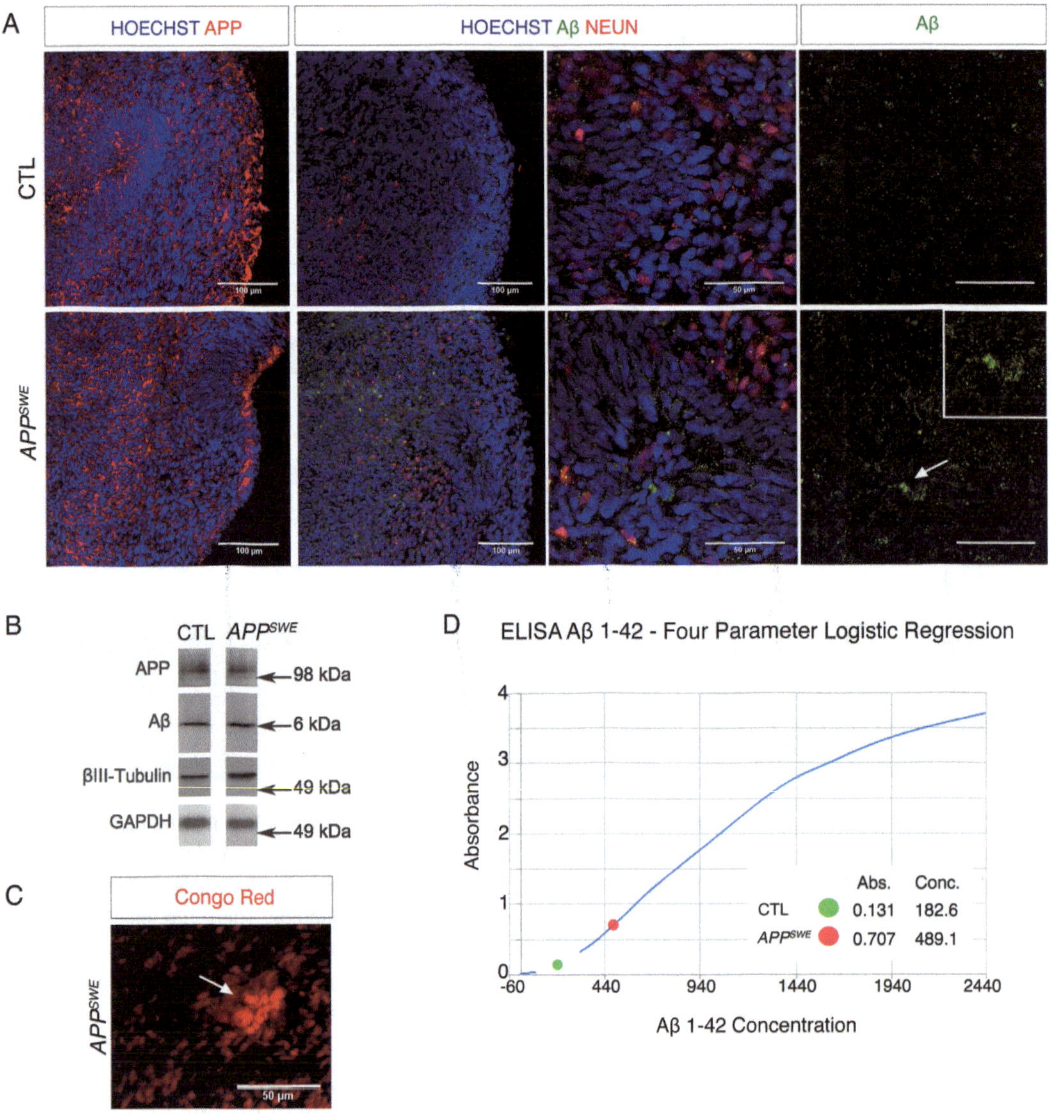

Fig. 3 Characterization of APP and Aβ in *APP*^{SWE} brain organoids at Day 90. (**a**) Representative micrographs depicting sections of organoids derived from cells taken from healthy donors (CTL) and from Swedish mutation (*APP*^{SWE}) patients. Organoid sections were immunostained for APP (*red*), Aβ (*green*), and NeuN (*red*). An increase in Aβ staining in *APP*^{SWE} organoids can be seen, as well as clear aggregates (*inset* and *white arrow*). (**b**) Representative examples of Western blots obtained using APP, Aβ, and βIII-Tubulin antibodies in CTL and *APP*^{SWE} samples, showing an increase in Aβ in *APP*^{SWE} organoids. Note the increase in βIII-Tubulin; this could be attributed to an increase in the ratio of neurons or of astrocyte progenitors. GAPDH is used as a loading control. (**c**) Representative image showing Congo Red staining (*red*) in *APP*^{SWE} organoids. Amyloid aggregates show higher fluorescence (*white arrow*). (**d**) A four-parameter logistic regression performed for the ELISA analysis of conditioned medium from CTL and *APP*^{SWE} organoids, showing an increase in Aβ 1–42 levels in *APP*^{SWE} samples

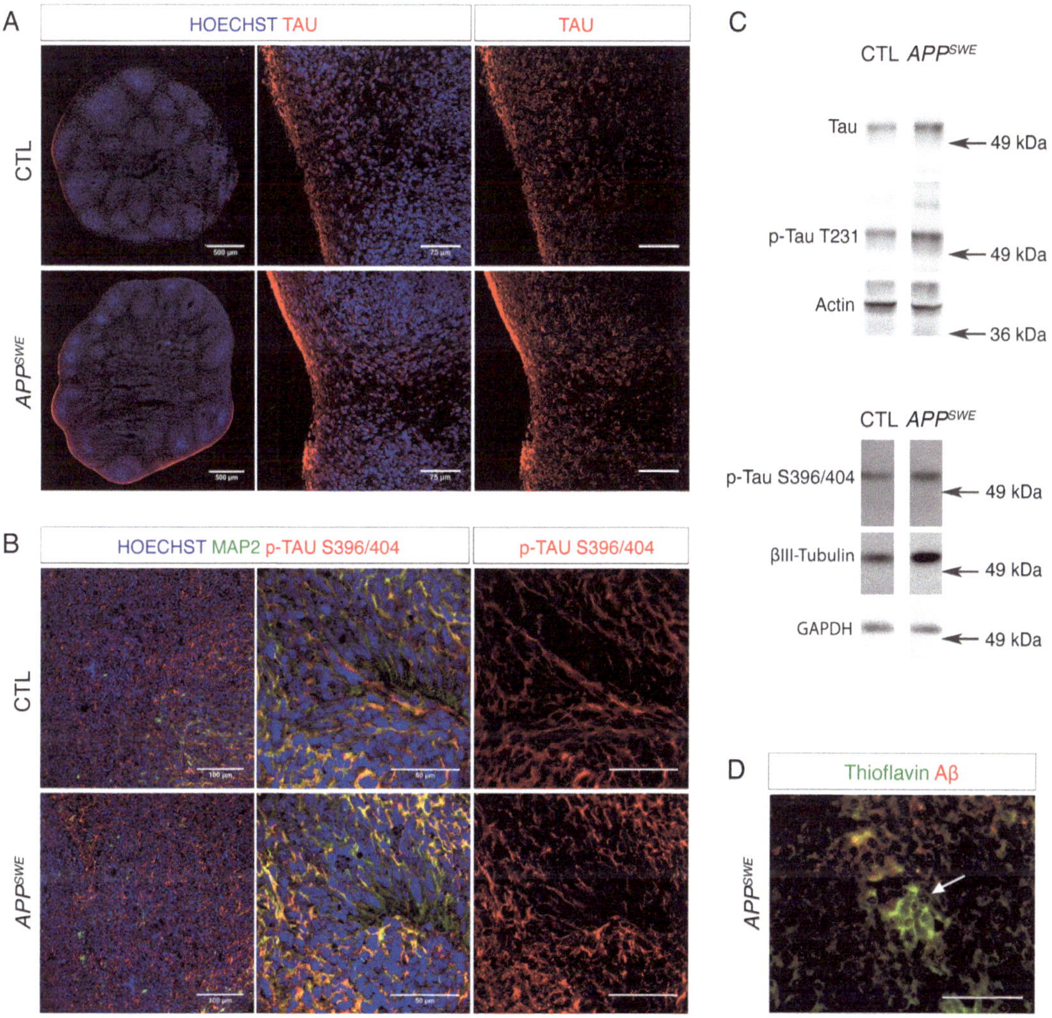

Fig. 4 Characterization of Tau and p-Tau in *APP*^{SWE} brain organoids at Day 90. (**a**) Representative micrographs showing Tau (*red*) with Hoechst (*blue*) nuclear counterstaining. The *APP*^{SWE} organoids exhibit an increase in Tau. (**b**) Micrographs depicting p-Tau S396/404 (*red*) and MAP2 (*green*) staining, with nuclei counterstained with Hoechst (*blue*). (**c**) Representative examples of Western blots obtained using Tau, p-Tau T231, p-Tau S396/404, and βIII-Tubulin, on samples from CTL and *APP*^{SWE} organoids, where an increase in tau forms and βIII-Tubulin are visible. GAPDH and actin are used as loading controls. (**d**) Representative image showing Thioflavin staining (*green*) and Aβ (*red*) immunolabeling in *APP*^{SWE} organoids. Note that some Thioflavin-positive staining does not colocalize with strong Aβ immunolabeling (*white arrow*)

13. Set the glass cast, with the solidified gels, into the running tank and fill it with Running buffer.
14. Load 20 μL of each sample, containing 40 μg of total protein, in each lane. Store any remaining sample at −80 °C for later analysis.
15. Load 5 μL of See Blue Plus2 prestained protein standard, or an equivalent molecular weight ladder, in one or more lanes.

16. Run at 80 V, until the buffer dye-front reaches the end of the stacking gel, then change the voltage to 100 V for the rest of the run.
17. Stop the run before the buffer dye-front reaches the end of the gel (*see* **Note 12**).
18. Transfer the proteins to a PVDF membrane, using a wet/tank blotting system, at 30 mA for 1 h and 45 min (*see* **Note 13**).
19. Immediately after the transference is over, place the membrane in a container with TBS. The subsequent steps are carried out in the container.
20. To stain proteins, remove the TBS and add the Ponceau Red solution. Incubate the membrane for 5 min, with gentle agitation, then rinse with 70% EtOH until the bands are visible (*see* **Notes 14** and **15**).
21. After documenting the Ponceau staining, wash away the remaining solution using TBS.
22. Incubate the membrane for 1 h in 5% BSA dissolved in TBS-T, to block nonspecific epitopes. Do not block with milk, since it contains phosphatases.
23. Dilute the corresponding primary antibodies in a 1% BSA solution, add the antibodies to the membrane and incubate overnight at 4 °C.
24. Wash the membrane 5 times using TBS-T (5 min each) at RT.
25. Incubate for 2 h with the corresponding secondary antibody.
26. Wash the membrane 5 times using TBS-T (5 min each) at RT.
27. Incubate membranes with ECL, or an equivalent reagent, and detect chemiluminescence using the ChemiDoc system or an equivalent method.
28. Quantify bands using ImageJ software.

    See Fig. 3b for representative images of APP and Aβ detection in Western blot analysis and Fig. 4c for Tau, p-Tau T231 and p-Tau S396/404 detection.

3.2.3 ELISA Assay for Aβ 1–42 Detection in Conditioned Medium

1. On Day 90 of brain organoid formation, collect 1.5 mL of conditioned medium from each of 8–10 brain organoids (that were cultured in 6-well plates), at least 4 days after the last medium change.
2. Transfer the conditioned medium to low-binding microtubes, using low-binding pipette tips, and centrifuge at 300 × *g* for 5 min, to remove cellular debris. Store the supernatant in 500 μL aliquots at −80 °C.
3. Keep the antibody-coated microplate at RT for 30 min before opening the protective wrapping.
4. Reconstitute calibrators and controls with 500 μL dH_2O for 10 min, before use.

References

1. Selkoe DJ (1991) The molecular pathology of Alzheimer's disease. Neuron 6(4):487–498. https://doi.org/10.1016/0896-6273(91)90052-2
2. Schellenberg GD, Montine TJ (2012) The genetics and neuropathology of Alzheimer's disease. Acta Neuropathol 124:305–323. https://doi.org/10.1007/s00401-012-0996-2
3. Zhang Y, Thompson R, Zhang H, Xu H (2011) APP processing in Alzheimer's disease. Mol Brain 4:3. https://doi.org/10.1186/1756-6606-4-3
4. Kakuda N (2017) Distinct deposition of amyloid-β species in brains with Alzheimer's disease pathology visualized with MALDI imaging mass spectrometry. Acta Neuropathol Commun 5:1–8
5. Lewczuk P, Lelental N, Spitzer P et al (2014) Amyloid-β 42/40 cerebrospinal fluid concentration ratio in the diagnostics of Alzheimer's disease: validation of two novel assays. JAD 43:183–191. https://doi.org/10.3233/JAD-140771
6. Long JM, Holtzman DM (2019) Alzheimer disease: an update on pathobiology and treatment strategies. Cell 179:312–339. https://doi.org/10.1016/j.cell.2019.09.001
7. Johnson GVW (2004) Tau phosphorylation in neuronal cell function and dysfunction. J Cell Sci 117:5721–5729. https://doi.org/10.1242/jcs.01558
8. Noble W, Hanger DP, Miller CCJ, Lovestone S (2013) The importance of Tau phosphorylation for neurodegenerative diseases. Front Neurol 4. https://doi.org/10.3389/fneur.2013.00083
9. Götz J, Ittner LM (2008) Animal models of Alzheimer's disease and frontotemporal dementia. Nat Rev Neurosci 9:532–544. https://doi.org/10.1038/nrn2420
10. Duyckaerts C, Potier M-C, Delatour B (2007) Alzheimer disease models and human neuropathology: similarities and differences. Acta Neuropathol 115:5–38. https://doi.org/10.1007/s00401-007-0312-8
11. Hall AM, Roberson ED (2012) Mouse models of Alzheimer's disease. Brain Res Bull 88:3–12. https://doi.org/10.1016/j.brainresbull.2011.11.017
12. Kitazawa M, Medeiros R, M LaFerla F (2012) Transgenic mouse models of Alzheimer disease: developing a better model as a tool for therapeutic interventions. CPD 18:1131–1147. https://doi.org/10.2174/138161212799315786
13. Sasaguri H, Nilsson P, Hashimoto S et al (2017) APP mouse models for Alzheimer's disease preclinical studies. EMBO J 36(17):2473–2487. https://doi.org/10.15252/embj.201797397
14. Takahashi K, Tanabe K, Ohnuki M et al (2007) Induction of pluripotent stem cells from adult human fibroblasts by defined factors. Cell 131:861–872. https://doi.org/10.1016/j.cell.2007.11.019
15. Essayan-Perez S, Zhou B, Nabet AM et al (2019) Modeling Alzheimer's disease with human iPS cells: advancements, lessons, and applications. Neurobiol Dis 130:104503. https://doi.org/10.1016/j.nbd.2019.104503
16. Zhang W, Jiao B, Zhou M et al (2016) Modeling Alzheimer's disease with induced pluripotent stem cells: current challenges and future concerns. Stem Cells Int 2016:1–12. https://doi.org/10.1155/2016/7828049
17. Oksanen M, Hyötyläinen I, Voutilainen J et al (2018) Generation of a human induced pluripotent stem cell line (LL008 1.4) from a familial Alzheimer's disease patient carrying a double KM670/671NL (Swedish) mutation in APP gene. Stem Cell Res 31:181–185. https://doi.org/10.1016/j.scr.2018.07.024
18. Li Y, Park C, Vellón L, Li X (2016) iPSCs: from bench to clinical bed. Stem Cells Int 2016:1–2. https://doi.org/10.1155/2016/8367587
19. Yin J, VanDongen AM (2021) Enhanced neuronal activity and asynchronous calcium transients revealed in a 3D organoid model of Alzheimer's disease. ACS Biomater Sci Eng 7(1):254–264. https://doi.org/10.1021/acsbiomaterials.0c01583
20. Raja WK, Mungenast AE, Lin Y-T et al (2016) Self-organizing 3D human neural tissue derived from induced pluripotent stem cells recapitulate Alzheimer's disease phenotypes. PLoS One 11:e0161969. https://doi.org/10.1371/journal.pone.0161969
21. Yagi T, Ito D, Okada Y et al (2011) Modeling familial Alzheimer's disease with induced pluripotent stem cells. Hum Mol Genet 20:4530–4539. https://doi.org/10.1093/hmg/ddr394
22. Chang K-H, Lee-Chen G-J, Huang C-C et al (2019) Modeling Alzheimer's disease by induced pluripotent stem cells carrying APP D678H mutation. Mol Neurobiol 56:

3972–3983. https://doi.org/10.1007/s12035-018-1336-x
23. Gerakis Y, Hetz C (2019) Brain organoids: a next step for humanized Alzheimer's disease models? Mol Psychiatry 24:474–478. https://doi.org/10.1038/s41380-018-0343-7
24. Lancaster MA, Renner M, Martin CA, Wenzel D, Bicknell LS, Hurles ME, Homfray T, Penninger JM, Jackson AP, Knoblich JA (2013) Cerebral organoids model human brain development and microcephaly. Nature 501(7467):373–379. https://doi.org/10.1038/nature12517
25. Kadoshima T, Sakaguchi H, Nakano T, Soen M, Ando S, Eiraku M, Sasai Y (2013) Selforganization of axial polarity inside-out layer pattern and species-specific progenitor dynamics in human ES cell–derived neocortex. Proc Natl Acad Sci 110(50):20284–20289.
26. Di Lullo E, Kriegstein AR (2017) The use of brain organoids to investigate neural development and disease. Nat Rev Neurosci 18: 573–584. https://doi.org/10.1038/nrn.2017.107
27. Fernandez I, Navarro BJ, Martinez E, Karmirian K, Holubiec M, Alloatti M, Goto-Silva L, Arnaiz Yepez C, Martins-de-Souza D, Nascimento JM, Bruno L, Saez TM, Rehen SK, Falzone TL (2022) DYRK1A regulates the bidirectional axonal transport of APP in human-derived neurons. J Neurosci JN-RM-2551-21-jneuro; JNEUROSCI. 2551-21.2022v1. https://doi.org/10.1523/JNEUROSCI.2551-21.2022
28. Amin ND, Paşca SP (2018) Building models of brain disorders with three-dimensional organoids. Neuron 100:389–405. https://doi.org/10.1016/j.neuron.2018.10.007
29. Paca SP (2018) The rise of three-dimensional human brain cultures. Nature 553:437–445. https://doi.org/10.1038/nature25032
30. Gordon A, Yoon S-J, Tran SS et al (2021) Long-term maturation of human cortical organoids matches key early postnatal transitions. Nat Neurosci. https://doi.org/10.1038/s41593-021-00802-y
31. Gonzalez C, Armijo E, Bravo-Alegria J et al (2018) Modeling amyloid beta and tau pathology in human cerebral organoids. Mol Psychiatry 23:2363–2374. https://doi.org/10.1038/s41380-018-0229-8
32. Haass C, Lemere CA, Capell A et al (1995) The Swedish mutation causes early-onset Alzheimer's disease by beta-secretase cleavage within the secretory pathway. Nat Med 1(12): 1291–1296. https://doi.org/10.1038/nm1295-1291
33. Tanaka Y, Cakir B, Xiang Y, Sullivan GJ, Park IH (2020) Synthetic analyses of single-cell transcriptomes from multiple brain organoids and fetal brain. Cell Rep 30(6):682–1689. https://doi.org/10.1016/j.celrep.2020.01.038

Chapter 8

Three-Dimensional Biohybrid StarPEG–Heparin Hydrogel Cultures for Modeling Human Neuronal Development and Alzheimer's Disease Pathology

Tohid Siddiqui, Hilal Celikkaya, Zeynep Tansu Atasavum, Stanislava Popova, Uwe Freudenberg, Carsten Werner, and Caghan Kizil

Abstract

In this chapter, we present the methodology currently used in our laboratory to generate a starPEG-MMP (starPEG)- and heparin maleimide HM06 (heparin)-based 3D cell culture system, in a hydrogel, that can be used to study human neuronal development and Alzheimer's disease (AD) pathology. A 3D cell culture system can mimic the in vivo cellular environment better than a 2D format, in which these cells exhibit neural network formation, electrophysiological activity, tissue-specific extracellular matrix (ECM) deposition, and neurotransmitter responsiveness. When treated with amyloid beta-42 (Aβ42) peptides, this system recapitulates many of the pathological effects of AD, including reduced neural stem cell proliferation, impaired neuronal network formation, dystrophic axonal ends, synaptic loss, failure to deposit ECM, elevated tau hyperphosphorylation, and formation of neurofibrillary tangles. Culturing human primary cortical astrocyte (pHA)- or induced pluripotent stem cell (iPSC)-derived human neural stem cells in this biohybrid hydrogel system has led to the discovery of novel regulatory pathways underlying neurodegenerative pathology in different phases of AD.

Key words Alzheimer's disease, Human neural development, 3D, Primary human astrocytes, iPSC-derived neural stem cells, Heparin, StarPEG, Biohybrid hydrogel cell culture, Amyloid beta-42 (Aβ42)

1 Introduction

Understanding the cellular physiology and network dynamics of human neurons and developing viable drug candidates against neurological disease require working with human cells in experimental settings that resemble the human nervous system. Although 2D cell culture formats are usually employed for these purposes, culturing human cells in a 3D system generates a microenvironment that is more physiologically relevant [1, 2].

Jerold Chun (ed.), *Alzheimer's Disease: Methods and Protocols*,
Methods in Molecular Biology, vol. 2561, https://doi.org/10.1007/978-1-0716-2655-9_8,

Here, we describe establishing a 3D biohybrid hydrogel culture, made of star-shaped polyethylene glycol with matrix metalloproteinase (MMP) cleavage sites (starPEG-MMP or *starPEG*) and end-functionalized glycosaminoglycan (GAG) heparin maleimide HM06 (*heparin*). This culture methodology is useful with various cell types, including, but not limited to, human primary neural stem cells and induced pluripotent stem cells (iPSCs) [3, 4]. In this protocol, we detail the maintenance of primary human astrocytes, the embedding procedure that employs 3D starPEG–heparin hydrogels, the treatment of cells with Aβ42 peptides, and optimal cell culture conditions.

There are numerous features of this culture system that are advantageous: its synthetic nature; its highly defined and reproducible composition; the possibility of customizing physicochemical properties, such as the degree of cross-linking or solid content; its transparency; its bio-modifiable nature (owing to the MMP cleavage sites); and its adaptability to high-throughput screening using automated liquid-handling systems. Cells cultured using this method exhibit a robust biological resemblance to nascent human brain tissues, in terms of neuronal subtype identity, patterns of gene expression, ECM composition, neuronal functionality, and successful recapitulation of neurodegenerative pathophysiology, such as the amyloid toxicity seen in Alzheimer's disease (AD) [5, 6] (Fig. 1).

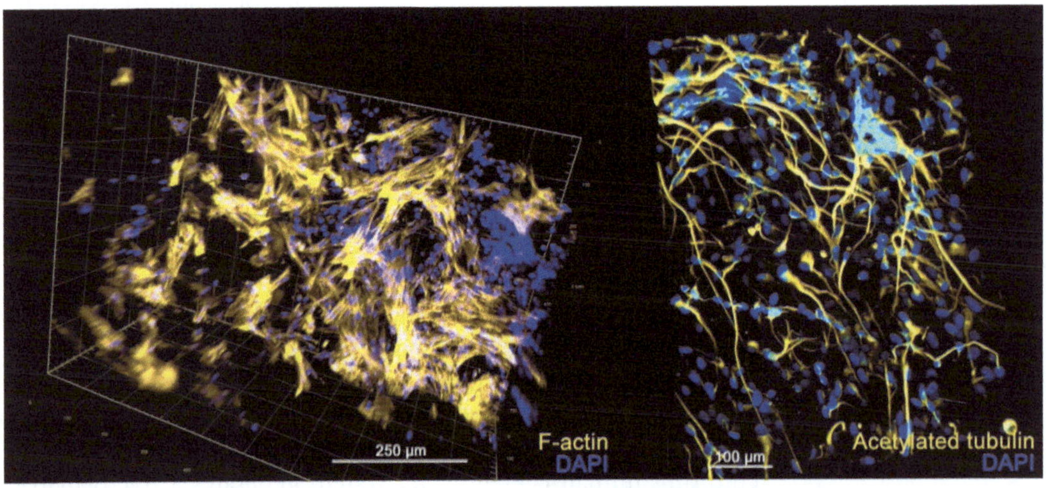

Fig. 1 Representative 3D rendered images. *Left*, a portion of the gel showing intricate networks in 3D. *Right*, high-power image showing neurons forming long extensions and connections

2 Materials

All materials used to prepare the hydrogels should be sterile and cell culture grade. Reagents should be stored at room temperature (RT), unless indicated otherwise.

2.1 Primary Human Astrocyte (pHA) Culture

1. Primary human astrocytes (*see* **Note 1**) kept at 4 °C (*see* Subheading 3.1, below).
2. *Complete astrocyte medium* (CAM): Astrocyte medium supplemented with fetal bovine serum, 2% (v/v); astrocyte growth supplement, 1% (v/v); and penicillin–streptomycin solution, 1% (v/v), and stored at 4 °C.
3. Cell culture incubator equipped with 95% air/5% CO_2 environment.
4. Accutase solution, sterile (stored at 4 °C temperature but brought to RT before use; *see* **Note 2**).
5. Double-distilled PBS (DPBS), sterile.
6. Parafilm or other hydrophobic sheet or surface, sterile.
7. Spatulas, disposable, sterile (*see* **Note 3**).
8. Two T25 cell culture flasks, sterile.
9. Serological pipettes; size: 2 mL, 5 mL, and 10 mL; sterile.
10. Microliter pipettors and tips; size: P1000, P100, and P10; sterile.
11. Centrifuge tubes; conical; size: 15 mL; sterile.
12. Non-cell-culture-treated plate, 24-well (*see* **Note 4**), sterile.
13. Tubes, 1.5 mL, sterile.
14. Cell culture vacuum aspiration system.
15. Manual cell-counting chamber or automated cell counter.
16. Inverted microscope with phase-contrast filter (*see* **Note 5**).
17. Vortex.

2.2 StarPEG-MMP and Heparin Maleimide HM06

1. Heparin, MW: 14.730 g/mol (manufacturer/supplier: ZetaScience GmbH; *see* **Note 6**).
2. StarPEG, MW: 15.800 g/mol (manufacturer/supplier: ZetaScience GmbH; *see* **Note 6**).

2.3 Amyloid Beta-42 Peptides

1. Aβ42 peptides (1 mM, prepared as in **Note** 7), in Milli-Q water with acetonitrile (1:1), sterile.
2. Automated solid-phase peptide synthesizer.
3. *Cleavage solution*: trifluoroacetic acid (TFA, 90 v/v), tri-isopropylsilane (TIS, 5 v/v), ddH_2O (2.5 v/v), dithiothreitol (DTT, 2.5 m/v).

4. Diethyl ether, ice-cold.
5. High-performance liquid chromatography (HPLC) column, semi-preparative, containing polystyrene divinylbenzene.
6. Reverse-phase ultra-HPLC, with an analytical C18 column (bead size, 1.7 μm).
7. Electrospray ionization-mass spectrometer (ESI-MS).

3 Methods

This protocol describes the formation of starPEG–heparin-based biohybrid hydrogels, encapsulating 20,000 pHAs in a 10 μL volume, with a Young's modulus of 0.5 kPa. It can be adapted to generate different disease-like conditions via the application of disease-specific cell culture treatments. Here, we describe pHAs treated with Aβ42 peptides to develop AD-like pathology. The protocol focuses on the preparation of 20 hydrogels; among them, 10 are for untreated pHAs, as controls, and the remaining 10 are for the Aβ42 peptide-treated pHAs. It is a 25-day procedure, without any halts. All the cell culture steps must be carried out using sterile cell culture techniques and, when applicable, under a biological safety hood, using appropriate safety gear.

3.1 Primary Human Cortical Astrocyte Culture (Day 1)

1. Remove the vial containing frozen pHAs from cryogenic storage, and immediately place it into a 37 °C water bath, but do not immerse the vial fully in the water. Let the contents thaw for about a minute, until only a small ice chip remains in the vial, and then remove it from the water bath.
2. Spray the exterior of the vial with 70% ethanol, wipe it dry, and transfer it to a biological safety hood.
3. Using a sterile, P1000 pipette, transfer the contents of the vial into a sterile, 15 mL conical tube.
4. Dispense 4 mL of pre-warmed CAM into the conical tube containing pHAs, dropwise (*see* **Note 8**), and gently triturate the cell suspension ~5×, using a sterile, 5 mL serological pipette.
5. Reserve 10 μL of the cell suspension for manual or automated cell counting (*see* **Note 9** and **Step 9**), and place on ice until use.
6. Determine whether it is recommended to remove the cryoprotectant from the cell type you are culturing (it is not recommended for pHAs). If your cells require the removal of cryoprotectant at this point, centrifuge the conical tube at 150–300 × *g* for 5–10 min and proceed to **Steps 7–13**. If not, then skip **Steps 7–8** and **13** and only perform **Steps 9–12**.

7. Bring the tube containing the cell pellet into the biological safety cabinet, and aspirate the supernatant carefully, without disturbing the pellet, using a vacuum aspiration system.
8. Resuspend the pellet in 5 mL of pre-warmed CAM by gently triturating with a sterile, 5 mL serological pipette. Save 10 μL of cells for manual or automated cell counting, and continue with **Step 9** (*see* **Note 10**).
9. Count the reserved volume of the cells from **Step 5** or **8**, using a manual counting chamber or an automated cell counter, per the manufacturer's instructions, and use this number to determine how many cells are in the pellet (*see* **Note 11**).
10. For the cell seeding, adjust the pHA concentration, in a 15 mL sterile conical tube, to 5000 cells/cm^2 (as per the cell provider's guidelines for pHAs or other cells), using CAM (*see* **Note 12**).
11. Using a sterile, 5 mL serological pipette, transfer one 5 mL aliquot of concentration-adjusted cell suspension into a T25 flask for control pHAs and another 5 mL aliquot into another T25 flask for Aβ42 peptide treatment of pHAs.
12. Incubate the flasks for 1 day at 37 °C, in 5% CO_2.
13. *Optional, for removing cryoprotectant*: Perform a total medium change 1 day after cell seeding (day 2), to remove any remaining cryoprotectant from the cultures, and incubate the cells under the culture conditions recommended by the supplier (including the medium change intervals), until encapsulation (day 4).
14. For disease-specific treatments (such as for AD tissues), continue with Subheading 3.2; otherwise, skip to Subheading 3.3.

3.2 Amyloid Beta-42 (Aβ42) Peptide Treatment (Day 2)

If investigating the effects of a certain agent on the cells—such as Aβ42 peptide—the mode, timing, and duration of the treatment must be determined a priori, taking into consideration the biochemical properties and potential effects of the agent. Treatment can occur before, during, or after encapsulation. The steps below describe the process for Aβ42 peptide treatment prior to encapsulation (Fig. 2).

1. Using a vacuum aspirator, remove the medium from the cell culture vessel, and replenish it with 5 mL CAM containing 10 μM Aβ42 peptides (*see* **Note 13**), using a sterile, 5 mL serological pipette. For the control vessel, replenish the CAM, but without adding any agents.
2. Incubate the control and treated cells for 2 days at 37 °C, in 5% CO_2 (*see* **Note 14**).

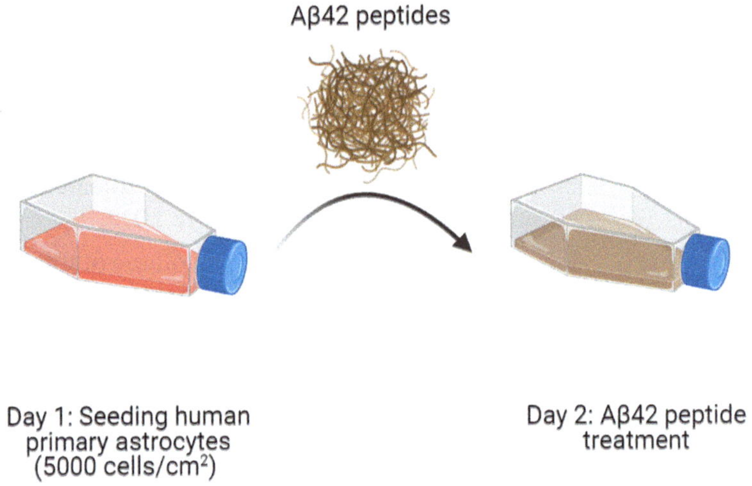

Fig. 2 Amyloid beta-42 (Aβ42) peptide treatment to induce Alzheimer-disease-like pathology. (This image was created with BioRender (https:/biorender.com/))

3.3 Preparation of Cells for Encapsulation (Day 4)

On day 4, the cell confluency for pHA should be around 40%, when the cells can be encapsulated.

1. Using a vacuum aspirator, remove the medium from the T25 flasks, and using a sterile, 5 mL serological pipette, wash both the control and treatment vessels with 5 mL of DPBS. Rock the vessels slowly, and then aspirate the DPBS.
2. Dissociate the cells by adding 1 mL of RT Accutase into each flask, using a sterile, 2 mL serological pipette.
3. Incubate at 37 °C, in 5% CO_2, for 5–10 min, examining the cells under a microscope every 5 min, to monitor cell detachment (*see* **Note 15**).
4. When the cells are fully detached, use a 5 mL serological pipette to add 4 mL CAM to each T25 flask, to quench the Accutase.
5. Using 5 mL serological pipettes, gently triturate each cell suspension ~8×, and then transfer them from the T25 flasks into sterile, 15 mL conical tubes.
6. Reserve a small aliquot (1–2 μL) of each cell suspension for counting, as in Subheading 3.1, **Step 5** (*see* **Note 16**).
7. Centrifuge the remainders of the cell suspensions at 250 × *g* for 8 min at RT.
8. Slowly aspirate the medium from the tubes, without disturbing the pellets, using the vacuum aspirator (*see* **Note 17**).
9. Based on the cell counts in **Step 6**, resuspend the pellets with enough DPBS to reach a final cell concentration of 8.0×10^6 cells/mL, for both (control and treated) conditions (*see* **Note 18**).

10. During the remaining steps, gently triturate (~10–12 times) the suspensions from time to time, to avoid cell clumping. Keep the tubes on ice.

3.4 Preparation of Heparin and StarPEG for Encapsulation (Day 4) (See Fig. 3)

For Heparin

1. Weigh 4.5 mg of heparin into a 1.5 mL tube, and then use a sterile P100 pipettor to add 50 μL of sterile DPBS, dropwise, to obtain a 6 mM final concentration (*see* **Note 19**).
2. Dissolve the heparin by triturating ~10 times, and then vortex for 5–10 s at medium speed, to ensure dissolution.
3. Using a P100 pipettor, aliquot 25 μL of the heparin solution into each of the two 1.5 mL tubes, for the control and treated conditions, keeping the tubes at RT.

For StarPEG

4. Weigh 3.6 mg of starPEG into a 1.5 mL tube, and then use a P100 pipettor to add 100 μL of DPBS, dropwise, to attain 2.25 mM (*see* **Note 20**). Vortex the starPEG for 5–10 s at medium speed, to ensure dissolution.
5. Aliquot 50 μL of the starPEG solution into each of the two 1.5-mL tubes, for the control and treated conditions. Keep the aliquots at RT.

3.5 Encapsulation of Cells in StarPEG–Heparin Hydrogels (Day 4)

1. Prepare a sterile, non-cell-culture-treated 24-well plate for the hydrogels, by adding 1 mL of CAM to each of 4 wells, 2 labeled for the control and 2 for the treated condition; there will eventually be 5 hydrogels per well.
2. Mix 25 μL of each cell suspension (from Subheading 3.3, **Step 9**) with 25 μL of heparin solution (from Subheading 3.4, **Step 3**), for each condition. This will result in 50 μL of 3 mM heparin for each cell suspension, at 4.0×10^6 cells/mL.
3. Gently triturate the suspensions 5–10 times to fully mix. Keep the tubes at RT.
4. Using a P10 pipettor, place 10 droplets of 5 μL volume of the control heparin-cell suspension onto the surface of a piece of sterile parafilm (or equivalent) (*see* **Note 21**).
5. For each of the 10 droplets, draw 5 μL of the control starPEG solution into a 10 μL pipette tip, and then eject it directly into the droplet, quickly triturating 3–5 times with the tip, to mix (*see* **Note 22**). The droplet will begin to show signs of gelation within a few seconds, resulting in the formation of a 10 μL hydrogel containing 20,000 cells (2×10^6 cells/mL) at a final concentration of 1.50 mM heparin and 1.12 mM starPEG (*see* **Note 23**).

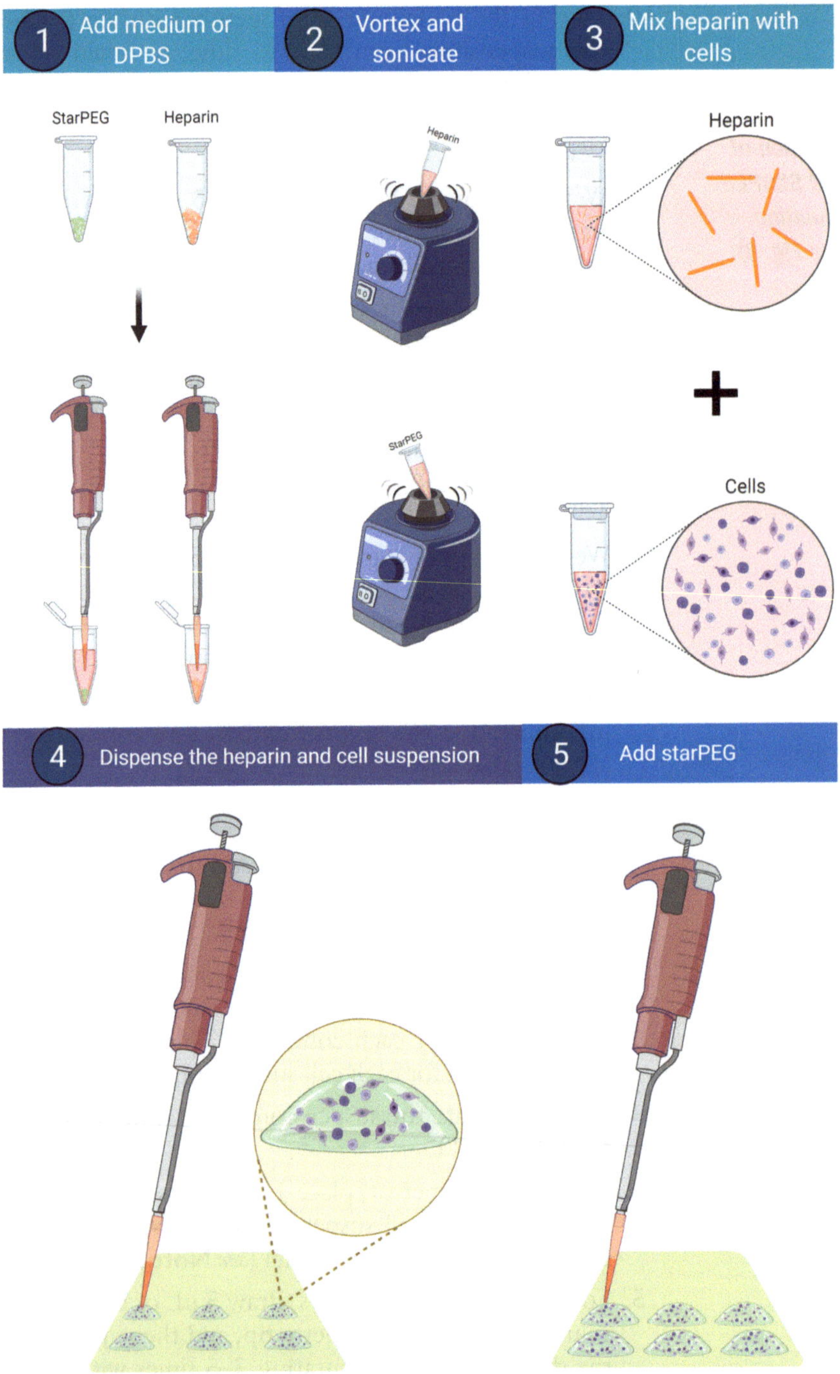

Fig. 3 Simplified preparation scheme of glycosaminoglycan (GAG)-, heparin-, and star-shaped polyethylene glycol (starPEG-MMP)-based hydrogels. (This image was created with BioRender (https:/biorender.com/))

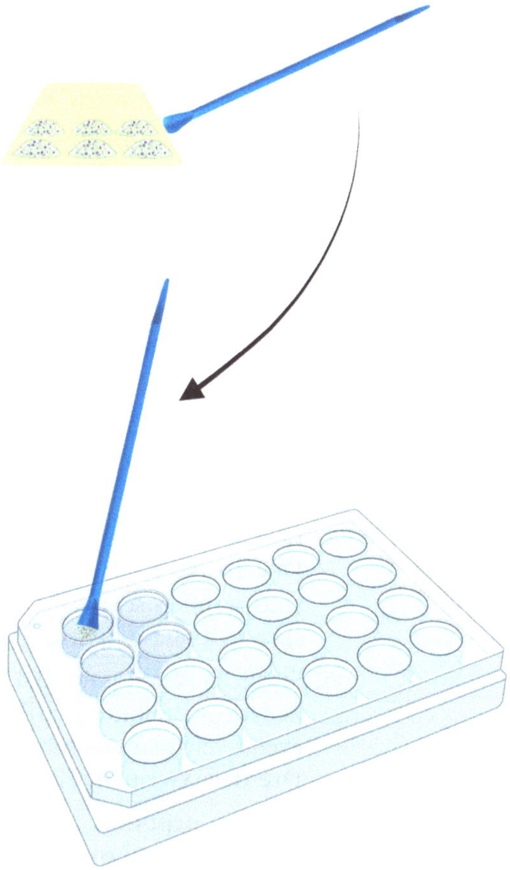

Fig. 4 Simplified preparation scheme of transferring the hydrogels to a 24-well plate. (This image was created with BioRender (https:/biorender.com/))

6. Repeats **Steps 4** and **5** for the second (treated) cell suspension.
7. After 4–5 min of formation, transfer the hydrogels into the prepared 24-well plate, at 5 hydrogels/well, using a sterile spatula (*see* **Note 24**, Fig. 4), and place the lid on the plate.
8. Submerge the hydrogels by gently tapping the plate on the sides, to ensure the hydrogels are fully immersed in the CAM (*see* **Note 25**).
9. Incubate the gels for 1 day at 37 °C, in 5% CO_2.

3.6 Medium Change and Hydrogel Treatment (Day 5 Through Day 24)

1. Aspirate the medium from each well using a P1000 pipettor, and then replenish the medium with 1 mL per well of either CAM (for controls) or with CAM plus treatment factors (*see* **Note 26**).
2. Incubate the hydrogels at 37 °C, in 5% CO_2, until the next medium change, in 2–3 days (3 times per week).
3. On *day 25*, incubation will be complete, so the hydrogels will be ready for further experimentation (e.g., immunofluorescence staining, RNA isolation, cell proliferation assays, etc.).

4 Notes

1. The primary human astrocytes used here were isolated from a male cerebral cortex at gestation week 21. We recommend using cells at a passage of 2–3, to avoid any effects of "aging." Also, we recommend using a cell-seeding concentration of 5000 cells/cm^2 for pHAs.
2. We do not recommend using a water bath to warm the Accutase solution. Instead, it should be brought to RT over a 2- to 3-h period, before use.
3. Disposable sterile spatulas are the best tools for picking up nascent hydrogels.
4. It is advisable to use a sterile, non-cell-culture-treated 24-well plate to incubate the hydrogels. It avoids the unnecessary attachment of hydrogels to the well bottoms.
5. Due to the very low intrinsic contrast of animal cells, a cell culture microscope should have a contrast-enhancing device, such as a phase-contrast filter.
6. StarPEG (catalog no. BB1: PEG-MMP, MW 15.8 g/mol) and heparin (catalog no. BB2: HEP-M6, MW 14.73 g/mol) are exclusively manufactured and supplied by ZetaScience GmbH, Dresden, Germany [7, 8].
7. Aβ42 peptides are prepared using a standard 9-fluorenylmethoxycarbonyl (Fmoc) chemistry with 2-(1H-benzotriazol-1-yl)-1,1,3,3-tetramethyluronoium hexafluorophosphate (HBTU) activation on an automated solid-phase peptide synthesizer. The peptide is cleaved from the resin with cleavage solution. Ice-cold diethyl ether is used for precipitation. Stock solutions of the peptides (1 mM) are prepared by dissolving the precipitate in 1:1 Milli-Q water–acetonitrile. For purification, use reverse-phase high-pressure liquid chromatography (HPLC), on a semi-preparative HPLC column containing polystyrene divinylbenzene. The purity can be confirmed using analytical reverse-phase ultra-HPLC with an analytical C18 column (bead size, 1.7 μm). The peptide products are analyzed by electrospray ionization-mass spectrometry (ESI-MS).
8. The volume of medium depends upon the initial cell concentration of the frozen vial. We recommend 100,000 cells/mL as a working cell concentration for this step.
9. Ideally, 10 μL of cell suspension should be enough to indicate the approximate number of cells. This volume can be altered if needed.

10. The amount of medium to add depends on the type of cell and the initial cell concentration. We recommend 100,000 cells/mL as a working cell concentration.
11. To seed pHAs in a T25 cell culture flask, the total number of pHAs required is 125,000, considering the required cell-seeding density of 5000 cells/cm^2. For this protocol, a total of 250,000 pHAs are required for two T25 flasks.
12. The areas of cell culture vessels are specified by the manufacturers, e.g., the area of the T25 flask is 25 cm^2. Cell-seeding concentrations should be adjusted according to the area of the vessel.
13. Aβ42 peptides can be quite sticky, so it is recommended to triturate the peptide solution 10–15 times, to bring them back into solution, before adding them to the culture.
14. Note that Aβ42 peptide treatment is only performed for 48 h, from day 2 to day 4; it is not applied for the full duration of the experiment.
15. Ideally, pHAs should be detached within 5–6 min.
16. Aβ42 peptides can hinder the counting of the treated pHAs. We recommend using a higher dilution (1:100) to avoid any errors.
17. Tilting the tubes slightly helps to aspirate the medium without interfering with the pellet.
18. Based on the cell count from Subheading 3.3, **Step 6**, calculate the volume of DPBS to be added to the pellet to reach a concentration of 8×10^6 cells/mL. The cells will be mixed with the heparin in a 1:1 (v:v) ratio, to reach a concentration of 4×10^6 cells/mL, followed by the addition of starPEG in a 1:1 (v:v) ratio, leading to a final concentration in each hydrogel of 2×10^6 cells/mL (20,000 cells per 10 μL hydrogel).
19. We advise calculating the volume of 6 mM heparin needed for all the hydrogels, and then adjusting the weight of heparin for the total volume needed. For example, if 100 hydrogels of 10 μL volume (a total volume of 1000 μL) is to be prepared, then the volume of 6 mM heparin required would be 250 μL. Using this volume and heparin with a molecular weight of 14.730 g/mol, the amount of heparin needed for 100 hydrogels would be 22.4 mg.
20. Calculate the volume of 2.25 mM starPEG needed for the total number of hydrogels, and then adjust the weight of starPEG for the volume needed. For example, if 100 hydrogels of 10 μL volume (a total volume of 1000 μL) is to be prepared, then the volume of 2.25 mM starPEG required would be 500 μL. Using this volume and starPEG with a molecular weight of 15.800 g/mol, the amount of starPEG needed for 100 hydrogels would be 18.02 mg.

21. Keep the pipette tip slightly above the parafilm while dispensing.
22. Strictly avoid the formation of air bubbles in the hydrogel, as this can alter the development of encapsulated cells. Air bubbles can be avoided by keeping the pipette tip in the hydrogel droplet at all times, while mixing. We do not recommend using hydrogels with entrapped air bubbles.
23. The progression of gelation can be tested by gently touching the hydrogels with a disposable spatula. Do not leave the hydrogels out of the medium for more than 10 min.
24. Gently scoop one hydrogel at a time and transfer it to the designated well containing the medium. Completely polymerized hydrogels will slide easily off the spatula, into the medium.
25. Make sure the medium doesn't splash on the lid.
26. Aspiration should be done carefully, without touching the hydrogels. Tilting the 24-well plate slightly during aspiration helps keep the P1000 tip away from the hydrogels. Make sure the hydrogels are fully immersed in the medium after replenishment.

References

1. Papadimitriou C, Celikkaya H, Cosacak MI et al (2018) 3D culture method for Alzheimer's disease modeling reveals Interleukin-4 rescues Aβ42-induced loss of human neural stem cell plasticity. Dev Cell 46:85–101.e8. https://doi.org/10.1016/j.devcel.2018.06.005
2. Celikkaya H, Cosacak MI, Papadimitriou C et al (2019) GATA3 promotes the neural progenitor state but not neurogenesis in 3D traumatic injury model of primary human cortical astrocytes. Front Cell Neurosci 13. https://doi.org/10.3389/fncel.2019.00023
3. Freudenberg U, Liang Y, Kiick KL, Werner C (2016) Glycosaminoglycan-based biohybrid hydrogels: a sweet and smart choice for multifunctional biomaterials. Adv Mater 28: 8861–8891. https://doi.org/10.1002/adma.201601908
4. Wieduwild R, Tsurkan M, Chwalek K et al (2013) Minimal peptide motif for non-covalent peptide-heparin hydrogels. J Am Chem Soc 135: 2919–2922. https://doi.org/10.1021/ja312022u
5. Bhattarai P, Thomas AK, Cosacak MI et al (2016) IL4/STAT6 signaling activates neural stem cell proliferation and neurogenesis upon amyloid-β42 aggregation in adult zebrafish brain. Cell Rep 17:941–948. https://doi.org/10.1016/j.celrep.2016.09.075
6. Kizil C, Iltzsche A, Thomas AK et al (2015) Efficient cargo delivery into adult brain tissue using short cell-penetrating peptides. PLoS One 10. https://doi.org/10.1371/journal.pone.0124073
7. Tsurkan M, Chwalek K, Levental KR et al (2016) Modular StarPEG-heparin gels with bifunctional peptide linkers. Macromol Rapid Commun 31:1529–1533. https://doi.org/10.1002/marc.201000155
8. Tsurkan M, Chwalek K, Prokoph S et al (2016) Defined polymer–peptide conjugates to form cell-instructive starPEG–heparin matrices in situ. Adv Mater 25:2606–2610. https://doi.org/10.1002/adma.201300691

Part III

Nucleic Acid Analyses of Human Brain

Chapter 9

Computational Approaches to Assess Abnormal Metabolism in Alzheimer's Disease Using Transcriptomics

Hatice Büşra Lüleci, Dilara Uzuner, Tunahan Çakır, and Madhav Thambisetty

Abstract

Transcriptome-integrated human genome-scale metabolic models (GEMs) have been used widely to assess alterations in metabolism in response to disease. Transcriptome integration leads to identification of metabolic reactions that are differentially inactivated in the tissue of interest. Among the methods available for mapping transcriptome data on GEMs, we focus here on an Integrative Metabolic Analysis Tool (iMAT), which we have recently applied to the analysis of Alzheimer's disease (AD). We provide a detailed protocol for applying iMAT to create models of personalized metabolic networks, which can be further processed to identify reactions associated with abnormal metabolism.

Key words Genome-scale metabolic models (GEMs), Transcriptomics, Reaction activity, Integrative Metabolic Analysis Tool (iMAT)

1 Introduction

Genome-scale metabolic models (GEMs) combine all known metabolic reactions in an organism or tissue with information on the genes and enzymes associated with those reactions. GEMs have become a popular tool for systems medicine and have been used in predicting molecular alterations, potential biomarkers, and plausible drug targets for several diseases [1, 2]. Transcriptome data can be mapped onto metabolic network models to identify disease-associated metabolic changes [3–5]. The Integrative Metabolic Analysis Tool (iMAT) [6] is an optimization algorithm that enables the integration of transcriptomic data with GEMs to convert an organism-specific GEM into a tissue- or condition-specific GEM, by predicting whether a metabolic reaction is active or inactive in healthy/diseased states [7]. Compared to other similar algorithms, iMAT is more suitable for mammalian cells since it has the advantage of not requiring the definition of a biological objective and

Jerold Chun (ed.), *Alzheimer's Disease: Methods and Protocols*,
Methods in Molecular Biology, vol. 2561, https://doi.org/10.1007/978-1-0716-2655-9_9,

measured uptake/secretion rates for its optimization-based predictions. In iMAT, reactions associated with highly expressed genes are forced to carry a non-zero flux in the metabolic network (i.e., active reaction), while the reactions associated with low-expression genes are forced to be inactive [6]. In other words, the algorithm aims to maximize the consistency between activity states of the reactions and gene expression levels.

Recently, iMAT was applied in a personalized manner by mapping transcriptome data for each sample separately on a genome-scale human metabolic network, to predict active/inactive reactions for each sample. These data were then used to assign a significance score to each reaction to identify significantly upregulated/downregulated reactions in the presence of Alzheimer's disease (AD) [8, 9]. In this chapter, the iMAT-based personalized approach is explained, step by step, as a computational tool to integrate transcriptome data with GEMs, for the prediction of significantly affected reactions in the diseased vs. non-diseased states.

2 Materials

All materials and specific steps of the method are demonstrated as a workflow in Fig. 1.

2.1 Computational Tools for Metabolic Modeling

All the tools listed below require MATLAB software, commonly used in programming and computing, which is available from the MathWorks website (https://www.mathworks.com/products/matlab.html) and requires a license (*see* **Note 1**). The additional software tools described below serve the overall purpose of implementing the iMAT approach.

1. The COnstraint-Based Reconstruction and Analysis (COBRA) toolbox is a General Public Licensed (GPL) MATLAB package [10] that provides functions for the reconstruction and analysis of GEMs. COBRA also enables data integration with GEMs to reconstruct context-specific models through built-in functions. A function to apply iMAT is also available in COBRA (*see* **Note 2**).
2. The Reconstruction, Analysis, and Visualization of Metabolic Networks (RAVEN) toolbox is another GPL MATLAB package for the reconstruction and analysis of GEMs. It also provides functions for model curation, constraint-based modeling, and simulation [11]. The structure of models reconstructed by RAVEN is different from COBRA models. To ensure compatibility with COBRA toolbox, RAVEN has a function to convert models to COBRA structure.

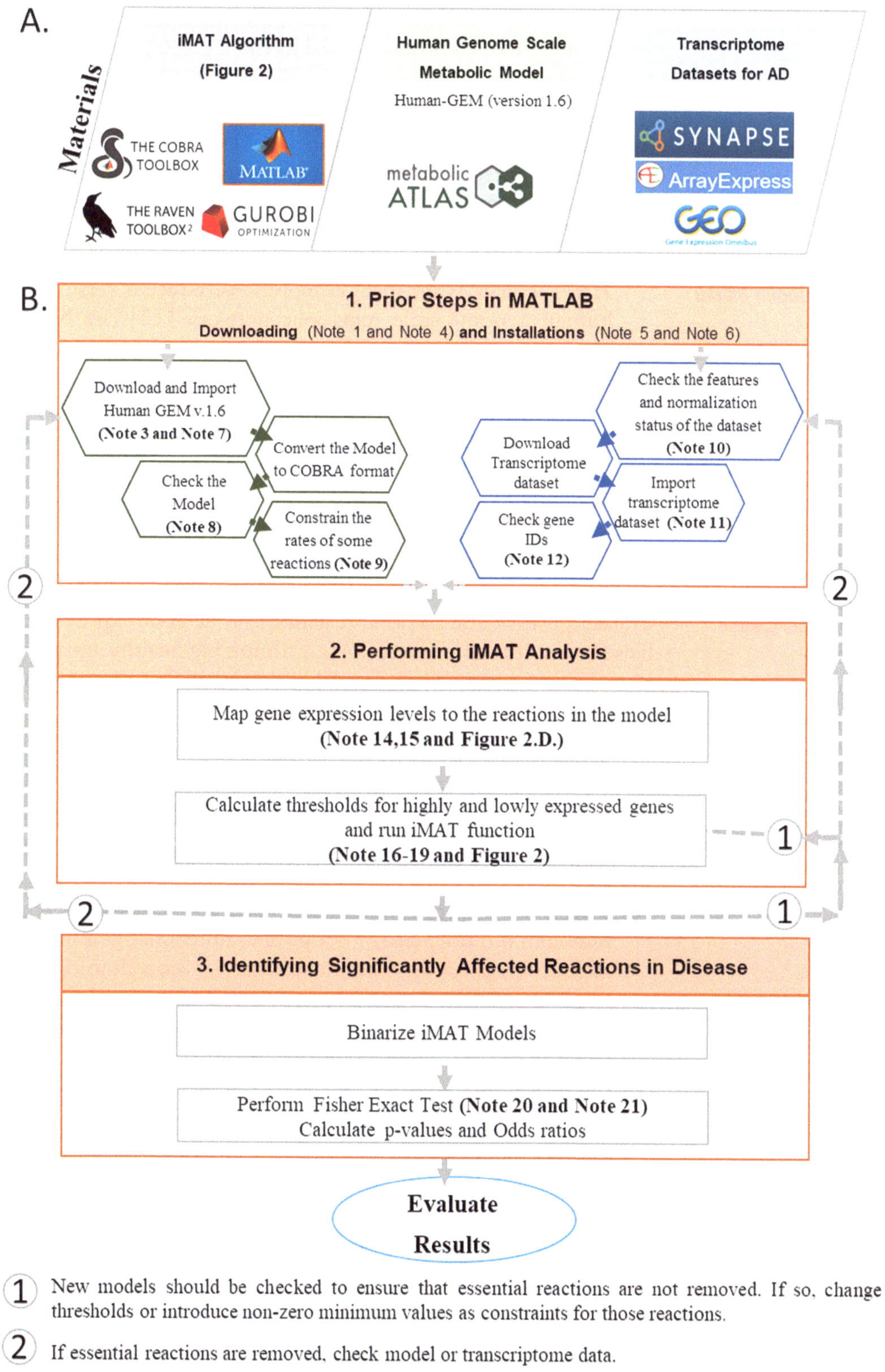

Fig. 1 Workflow for Integrative Metabolic Analysis Tool (iMAT) analysis. (**a**) All materials and (**b**) all steps of iMAT analysis are summarized as a workflow

3. The iMAT algorithm is a tool to reconstruct context-specific models, and it is implemented in the COBRA toolbox [6, 7].
4. The Gurobi Optimizer (www.gurobi.com) is a fast and powerful tool to solve linear, quadratic, and mixed-integer programming problems. It has Python and MATLAB interfaces. It can be used to solve constraint-based modeling problems in systems biology.

2.2 Human Genome-Scale Metabolic Model

1. Human GEM, the generic genome-scale metabolic model of *Homo sapiens*, is the most comprehensive recent version of the human genome-scale metabolic network [12] (*see* **Note 3**). Version 1.12, the latest version, includes 13,070 reactions, 8369 metabolites, and 3067 genes. It can be downloaded from GitHub (https://github.com/SysBioChalmers/Human-GEM) in commonly used file formats such as MAT-file and SBML.
2. Metabolic Atlas [13] (https://metabolicatlas.org/) is an online resource for human metabolism, which provides a search-and-browse interface for the human GEM.

2.3 Transcriptome Databases

Transcriptome datasets provide genome-wide gene expression profiles for conditions of interest, e.g., comparing healthy and diseased states. There are publicly accessible databases that store transcriptome datasets, such as those related to AD.

1. Synapse [14] (https://www.synapse.org/) is an open-source research platform developed by Sage Bionetworks. It enables the storage and sharing of data, codes, and results. It includes data for AD and other neurological diseases from different projects.
2. The Gene Expression Omnibus (GEO) [15] (https://www.ncbi.nlm.nih.gov/geo/) is a public, functional genomics data repository under the National Center for Biotechnology Information (NCBI). It includes array and sequence-based transcriptomic datasets for various organisms. A keyword search of GEO for "Alzheimer's disease" leads to 339 human datasets in GEO.
3. ArrayExpress [16] (https://www.ebi.ac.uk/arrayexpress/) is another publicly available database of transcriptome datasets generated by microarray and sequencing technology.

3 Methods

The iMAT algorithm can be run in MATLAB using the Gurobi Optimizer and COBRA toolbox [10] to identify the activity/inactivity of the reactions/pathways in healthy and diseased states. The main advantage of the iMAT optimization algorithm is that it does not require objective function nor any measured values specific to the condition of interest [6]. Thus, it does not require additional information about the disease studied. For instance, it only requires a transcriptome dataset covering AD and control samples for the analysis.

3.1 Preparation of MATLAB

1. Download MATLAB from the MathWorks website (*see* **Notes 1** and **4**).
2. Install the RAVEN toolbox, COBRA toolbox, and Gurobi Optimizer in MATLAB (*see* **Notes 5** and **6**).
3. Start MATLAB and run this code to check if the installation is successful:

```
>>initCobraToolbox
```

4. Set the working directory to the folder where RAVEN is installed and run the code below:

```
>>checkInstallation
```

5. To make Gurobi the default solver in COBRA, run the code below:

```
>>changeCobraSolver('gurobi','all')
```

3.2 Downloading and Importing the Human Metabolic Model on MATLAB

1. Download the most recent version of the human genome-scale metabolic model, Human GEM v.1.12, from the SysBioChalmers/Human GEM GitHub page (https://github.com/SysBioChalmers/Human-GEM). It can be downloaded in SBML format (.xml), MAT-file format (.mat), Excel format (.xlsx), or text format (.txt) (*see* **Note 7**).
2. Set the working directory to the folder in which the model file is located. The MAT-file format of the model can be loaded by the function below:

```
>>load('HumanGEM.mat')
```

3. The MAT-file is compatible with RAVEN, but not compatible with COBRA. For further analyses, convert the model to the COBRA format by using the function available in the RAVEN toolbox:

```
>>CobraModel = ravenCobraWrapper(RavenModel)
```

4. After reading the model on MATLAB, double-check the number of reactions, metabolites, genes, gene–protein–reaction (GPR) rules and other information about the model (*see* **Note 8**).
5. Before the iMAT analysis, constrain the rates of reactions essential for cell viability to be non-zero, to prevent iMAT from removing these reactions from the model (*see* **Note 9**). Such reactions include, but are not limited to, glucose uptake, oxygen uptake, and biomass (macromolecule) production.

3.3 Downloading Transcriptome Data and Importing Into MATLAB

1. Search transcriptome datasets with keywords of interest on GEO, ArrayExpress, and Synapse, and retrieve the relevant data.
2. Check the dataset for the types of cases, number of samples, and normalization of the data (*see* **Note 10**).
3. Download the dataset to the folder that was set as the working directory.
4. Read transcriptome data into MATLAB (*see* **Note 11**).
5. Check the IDs of the genes; they should be compatible with human GEM (*see* **Note 12**).

3.4 Performing the iMAT Analysis

There are two main steps for the iMAT simulations (*see* Fig. 2 and **Note 13**). The first is mapping gene expression levels to the reactions, based on gene–protein–reaction rules. The second is running the iMAT algorithm to predict a condition-specific metabolic model. To perform iMAT for each sample separately [8], use "for loop" (*see* **Note 14**).

1. Mapping gene expression levels to the reactions in the model.

   The relationships between genes and reactions are available in the model *.mat* files as logical rules. Gene–reaction rules define which reaction is controlled by a specific gene. Documenting gene–reaction rules is important since there can be more than one gene controlling a reaction, or a single gene can control multiple reactions (*see* **Note 15**). Use the code below to map the gene expression levels onto the genome-scale metabolic model:

```
>>[exp_rxns parsedGPR] = mapExpressionToReactions (model,
expressionData)
```

   There are two inputs to the function; one is the metabolic model used, called "model" above. The other is expression data, called "expressionData" (*see* **Note 16**).

2. Run the iMAT function using the following code:

```
>>new_model = iMAT(model, exp_rxns, lower, higher)
```

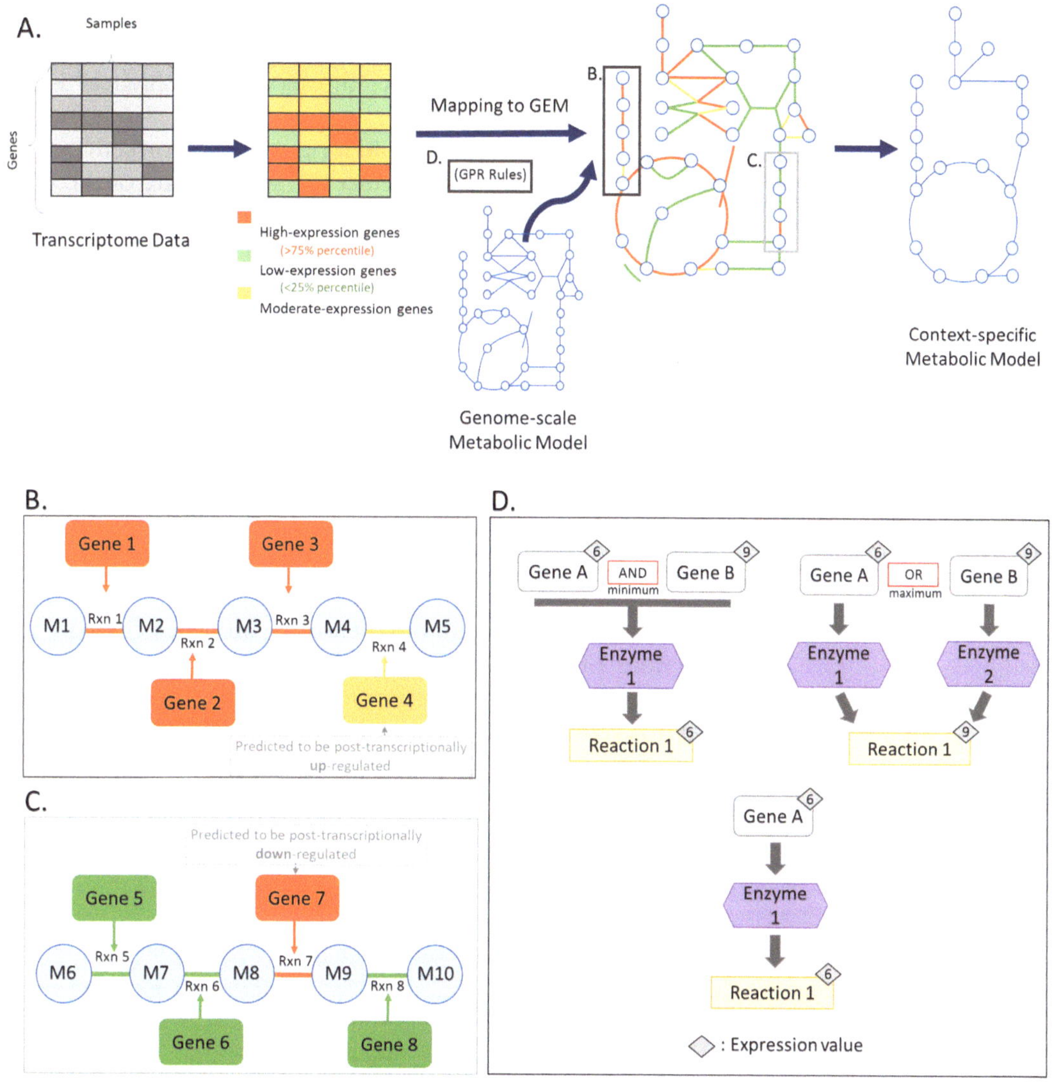

Fig. 2 Integrative Metabolic Analysis Tool (iMAT) analysis. (**a**) Genes in transcriptome data are divided into three categories based on their expression levels (high, low, and moderate). Then, the genes are mapped onto a genome-scale metabolic model via gene–protein–reaction (GPR) rules. In this way, gene scores are converted into reaction scores. The iMAT algorithm can construct sample-specific metabolic models by using transcriptome-mapped genome-scale metabolic model (GEM). iMAT ensures maximum consistency with the gene expression data while predicting metabolite routes that obey mass balancing. Therefore, iMAT can predict moderate- or low-expression genes as post-transcriptionally upregulated (**b**) or post-transcriptionally downregulated (**c**). (**d**) GPR rules of human genome-scale metabolic models contain reactions whose associated genes are linked with "AND" or "OR." Expression values of genes are transferred to reactions via these rules

There are four inputs: "model" is the metabolic model, "exp_rxns" is the mapped scores of the reactions (the output of **Step 1**), "lower" is the lower threshold, and "higher" is the upper threshold.

Be sure to set the threshold for gene levels correctly, as this is important for iMAT to predict the activity state of the reactions. The 25th percentile and 75th percentile (based on the average value of the expression values of all the genes, across samples) should be used as thresholds for low and high levels of gene expression, respectively, in the iMAT analysis (*see* **Note 17**).

After the iMAT simulation, a new model is predicted, where some reactions in the input model that were predicted to be inactive are removed (*see* **Notes 18** and **19**).

3.5 Performing Fisher's Exact Test for iMAT Models

The last step in the iMAT-based comparison of control and diseased states (to identify significantly affected reactions) is the use of statistical analyses [8] (*see* Fig. 3).

1. Represent the models in binary format, and then compare the reactions of each sample-specific model with the reactions in human GEM. Set the reaction value to 0 if it is inactive and to 1 if it is active. Apply binarization to all models.
2. Construct a contingency table for each reaction in binary models (*see* **Note 20**).
3. Apply Fisher's exact test [17] on contingency tables (*see* **Note 21**), using the function below in MATLAB:

```
>>[H, P, STATS] = fishertest (contingency_table)
```

The outputs of this function are:

- H: A logical value. It is TRUE if the p-value is <0.05 and FALSE if it is >0.05.
- P: The p-value of the Fisher's exact test.
- STATS: A structure. STATS.OddsRatio includes the odds ratio (OR) of the reaction. The OR is calculated as $OR = (AD1/C1)/(AD0/C0)$, via a contingency table. If OR is >1, the reaction is *active* in the AD state; otherwise, it is *active* in the healthy state.

4. After performing Fisher's exact test for all reactions, identify significantly affected (active or inactive in disease) reactions based on a chosen p-value cutoff [8] (*see* **Note 22**).

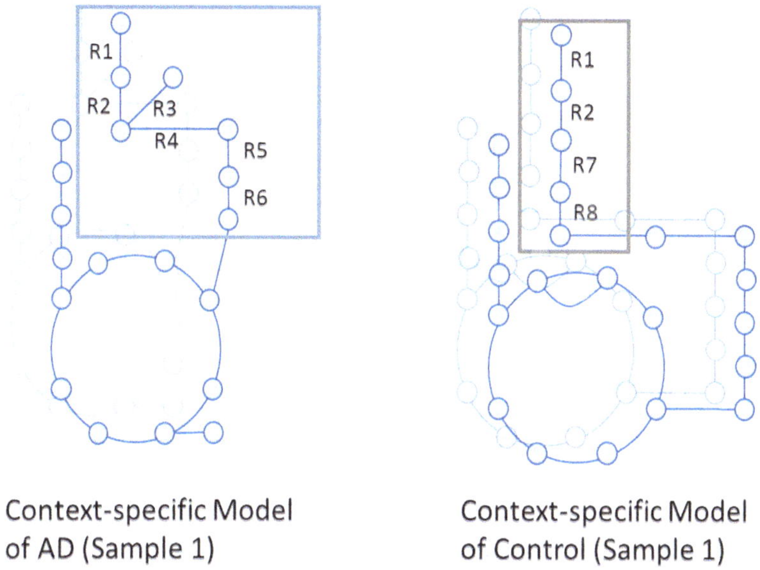

Binary Models

	Alzheimer'S Disease			Control		
Reaction	Sample 1	Sample 2	Sample 3	Sample 1	Sample 2	Sample 3
R1	1	1	1	1	0	0
R2	1	0	1	1	1	0
R3	1	0	1	0	0	1
R4	1	1	0	0	1	1
R5	1	0	0	0	0	1
R6	1	1	1	0	0	0
R7	0	0	0	1	1	1
R8	0	0	1	1	0	1

Contingency Tables for Each Reaction

R1	AD	Control
1 (active)	3	1
0 (inactive)	0	2

R2	AD	Control
1 (active)	2	2
0 (inactive)	1	1

...

Fisher's Exact Test

Reactions Significantly Affected in AD

Fig. 3 Fisher's exact test for Integrative Metabolic Analysis Tool (iMAT) models. First, reconstructed sample-specific models should be converted to a binary format to perform Fisher's exact test. In this example, the first

4 Notes

1. MATLAB requires a license. Some universities have an institution-specific license; check https://www.mathworks.com/academia/tah-support-program/eligibility.html.
2. The COBRA toolbox also provides handy functions for flux balance analysis (FBA), flux variability analysis (FVA), and other constraint-based analysis algorithms. A Python version, called COBRApy, is also available.
3. Human genome-scale metabolic models include all known reactions in an organism with no tissue specificity and the corresponding gene–protein–reaction associations. Here, nutrients such as glucose enter the human metabolic network and are converted to amino acids, nucleic acids, lipids, carbohydrates, and several by-products, through thousands of reactions. Human GEMs can be used as an input to an FBA framework for the prediction of fluxes through metabolic reactions for a given condition, using optimization techniques. Human GEMs also serve as a platform for the integration of omics data [18]. They have been employed to study disease effects on human metabolism [19].

   Different versions of GEMs for human metabolism are publicly available. Some examples are human metabolic reconstruction (HMR) [20] and its improved versions, HMR2 [21] and iHsa [22], and Recon1 [23] and its improved versions, Recon2 and Recon3D [24].
4. To perform this method with human GEM, MATLAB version R2018 or later is required, and the parallel computing toolbox of MATLAB is optional. It is only required if the *parfor* function will be used (*see* **Note 11**).
5. The COBRA toolbox can be installed by following the instructions at https://opencobra.github.io/cobratoolbox/latest/installation, and the RAVEN toolbox can be installed by following the instructions on https://github.com/SysBioChalmers/RAVEN. The RAVEN toolbox requires version R2016b, or later, of MATLAB. The Gurobi Optimizer can be downloaded from https://www.gurobi.com/downloads.

Fig. 3 (continued) 8 reactions of Alzheimer's disease (AD) and control models are illustrated. For Sample 1, since R7 and R8 are inactive in the AD model, they are represented by 0 in the binary model. Similarly, R3–R6 are inactive in the control model, and they are represented by 0 in the binary model. Active reactions are represented by 1. Then, contingency tables for each reaction are constructed. Finally, Fisher's exact test is performed for each contingency table to calculate the *p*-value and odds ratio for each reaction

6. Gurobi requires a license but has a free license option for academic use. Academic licenses can be obtained and installed from the https://www.gurobi.com/downloads/end-user-license-agreement-academic/ webpage. Compared to free alternatives, such as *G*NU *L*inear *P*rogramming *K*it (GLPK), it is more stable and should be the choice of optimizer for large systems with over 1000 unknowns (reactions).
7. The most practical way to access the model in MATLAB is via the MAT-file. The Excel format is useful to go through the reactions or genes in the model and their associated pathways.
8. Sometimes a model cannot be imported/loaded properly. To assess this, the number of reactions, metabolites, and genes should be checked. The expected numbers for the most recent version of the model can be found on the GitHub page for the model.
9. We recommend setting the upper-bound values of glucose uptake and oxygen uptake reactions to −0.01 and the lower-bound value of the biomass reaction to 0.0001, to prevent the removal of those reactions by iMAT. This ensures that those reactions will always be active in iMAT predictions.
10. Information about the normalization of data is explained in the databases that are used to download transcriptome data. Before the iMAT analysis, the retrieved data may have to be normalized. For instance, for microarray-based gene expression data, the CELL file of each sample should be downloaded from the database (for example, GEO), and the Robust Multi-array Average (RMA) normalization [25] should be performed in the R environment.
11. Gene expression data must be in numeric format for analysis in MATLAB. Because of MATLAB's data-reading functions, numeric data can be erroneously read as a string. To avoid such problems, the optimal method is to read the data from an Excel file. Always check variable types after reading data from a file.
12. The IDs of genes in the data and the model should be of the same type. The gene IDs in human GEM are Ensembl gene IDs. If the IDs are not in the same format, mapping is not possible. In the case of different ID types, the code below in the RAVEN toolbox can be used to change the IDs of genes in the model before changing the model format to COBRA:

```
>> translateGrRules()
```

For example:

```
%Generate a model that uses Entrez IDs in the gene-related fields
>>[grRules, genes, rxnGeneMat] = translateGrRules (model.grRules, 'Entrez');
%Update the model fields with the new gene ID type
>>model.grRules = grRules;
>>model.genes = genes;
>>model.rxnGeneMat = rxnGeneMat;
```

13. iMAT integrates transcriptomic or proteomic data with GEMs by using gene–protein–reaction (GPR) rules in GEMs (*see* Fig. 2d). Then, it uses constraint-based modeling and optimization to remove reactions associated with low-expression (i.e., inactive) genes from the input GEM. The modeling here is based on making the mass balances around intracellular metabolites equal to zero, by assuming that they do not accumulate for a sufficiently long time [9, 26].

14. If the dataset has a large number of samples, such as for single-cell RNAseq data, iMAT analysis can take too long to run. To avoid this, the *parfor* function of MATLAB can be used. This function shortens the working time by running the code in parallel. It needs the Parallel Computing Toolbox of MATLAB. iMAT can be run with *parfor* using the code below:

```
>>parpool(X); % X is the number of threads (workers), the maximum possible value is the number of available threads in your computer
>>parfor i=1:size(gene_expression_value,2)
%% Map the expression data to model %%
expressionData =[];
>>expressionData.gene = gene_ids;
>>expressionData.value = gene_expression_value;
>>[exp_rxns parsedGPR] = mapExpressionToReactions(model, expressionData);
%% iMAT %%
>>new_model=iMAT(model, exp_rxns, lower, higher);
>>end
```

You should remove the *switch* statement and all the *case* statements in SolveCobraMILP.m file except the *case* statement for Gurobi to properly run the *parfor* code.

15. Some reactions are catalyzed by enzyme complexes. In this case, all enzymes in the complex must be available for the corresponding reaction to proceed. In the gene–reaction rule, this condition is defined with an "*and*" relation between all the

genes controlling the enzymes in a complex. This means that the activity of such a reaction is limited by the enzyme in the complex with the lowest concentration. If gene expression is assumed to be correlated with enzyme levels, a minimum value for the mRNA levels of the genes controlling enzymes of a complex must be used to score the corresponding reaction.

Other types of enzymes are isoenzymes, which can catalyze a reaction independently. This condition is defined with an "*or*" relation in the gene–reaction rule. This means that availability of any enzyme in the rule is sufficient for the reaction to proceed. There are two alternatives in the literature about how to map gene expression data on the reactions for isoenzyme-catalyzed reactions. The reaction score can be calculated by summing the mRNA levels of all its isoenzymes or by considering the maximum of the mRNA levels. The default option in the related COBRA function is to use the maximum of the gene expression levels for isoenzymes.

For reactions associated with isoenzymes, the default of the function *mapExpressionToReactions* is to take the maximum of the expression levels of associated genes in mapping. However, you can calculate reaction scores by summing the gene levels of all its isoenzymes. For this, you should add a new parameter to the code, shown below:

```
>>mapExpressionToReactions (model, expressionData, minSum)
```

16. Expression data should be prepared as a structure on MATLAB. This structure includes IDs of genes and gene expression levels for all samples:

```
 expressionData.gene = gene_ids
expressionData.value = gene_expression_value
```

17. Thresholds can be adjusted based on the data. For this reason, the calculated threshold should be checked and compared to overall gene expression levels. If the 25th and 75th percentiles are not suitable, they can be changed based on your specific analyses. If RNAseq count data are used, the lower threshold should be set to 1, because count data show the number of reads for a gene.
18. The new model should be checked in terms of reactions, to make sure that essential reactions were not removed. For example, a non-zero glucose uptake rate and oxygen uptake rate are important for brain metabolism. If the predicted model does not include these reactions of interest, a lower-bound constraint (or an upper-bound constraint in the case of uptake reactions) should be added to the model to prevent the removal of these reactions.

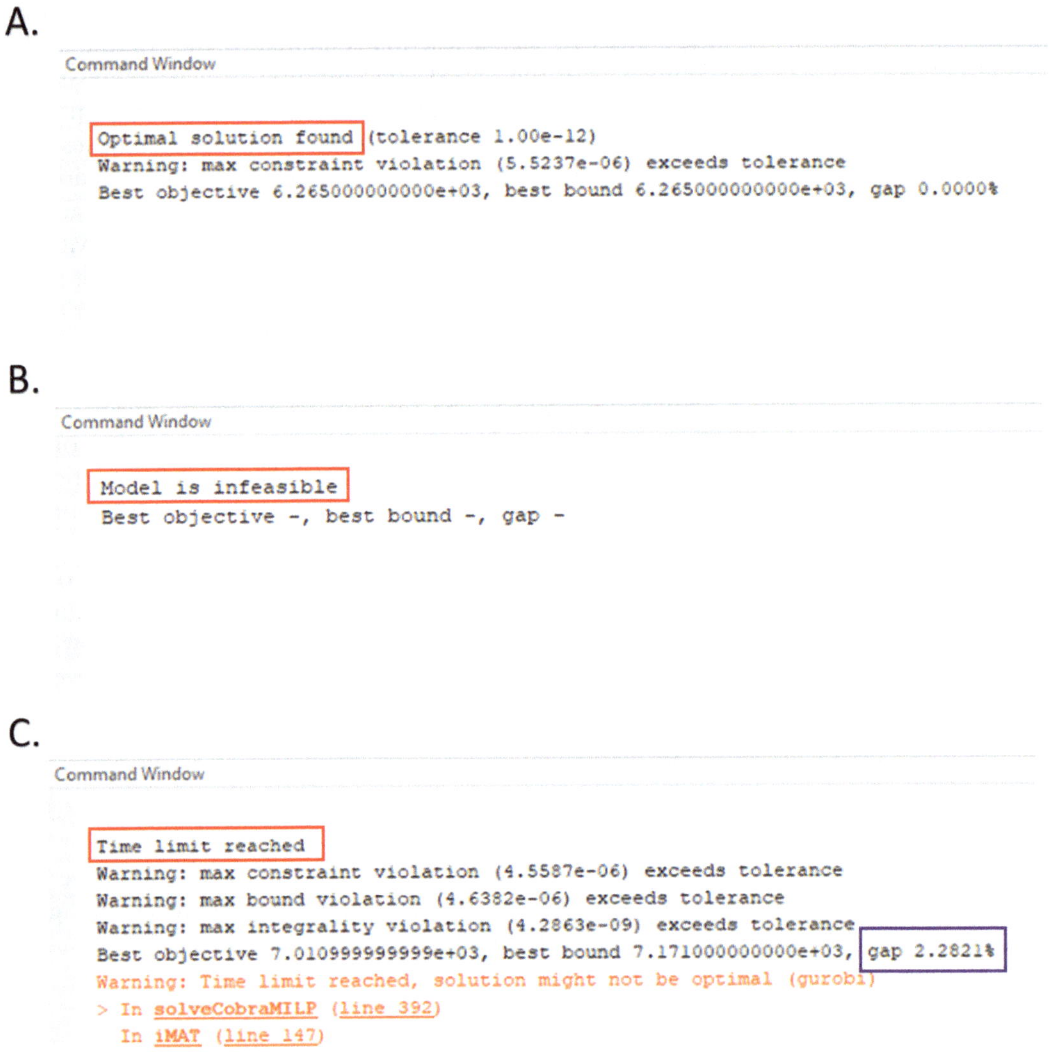

Fig. 4 Command window outputs of Integrative Metabolic Analysis Tool (iMAT). (**a**) Optimal model found, (**b**) infeasible model, and (**c**) time limit reached. *Red boxes* show status of the model. *Purple box* shows the gap value of the model when the status is "time limit reached"

19. When an iMAT run is completed, information about the optimization status of the reduced model is shown in the command window of MATLAB. There are three possible outcomes:

    (i) "Optimal model found" (*see* Fig. 4a): This indicates that iMAT found an optimal solution and produced a reduced model.

    (ii) "Infeasible model" (*see* Fig. 4b): This indicates that iMAT could not produce a feasible model. In this case, you should check the input genome-scale metabolic model. It is possible that some of the constraints chosen resulted in iMAT not being able solve the optimization problem or find an optimal solution.

(iii) "Time limit reached" (*see* Fig. 4c): This indicates that iMAT cannot find an optimal solution within the default runtime (60 s) of the algorithm. In this case, check the calculated gap value. If the gap value is <0.01, the created model is acceptable. If the gap value is >0.01, increase the "runtime" value, which is one of the optional parameters of the iMAT algorithm, and rerun the iMAT algorithm.

20. Contingency tables are frequency distribution tables that show relationships between categorical data. In the scope of this chapter, the reactions and samples are categorical variables. Reactions have two groups: *active* (represented by 1) and *inactive* (represented by 0). Samples also have two groups: for example, *control* and *AD* groups. An example of a contingency table is shown in Table 1. In Table 1, for a given reaction, C0 (or AD0) is the sum of the number of control (or AD) models in which the reaction was predicted to be *inactive*, and C1 (or AD1) is the sum of the number of control (or AD) models in which the reaction was predicted to be *active*. To perform Fisher's exact test for each reaction separately, use the "for loop."

21. If the number of samples is low (for example, lower than 4 per condition), Fisher's exact test may not give reliable results. In this case, Barnard's test [27] can be used.

22. In a recent study [9], we used the iMAT algorithm to predict metabolic alterations in the brain in AD. The iMAT-based *significantly affected reaction* list showed that several reactions related to de novo cholesterol biosynthesis were altered in the AD brain, relative to controls. We also observed regional differences in the distribution of alterations in these pathways, with iMAT predicting a more pronounced downregulation of reactions in the pre-squalene biosynthetic pathway of cholesterol in the hippocampus, whereas the post-squalene biosynthetic pathway was more significantly affected in the entorhinal cortex. Vitamin D metabolism was another important pathway affected in the AD brain. In this study, we also used iMAT to test whether metabolic reactions predicted to be altered in AD were similarly affected in another neurodegenerative disease,

Table 1
An example of a contingency table

	AD	Control
1 (Active)	AD1	C1
0 (Inactive)	AD0	C0

such as Parkinson's disease (PD), in the substantia nigra. Genome-scale metabolic network modeling in PD samples relative to controls showed that reactions predicted to be altered in the AD hippocampus/entorhinal cortex were unaltered in the substantia nigra, suggesting that this approach may also be useful to examine the disease specificity of altered metabolic reactions. The original study that used iMAT-derived personalized metabolic networks to identify affected reactions 8 reported alterations in bile acid metabolism in AD in frontal cortex, temporal cortex and parahippocampal gyrus.

Acknowledgments

This work was supported in part by the intramural program of the National Institute on Aging (NIA), of the National Institutes of Health (NIH), and by the Scientific and Technological Research Council of Turkey, TUBITAK (Project Code: 120S824)

References

1. Chowdhury S, Fong SS (2020) Leveraging genome-scale metabolic models for human health applications. Curr Opin Biotechnol 66: 267–276
2. Zhang C, Hua Q (2016) Applications of genome-scale metabolic models in biotechnology and systems medicine. Front Physiol 6:413
3. Kim MK, Lun DS (2014) Methods for integration of transcriptomic data in genome-scale metabolic models. Comput Struct Biotechnol J 11:59–65
4. Bidkhori G, Benfeitas R, Klevstig M et al (2018) Metabolic network-based stratification of hepatocellular carcinoma reveals three distinct tumor subtypes. Proc Natl Acad Sci 115: E11874–E11883
5. Cho JS, Gu C, Han TH et al (2019) Reconstruction of context-specific genome-scale metabolic models using multiomics data to study metabolic rewiring. Curr Opin Syst Biol 15:1–11
6. Shlomi T, Cabili MN, Herrgård MJ, Palsson BØ, Ruppin E (2008) Network-based prediction of human tissue-specific metabolism. Nat Biotechnol 26:1003–1010
7. Zur H, Ruppin E, Shlomi T (2010) iMAT: an integrative metabolic analysis tool. Bioinformatics 26:3140–3142
8. Baloni P, Funk CC, Yan J et al (2020) Identifying differences in bile acid pathways for cholesterol clearance in Alzheimer's disease using metabolic networks of human brain regions. Cell Rep Med 1:17
9. Varma VR, Lüleci HB, Oommen AM et al (2021) Abnormal brain cholesterol homeostasis in Alzheimer's disease–a targeted metabolomic and transcriptomic study. NPJ Aging Mech Dis 7:11
10. Heirendt L, Arreckx S, Pfau T et al (2019) Creation and analysis of biochemical constraint-based models using the COBRA Toolbox v. 3.0. Nat Protoc 14:639–702
11. Wang H, Marcišauskas S, Sánchez BJ et al (2018) RAVEN 2.0: a versatile toolbox for metabolic network reconstruction and a case study on Streptomyces coelicolor. PLoS Comput Biol 14:e1006541
12. Robinson JL, Kocabaş P, Wang H et al (2020) An atlas of human metabolism. Sci Signal 13(624):eaaz1482
13. Pornputtapong N, Nookaew I, Nielsen J (2015) Human metabolic atlas: an online resource for human metabolism. Database 2015:bav068
14. Viennot N, Lécuyer M, Bell J, et al (2015) Synapse: a microservices architecture for heterogeneous-database web applications. In: Proceedings of the tenth european conference on computer systems, pp 1–16
15. Barrett T, Troup DB, Wilhite SE et al (2010) NCBI GEO: archive for functional genomics data sets—10 years on. Nucleic Acids Res 39: D1005–D1010

16. Kolesnikov N, Hastings E, Keays M et al (2015) ArrayExpress update—simplifying data submissions. Nucleic Acids Res 43: D1113–D1116
17. Fisher RA (1958) The genetical theory of natural selection. Рипол лассик
18. Gu C, Kim GB, Kim WJ et al (2019) Current status and applications of genome-scale metabolic models. Genome Biol 20:1–18
19. Cook DJ, Nielsen J (2017) Genome-scale metabolic models applied to human health and disease. Wiley Interdiscip Rev Syst Biol Med 9:e1393
20. Mardinoglu A, Agren R, Kampf C et al (2014) Genome-scale metabolic modelling of hepatocytes reveals serine deficiency in patients with non-alcoholic fatty liver disease. Nat Commun 5:1–11
21. Björnson E, Mukhopadhyay B, Asplund A et al (2015) Stratification of hepatocellular carcinoma patients based on acetate utilization. Cell Rep 13:2014–2026
22. Blais EM, Rawls KD, Dougherty BV et al (2017) Reconciled rat and human metabolic networks for comparative toxicogenomics and biomarker predictions. Nat Commun 8:1–15
23. Ryu JY, Kim HU, Lee SY (2015) Reconstruction of genome-scale human metabolic models using omics data. Integr Biol 7:859–868
24. Brunk E, Sahoo S, Zielinski DC et al (2018) Recon3D enables a three-dimensional view of gene variation in human metabolism. Nat Biotechnol 36:272
25. Irizarry RA, Hobbs B, Collin F et al (2003) Exploration, normalization, and summaries of high density oligonucleotide array probe level data. Biostatistics 4:249–264
26. Orth JD, Thiele I, Palsson BØ (2010) What is flux balance analysis? Nat Biotechnol 28: 245–248
27. Barnard GA (1947) Significance tests for 2×2 tables. Biometrika 34:123–138

Chapter 10

FISHing for Chromosome Instability and Aneuploidy in the Alzheimer's Disease Brain

Yuri B. Yurov, Svetlana G. Vorsanova, and Ivan Y. Iourov

Abstract

Fluorescence in situ hybridization (FISH) is the method of choice for visualizing chromosomal DNA in post-mitotic cells. The availability of chromosome-enumeration (centromeric), site-specific, and multicolor-banding DNA probes offers opportunities to uncover genomic changes, at the chromosomal level, in single interphase nuclei. Alzheimer's disease (AD) has been associated repeatedly with (sub)-chromosome instability and aneuploidy, likely affecting the brain. Although the types and rates of chromosome instability in the AD brain remain a matter of debate, molecular cytogenetic analysis of brain cells appears to be important for uncovering mechanisms of neurodegeneration. Here, we describe a FISH protocol for studying chromosome instability and aneuploidy in the AD brain.

Key words Alzheimer's disease, Aneuploidy, Brain, Chromosome instability, DNA probes, Fluorescence in situ hybridization, FISH, Interphase nuclei, Molecular neurocytogenetic analysis, Interphase chromosome-specific multicolor banding, ICS-MCB

1 Introduction

Molecular neurocytogenetic analysis of the human brain has been long considered a promising way to unravel genetic/genomic mechanisms of neuronal diversity and brain disease [1, 2]. The visualization of chromosomes represents an efficient approach to elucidating chromosomal behavior and structure in the brain. In particular, interphase fluorescence in situ hybridization (FISH) is an essential molecular cytogenetic platform for visualizing chromosomal DNA and whole interphase chromosomes in post-mitotic cells [3–5]. Although, in Alzheimer's disease (AD) studies, FISH-based molecular cytogenetic analysis of chromosomal imbalances in the diseased brain has had a variety of technical problems [6, 7], it has proven invaluable for discovering chromosome instability and aneuploidy [8–10], as well as sub-chromosomal instability, involving the *APP* gene [11] in the AD brain. Indeed, these technologies

Jerold Chun (ed.), *Alzheimer's Disease: Methods and Protocols*,
Methods in Molecular Biology, vol. 2561, https://doi.org/10.1007/978-1-0716-2655-9_10,

have proven essential for the study of possible roles for interphase chromosomes in disease- or age-related neurodegeneration (e.g., chromosome/genome instability, DNA replication/reparation stress, cohesion, etc.) [12–16]; thus, one cannot overestimate the usefulness of molecular neurocytogenetic methods in studying brain genetics and the neurobiology of normal and accelerated aging.

The use of FISH-based techniques has also advanced studies of non-AD brain disease [1, 17–23], although certain drawbacks may diminish the efficiency of using FISH for molecular neurocytogenetic analysis [24–26]. Fortunately, there is a wide spectrum of FISH modifications that can solve some of these problems (for an overview, *see* ref. 27). Moreover, the ***interphase chromosome-specific multicolor-banding*** (ICS-MCB) technique can detect whole interphase chromosomes in their integrity at molecular resolutions and at any stage of the cell cycle [28]. ICS-MCB enhances FISH by using micro-dissected DNA probes (for technical details, *see* ref. 29), which generate multicolored banding of homologous interphase chromosomes. ICS-MCB has been shown repeatedly to be efficient in molecular cytogenetic studies of chromosome behavior and variability/instability in non-dividing cells [30–33].

Alzheimer's disease is associated with many types of chromosomal mosaicism and genomic/chromosomal instability [34–36]; these processes have dynamic natures, suggesting a need for molecular cytogenetic monitoring of the available mitotic tissues [37]. It is thought that these dynamic changes in cellular genomes result from abnormal cell cycle regulation (i.e., endomitosis), mitotic defects, DNA damage, and microtubule dysfunction [38–42], and the origins and consequences of these can be discovered using FISH-based chromosomal analysis during interphase. Furthermore, chromosomal variations are likely to underlie the uniqueness of each neuron in the human brain [25, 43, 44]; thus, neuronal diversity is another phenomenon that could be analyzed using FISH-based techniques [1, 2, 17]. Other mechanisms associated with AD pathogenesis probably involve age-related accumulation of chromosomal changes [12, 20, 45]; indeed, the AD brain exhibits increased rates of X-chromosome loss (aneuploidy), a cytogenetic hallmark of human aging [46, 47].

Although the use of FISH with chromosome enumeration probes [48] is effective in detecting mosaic aneuploidy and chromosome instability, there are three difficulties related to detecting aneuploidy confined to the brain [20, 49–51]: (i) the lack of the integral view of whole chromosomes (canonical FISH techniques detect specific chromosomal regions only, i.e., aneuploidy is suggested but not proved), (ii) the tissue specificity of cell suspension preparations for FISH-based analyses, and (iii) the need to differentiate between associations (pairing) of chromosomal regions and chromosomal loss in post-mitotic cells with transcriptionally active genomes. The first difficulty is reduced by using multiprobe FISH [52–54] but requires ICS-MCB to visualize whole, individual chromosomes [28, 30, 32, 46]; the second has been addressed with a number of specific protocols [55, 56]; and the third has been solved using quantitative FISH [57–60].

Finally, an important issue in human neurocytogenetics is the discrepancy between data on intercellular genomic variations in the AD brain obtained by single-cell whole-genome analysis (e.g., single-cell sequencing) and FISH-based methods [18, 19, 61]. The determination of the intrinsic rates of chromosome instability and somatic mosaicism in the AD brain seems to require the use of FISH-based techniques [12, 19, 20, 24], the success of which requires determining the appropriate set of DNA probes, the optimal cell suspension preparation, and robust image-processing protocols. When FISH is combined with whole-genome analysis in the human AD brain, single-neuron, individual genomic profiling will be feasible [62].

Here, we describe a FISH-based protocol applicable for the study of chromosome instability and mosaicism in the AD brain. We describe the appropriate DNA probe set, basic FISH/ICS-MCB methodology, and quantitative FISH protocols (*see* Fig. 1), to facilitate AD research.

2 Materials

2.1 Preparation of Brain Cell Suspensions

1. *Carnoy's fixative* (5–15 mL per sample): methanol/glacial acetic acid, 3:1 (v/v), freshly prepared at 4 °C and stored at −20 °C (*see* **Note 1**).
2. Ethanol (EtOH) dilutions (50–70 mL per Coplin jar): 100%, 96%, and 70%, in ddH_2O.
3. *Earle's buffered saline solution (EBBS)* (250 mL): calcium chloride ($CaCl_2$) dihydrate, 66.3 mg; anhydrous magnesium sulfate ($MgSO_4$), 24.4 mg; potassium chloride (KCl), 100.0 mg; sodium bicarbonate ($NaHCO_3$), 550.0 mg; sodium chloride (NaCl), 1.7 g; anhydrous sodium phosphate dibasic (Na_2HPO_4), 30.5 mg; glucose, 250 mg; phenol red (5 mg/mL), 550.0 μL.
4. *10× Phosphate-buffered saline (PBS)* (100 mL): NaCl, 1.37 M; KCl, 27 mM; Na_2HPO_4, 100 mM; and potassium phosphate, monobasic (KH_2PO_4), 18 mM, dissolved in 80 mL ddH_2O, pH adjusted to 7.4 with 1 N hydrochloric acid (HCl), diluted to a final volume of 100 mL and stored at room temperature (RT).
5. 1× PBS, pH 7.3, with 0.1% (w/v) Nonidet P-40 (100 mL) stored at RT.
6. Magnesium chloride ($MgCl_2$), 2 M in ddH_2O (10 mL).

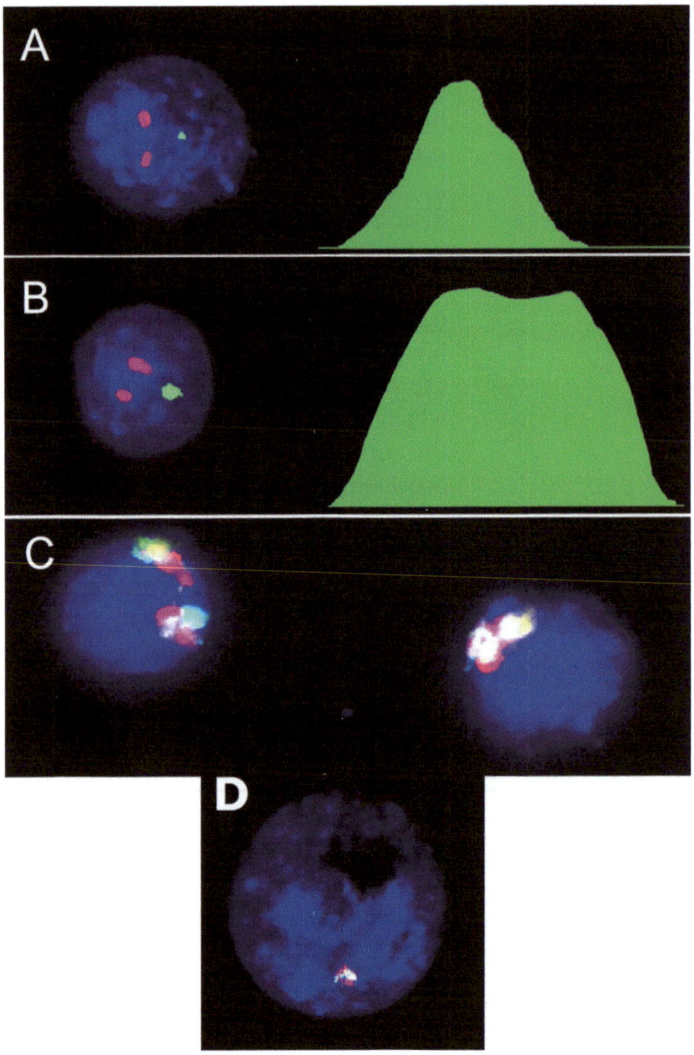

Fig. 1 Molecular neurocytogenetic analysis of the Alzheimer's disease brain. (**a**) Multiprobe (two-probe) and quantitative FISH using DNA probes for chromosomes 1 (two *red* signals, D1Z1) and X (one *green* signal, DXZ1; relative intensity is 2120 pixels), demonstrating true X-chromosome monosomy; (**b**) multiprobe (two-probe) and quantitative FISH using DNA probes for chromosomes 1 (two *red* signals, D1Z1) and X (one *green* signal, DXZ1; relative intensity is 4800 pixels), demonstrating overlap of two X-chromosome signals, but not a chromosome loss; (**c**) ICS-MCB with a probe set for chromosome X, showing one nucleus bearing two chromosomes X and another nucleus bearing a single chromosome X (from ref. 46); (**d**) ICS-MCB with a chromosome 21–specific probe—monosomy (loss) of chromosome 21 in a nucleus isolated from the Alzheimer's disease brain (from ref. 3)

7. *PBS/$MgCl_2$ solution* (100 mL): PBS, 1×, and $MgCl_2$, 2 M, 38: 1 (v/v).
8. Formaldehyde, 37% in ddH_2O (5–6 mL).
9. *1% Formaldehyde/PBS/$MgCl_2$ solution* (100 mL): 37% formaldehyde, 2.7 mL, diluted to 100 mL with PBS/$MgCl_2$ solution.
10. Glacial acetic acid, 60% in ddH_2O (w/v) (20 mL).
11. *20× saline sodium citrate (SSC) solution* (make 300 mL for the entire protocol): NaCl, 3 M; trisodium citrate ($Na_3C_6H_5O_7$), 0.3 M; and Tween-20, 0.5% (v/v), diluted to 300 mL with ddH_2O and stored at RT.
12. *2% Pepsin solution* (5 mL): Preheat 5 mL of 0.2 M HCl to 37 °C, and then add 100 mg pepsin (final pepsin concentration = 2% w/v).
13. *Sudan Black solution* (150 mL): Sudan Black, 0.7 g, mixed in 100 mL of 96% EtOH and then brought to a 150 mL final volume with ddH_2O; stir and then store at RT.
14. RNase, 0.5% in ddH_2O (w/v) (0.5 mL).
15. Forceps, sterile.
16. Dounce homogenizer, consisting of a Teflon pestle (shaft diameter, 24.5 mm) with a glass tube (40 mL, length ~150 mm).
17. Pipet, 5 mL, adjustable.
18. Petri dish (100 mm × 20 mm).
19. Coplin jars, 50–70 mL.
20. Screw-top tubes, sterile, glass or plastic,15 mL.
21. Microscope slides, 25 × 75 × 1 mm, plain.
22. Microscope coverslips, 20 × 40 mm and 24 × 24 mm.
23. Light microscope equipped with phase contrast.
24. Centrifuge (≥3500 × *g*).

2.2 Preparation of Fixed and Embedded Sections

1. Xylene (100%) (50 mL).
2. Ethanol dilutions (50–70 mL per Coplin jar): 100%, 96%, and 70%, in ddH_2O.
3. 20× SSC solution (100 mL).
4. Sodium isothiocyanate (NaSCN), 1 M (50–70 mL per Coplin jar) (*see* **Note 1**).
5. RNase, 0.5% in ddH_2O (w/v) (0.5 mL).
6. 2% Pepsin solution.
7. 1× PBS, pH 7.3, with Nonidet P-40, 0.1% (w/v) (100 mL), stored at RT.
8. $MgCl_2$, 2 M in 1× PBS (10 mL).
9. *PBS/$MgCl_2$ solution:* PBS, 1×, and $MgCl_2$, 2 M, 38:1 (v/v) (100 mL).

Table 1
DNA probes for FISH-based analysis of the Alzheimer's disease brain

Method	DNA probes (*see* Note 2)	Chromosomes
Multiprobe FISH + quantitative FISH [32, 46, 48, 52, 53, 58]	D1Z1, D7Z1, D8Z2, D9Z1, D10Z1, D11Z1, D12Z3, D15Z4, D16Z3, D17Z1, D18Z1, DXZ1, DYZ3	1, 7, 8, 9, 10, 11, 12, 15,16, 17, 18, X, and Y
FISH with site-specific DNA probes [11]	*APP* gene probe	21q21.3
ICS-MCB (for more technical details, *see* refs. 28–30, 32, 46)	MCB probes for chromosomes 21 and X	21 and X

10. Formaldehyde, 37% in ddH_2O (5–6 mL).
11. 1% Formaldehyde/PBS/$MgCl_2$ solution (100 mL).
12. Coplin jars, 50–70 mL.
13. Microscope slides, 25 × 75 × 1 mm, plain.
14. Microscope coverslips, 20 × 40 mm and 24 × 24 mm.

2.3 FISH

1. DNA probes (*see* Table 1 for a list of DNA probes recommended for FISH-based AD studies).
2. Rubber cement.
3. 20× SSC solution, 100 mL.
4. *Formamide solution* (150 mL): formamide, 50% in 2× SSC (adjusted to pH ~7 with 1 N HCl, if needed) (*see* **Note 1**).
5. 4,6-Diamidino-2-phenylindole (DAPI) solution, 300 nM in ddH_2O (1 mL).
6. Warming plate with metal surface (e.g., metal block of a thermocycler).
7. Humid chamber (a plastic box containing a wet piece of filter paper).
8. Opaque Coplin jars, 50–70 mL.
9. Microscope slides, 25 × 75 × 1 mm, plain.
10. Microscope coverslips, 20 × 40 mm and 24 × 24 mm.

2.4 Quantitative FISH

1. Fluorescence microscope equipped with a set of fluorescent filters (at least for DAPI, FITC, and Cy3) and 100× or 63× objectives.
2. Charge-coupled device (CCD) camera mounted on the fluorescence microscope.
3. Computer connected to the CCD camera.
4. *ImageJ* software (freeware: https://imagej.nih.gov/ij/).

3 Methods

3.1 Preparation of Brain Cell Suspensions from Fresh-Frozen Tissues [55, 56]

1. Rinse the brain tissue (*see* **Note 3**) in a Petri dish with 2 mL of EBSS.
2. Using a pair of sterile forceps, take a piece of the tissue (~3.5 mm^3 or ~1.5 × 1.5 × 1.5 mm), and place it into the bottom of the glass homogenization tube. Using the Teflon pestle, homogenize the tissue by hand until it resembles a liquid at RT.
3. Add 2 mL of 1× PBS into the glass tube and homogenize the contents again, until it becomes a homogeneous suspension.
4. Using a 5 mL adjustable pipet, transfer the suspension into a 15 mL sterile glass or plastic tube, then add 1 mL of 60% (w/v) glacial acetic acid, and let it stand for 3–5 min at RT.
5. Add 9 mL of Carnoy's fixative solution and centrifuge for 5 min at 1000 ×*g*.
6. Remove supernatant (an upper liquid phase, which is clearer than the lower liquid phase, which is turbid) and discard it.
7. Bring the total volume of the remaining liquid to 10 mL with Carnoy's fixative, and centrifuge for 8 min at 3000 ×*g*. Discard supernatant.
8. Repeat **Step** 7 three more times.
9. Discard all but ~2 mL of the lower phase (the cell suspension), which will later be used for FISH (*see* **Note 4**).
10. Place 100 μL of the suspension on a microscope slide, and let it air-dry for 15–20 min.
11. Add enough of the 2% pepsin solution to cover the dried suspension on the slide (about 20–100 μL), and let it stand for 3–5 min.
12. Place slide into Coplin jar with 1× PBS for 5 min.
13. Dehydrate the cells using an EtOH series (70%, 96%, and 100% for 3 min each), and let them air-dry.
14. To assess the quality of the cell suspension, drop 5–12 μL (depending on the amount of the whole suspension obtained) of the suspension onto a new microscope slide, and let it air-dry.
15. Check the distribution of nuclei using phase contrast under a light microscope: if the nuclei are overlapping, it is considered "extremely dense"; if nuclei are only occasionally seen in the view field, it is considered "extremely low."
15. If the nuclei are distributed satisfactorily (e.g., with a dense distribution but without large blocks of overlapped nuclei), proceed to Subheading 3.3 (FISH).

16. If there is a low distribution of nuclei:
    (a) Centrifuge the cell suspension obtained in **Step 9** for 7 min at 3500 ×*g*.
    (b) Remove and discard the upper 50% of the cell suspension, mix by inversion, and repeat **Step 4**.
17. If there is an extremely dense distribution of nuclei:
    (a) Centrifuge the cell suspension from **Step 9** for 7 min at 3500 ×*g*.
    (b) Remove and discard the upper 50% of the suspension.
    (c) Add 0.3–0.7 mL of Carnoy's fixative solution to adjust the volume to ~2 mL.
    (d) Repeat **Steps 10–13** and proceed to use the slides for Subheading 3.3 (FISH).

3.2 Using Formalin-Fixed and Paraffin-Embedded Sections

1. Mount the sections (*see* **Note 5**) on a microscopic slide.
2. Dry the slide at 60 °C overnight.
3. Immerse the slide in 100% xylene at RT for 5 min.
4. Remove and refresh the xylene, and let it stand for another 5 min at RT.
5. Rehydrate the sections in an EtOH series: 100%, 96%, and 70%, for 2 min each.
6. For disruption of DNA–protein complexes, place the slides into a Coplin jar, and cover with 1-M NaSCN for 3–5 h at 75 °C (*see* **Note 1**).
7. Rinse slides in another Coplin jar, filled with ddH_2O, for 3–5 s (*see* **Note 6**).
8. Add 100 μL of RNase.
9. Mount a coverslip over the sections.
10. Keep the slides at 37 °C for 15–30 min.
11. Remove the coverslip.
12. Add 100 μL of 2% pepsin solution to the sections, and let it stand for 3–5 min.
13. Rinse slides in a Coplin jar with 2× PBS for 3–5 s.
14. Rinse slides in a Coplin jar with PBS/$MgCl_2$ solution for 3–5 s.
15. Rinse slides in a Coplin jar with formaldehyde/PBS/$MgCl_2$ solution for 3–5 s.
16. Place slides in a Coplin jar containing 1× PBS for 5 min.
17. Dehydrate through an EtOH series: 70%, 96%, and 100%, for 3 min each), and let the slides air-dry.
18. Use the slides for FISH.

3.3 FISH (See Note 7)

1. Put 5 μL of the desired DNA probe over the suspension, and cover it with a coverslip.
2. Place the slide on a warming plate for 2–7 min at 72–76 °C.
3. Seal the edges of the coverslip with rubber cement.
4. Transfer the slide into a humid chamber and incubate at 37 °C overnight (*see* **Note 8**).
5. Remove the coverslip by dispensing ddH_2O or 2× SSC around the coverslip's perimeter.
6. Place the slide in a Coplin jar with formamide solution for 5–15 min at 42–45 °C.
7. Place the slide in a Coplin jar with 2× SSC for 5–15 min at 42–45 °C.
8. Dehydrate through an EtOH series: 70%, 96%, and 100%, for 3 min each.
9. Put 24 μL of DAPI solution on the sections.
10. Cover the sections with a coverslip (20 × 40 mm to cover the entire working surface of the slide or 24 × 24 mm to cover only a part of it).

3.4 Quantitative FISH [57, 59]

1. Capture images of FISH results (*see* Fig. 1) using the CCD camera mounted on a fluorescence microscope equipped with a set of fluorescent filters, 100× or 63× objectives, and analysis software provided with the imaging system or specifically designed for quantitative imaging. Use an appropriate filter for each fluorochrome.
2. Save image files separately, as 8-bit black-and-white images, to the computer connected to the CCD camera.
3. Load each FISH image into ImageJ software (*see* **Note 9**).
4. Select FISH signal area to be quantified using the "Rectangular" selection tool.
5. Assign the area to the first lane using "Analyze/Gels/Select First Lane"; alternatively, press Ctrl + 1.
6. Obtain an image plot containing the graph depicting intensity profiles (*see* **Note 10**) by pressing Ctrl + K ("Analyze/Plot Profile").
7. Use "Image/Adjust/Threshold" (or Ctrl + Shift + T) to remove the grid from the image plot.
8. Draw a line to define the borders of the graph corresponding to the desired FISH signal (which determines the area to be measured) using "Edit/Draw" (or Ctrl + D).

9. Select the area of the graph to measure using the "Wand" (tracing) tool.
10. Quantify the selection with "Analyze/Measure" (or Ctrl + M). Numerical values are outputted in a separate window.
11. Compare the numerical values of different signals from the same image as shown in Fig. 1a, b (*see* **Note 11**).

4 Notes

1. These reagents are generally toxic, so these should be used with caution, under a fume hood, and handled with gloves. All excess reagents and any materials that come into contact with them should be collected and treated as hazardous waste after use.
2. FISH DNA probes either are made in the laboratory or are commercially available.
3. Fresh-frozen brain tissue must be stored at −80 °C; it is recommended that the post-mortem interval (PMI) be less than 100 h (less is better); the samples should be processed with sterile forceps under a fume hood and handled with gloves.
4. These suspensions may be stored at −20 °C for at least 6–12 months.
5. The thickness of a specimen is generally 3–12 μm.
6. Avoid letting the slides dry before the next step.
7. The basic FISH procedure underlies all the FISH-based methods described in this chapter (i.e., multiprobe FISH, quantitative FISH, FISH with site-specific DNA probes, and ICS-MCB) (also *see* refs. 27, 29, 43, 56, 59).
8. Doing FISH with DNA probes that stain large chromosomal regions (e.g., ICS-MCB) requires longer incubation periods (i.e., 2–3 nights).
9. We have repeatedly/successfully used ImageJ freeware (https://imagej.nih.gov/ij/) (1997–2018) for analyses of digital microscopic images (e.g., measuring FISH signal intensities); for more details, *see* ref. 59.
10. FISH signal intensities are proportional to the content of DNA located within the stained chromosomal loci [57, 58].
11. Relative intensities are measured in pixels and compared.

Chapter 11

Somatic CNV Detection by Single-Cell Whole-Genome Sequencing in Postmortem Human Brain

Diego Perez-Rodriguez, Maria Kalyva, Catherine Santucci, and Christos Proukakis

Abstract

The evidence for a role of somatic mutations, including copy-number variants (CNVs), in neurodegeneration has increased in the last decade. However, the understanding of the types and origins of these mutations, and their exact contributions to disease onset and progression, is still in its infancy. The use of single-cell (or nuclear) whole-genome sequencing (scWGS) has emerged as a powerful tool to answer these questions. In the present chapter, we provide laboratory and bioinformatic protocols used successfully in our lab to detect megabase-scale CNVs in single cells from multiple system atrophy (MSA) human postmortem brains, using immunolabeling prior to selection of nuclei for whole-genome amplification (WGA). We also present an unpublished comparison of scWGS generated from the same control substantia nigra (SN) sample, using the latest versions of popular WGA chemistries, MDA and *PicoPLEX*. We have used this protocol to focus on brain cell types most relevant to synucleinopathies (dopaminergic [DA] neurons in Parkinson's disease [PD] and oligodendrocytes in MSA), but it can be applied to any tissue and/or cell type with appropriate markers.

Key words Parkinson's disease, Multiple system atrophy, Single-cell sequencing, Whole-genome amplification, Somatic mutation, CNV, Mosaicism, Substantia nigra

1 Introduction

Recent technical developments have increased the appreciation of mosaicism arising from somatic mutations in the human brain, which has likely roles in neurodevelopmental and neurodegenerative diseases [1–6]. Since the "signal" of a low-level somatic mutation can be lost in sequencing from "bulk" tissue homogenates, single-cell sequencing after whole-genome amplification (WGA) is increasingly utilized [7]. A range of WGA options now exist, with isothermal multiple-displacement amplification (MDA), PCR-based methods and hybrid methods available commercially for the past few years, and newer techniques emerging [8, 9]. Single-cell whole-genome amplification (scWGA) and sequencing are

Jerold Chun (ed.), *Alzheimer's Disease: Methods and Protocols*,
Methods in Molecular Biology, vol. 2561, https://doi.org/10.1007/978-1-0716-2655-9_11,

giving unprecedented insight into neuronal somatic mutations, although numerous technical and analytical challenges remain [10]. Several studies of human single neurons have reported megabase-scale somatic copy-number variations (CNVs), although the precise frequency of these remains unclear [11–14]. These studies analyzed frontal cortex or hippocampal neurons, with likely glial cells also included in one study [12]. Somatic CNVs may be more common in young vs. aged healthy brains [12], with evidence for an origin in embryonic neurogenesis in mouse [15] and analysis of clonally expanded human fetal neuronal precursors showing that complex genomic structural variants can arise in early human neurodevelopment [16].

The contribution of somatic CNVs, or somatic mutations in general, to common sporadic neurodegenerative diseases is only just beginning to be explored. Somatic recombination of the amyloid precursor protein gene (*APP*) is associated with Alzheimer's disease, with a novel mutation mechanism generating "genomic cDNAs" (gencDNAs) [17]. We have detected somatic CNVs of *SNCA*, encoding alpha-synuclein, in the substantia nigra (SN) and cingulate cortex in two synucleinopathies, Parkinson's disease (PD), and multiple system atrophy (MSA), using fluorescence in situ hybridization (FISH), with preliminary evidence of clinical correlations and increased likelihood for the presence of alpha-synuclein inclusions in the same cells [18, 19].

Single-neuron genomic studies have mostly relied on fluorescence-activated nuclear sorting (FANS) after nuclear isolation and staining for the neuronal nuclear marker NeuN [12, 20, 21]. These studies have not investigated the main cell types of interest in synucleinopathies: SN dopaminergic (DA) neurons in PD and glial oligodendrocytes in MSA. The only studies we are aware of using FANS for human SN DA neurons relied on the combination of NeuN staining and size gating [22] or size only [23]. However, NeuN may not be reliably expressed in SN DA neurons in humans [24] and other species [25, 26]. Indeed, NeuN has been used to identify non-DA neurons in the SN [27]. Transcription factors such as Sox6, highly expressed in SN DA neurons, especially the pars compacta, may be better suited to identifying their nuclei [28–30], combined with other morphological features, such as size and a prominent nucleolus. Further work to validate a selection strategy should include single-nucleus RNA sequencing of sorted nuclear fractions using this and other DA neuron transcription factors (e.g., LMX1a, Nurr1) and include PD cases, since Sox6 may be reduced in these [31]. Oligodendrocyte nuclei can be selected by Olig2 [32], which is expressed throughout the oligodendrocyte lineage [33] but may be low in mature oligodendrocytes [34]; Sox10 has also been used [35] so, once again, comparative validation is required. Selection of markers may be aided by analysis of publicly available RNA data (*see* **Note 1**).

In this chapter, we present and discuss in detail the laboratory and bioinformatic protocols used for our recent single-cell analysis in MSA, which was based on WGA using *PicoPLEX Gold* [19]. We also present the protocol for a WGA method based on MDA, *REPLI-g Advanced,* and an unpublished comparison of the two. Our protocol was used for brain regions (substantia nigra, pons, putamen) and cell types (DA neurons, oligodendrocytes) relevant to synucleinopathies but can be used for any cell type.

2 Materials

2.1 Tissue Dissociation, Nuclei Isolation, and Immunostaining

All buffer solutions for isolation should be freshly prepared on the day of the protocol, sterilized by filtration, and kept at 4 °C unless otherwise indicated.

1. *Nuclear isolation medium* (NIM; 2 mL per sample): potassium chloride (KCl), 25 mM; magnesium chloride ($MgCl_2$), 5 mM; Tris–HCl, 10 mM; pH 8.8; sucrose, 250 mM; dithiothreitol (DTT), 1 mM.
2. *Optimal diluent for nuclei* (ODN) (2 mL per sample): KCl, 150 mM; $MgCl_2$, 30 mM; Tris-HCl, 60 mM; pH 8.8; sucrose, 250 mM.
3. Protease inhibitor cocktail, EDTA free (Roche, or equivalent).
4. 10% Triton X-100 (stock).
5. Optiprep Density Gradient Medium (ODGN).
6. *25% Iodixanol solution* (1 mL per sample): NIM–ODGN–ODN at a 6:5:1 ratio.
7. *29% Iodixanol solution* (1 mL per sample): ODGN–ODN at a 29:31 ratio.
8. Rabbit polyclonal antibody against Sox6 (Sigma Aldrich #HPA001923, or equivalent), 0.5 μg/mL.
9. Rabbit monoclonal antibody against Olig2 (Abcam #Ab109186, or equivalent), 1 μg/mL.
10. Mouse monoclonal antibody against αSyn (Santa Cruz Biotechnology #sc-12767, or equivalent), 1 μg/mL.
11. Goat anti-mouse IgG antibody conjugated with Alexa 568 (Life Technologies, or equivalent), 2 μg/mL.
12. Goat anti-mouse IgG antibody conjugated with Alexa 488 (Life Technologies, or equivalent), 2 μg/mL.
13. Goat anti-rabbit IgG antibody conjugated with Alexa 568 (Life Technologies, or equivalent), 2 μg/mL.
14. Goat anti-rabbit IgG antibody conjugated with Alexa 488 (Life Technologies, or equivalent), 2 μg/mL.

15. *Phosphate-buffered saline* (PBS): sodium chloride (NaCl), 137 mM; KCl, 2.7 mM; sodium phosphate, dibasic (Na_2HPO_4), 10 mM; and potassium phosphate, monobasic (KH_2PO_4), 1.8 mM; pH 7.4 (prepare at least 100 mL per experiment).
16. Goat serum, 10% (v/v) in PBS.
17. 4′,6-diamidino-2-phenylindole (DAPI), 1 mg/mL in PBS (1000× stock solution).
18. Dounce tissue homogenizer set, 2 mL, including a large-clearance and small-clearance pestle (all autoclaved).
19. Pipette filter tips, sized 1 mL, 200 μL, and 10 μL, sterile.
20. Microcentrifuge tubes, 1.5 mL, sterile.
21. Microcentrifuge.
22. Orbital shaker.
23. Hemocytometer.
24. Inverted microscope.

2.2 Manual Isolation of Single Nuclei

1. *QIAscout* (Qiagen) or *CellRaft* (Cell Microsystems) (*see* **Note 2**).
2. Inverted microscope with fluorescence filters (*see* **Note 2**).
3. Motorized microscope stage (*see* **Note 2**).
4. CCD microscope camera.
5. Sterile, nuclease-free double-distilled water (ddH_2O).
6. Sterile PBS.
7. Cell-Tak tissue adhesive or similar (refer to isolation device manufacturer for other options).
8. DNase solution, 200 U/mL.
9. Ethanol (EtOH), 70%.
10. Retrieval wand supplied with *QIAscout* or *CellRaft*, washed with DNase I solution and 70% EtOH (*see* **Note 3**).
11. Low-EDTA Tris-EDTA (TE) buffer (1×), pH 8.0.
12. Pipette filter tips, sized 1 mL, 200 μL, and 10 μL, sterile.
13. Microcentrifuge tubes, 0.2 mL, sterile.

2.3 Whole-Genome Amplification (WGA), Library Construction, and Sequencing

1. *REPLI-g* Advanced single-cell kit (*REPLI-g*, *Qiagen*; *see* **Note 4**).
2. *QIAseq FX* single-cell DNA library kit (Qiagen; *see* **Note 4**).
3. *SMARTer PicoPLEX Gold* single-cell DNA-seq kit (PicoPLEX, Takara Bio; *see* **Note 4**).
4. *SMARTer DNA HT Dual Index* kit (Takara Bio; *see* **Note 4**).
5. PCR cabinet capable of preventing contamination.

6. MiniAmp thermal cycler (Applied Biosystems) or similar, compatible with 0.2 mL tubes.
7. 7.Real-time PCR system (*StepOne*, Applied Biosystems), or similar, capable of real-time monitoring.
8. EvaGreen dye, 20× (Biotium).
9. Fluorescein calibration dye (Bio-Rad).
10. *Dye stock solution*: a 9:1 ratio of 20× EvaGreen dye and 1:500 fluorescein calibration dye (diluted in PCR-grade water).
11. Qubit fluorometer (Thermo Fisher).
12. A broad-range or high-sensitivity dsDNA assay kit.
13. *QIAseq* Library Quant assay kit (Qiagen).
14. Agilent 2100 Bioanalyzer or Agilent 4200 Tapestation system.
15. Agencourt AMPure XP Beads (Beckman Coulter).
16. DynaMag-96 side magnet (Thermo Fisher) or equivalent.
17. MiSeq sequencing system (Illumina).
18. NextSeq sequencing system (Illumina).
19. Pipette filter tips, sized 1 mL, 200 μL, and 10 μL, sterile.
20. Microcentrifuge tubes, 0.2 mL, sterile.

2.4 Bioinformatic Analyses

Use the latest versions of the software tools listed below, whenever possible.

1. *Trimmomatic* software.
2. *FastQC* software.
3. *Bowtie2* software.
4. *Samtools* software.
5. *Picard tools* software.
6. *GATK* software.
7. *IGV* software.
8. *BEDtools* software.
9. *Ginkgo* software.
10. *R* software.
11. *Mixtools* software.

3 Methods

The methods described below have been used successfully in the human cerebral cortex, substantia nigra, pons, and putamen but could be potentially applied to any brain region.

3.1 Isolation of Individual Nuclei (See Note 5)

Unless otherwise indicated, all steps must be performed at 4 °C, keeping the reagents at 4 °C throughout.

3.1.1 Nuclear Suspension Preparation

1. Transfer 20–50 mg of frozen tissue (*see* **Note 6**) to a microcentrifuge tube containing 500 μL of NIM, and gently triturate it with a cut 1 mL pipette tip.
2. Repeat this procedure several times with 1 mL pipette tips cut to a progressively smaller size to further dissociate the tissue, until the tissue suspension can pass through an uncut 1 mL pipette tip without clogging.
3. Add Triton X-100 from the 10% stock solution to achieve a final concentration of 0.1%.
4. Homogenize the tissue in a 2 mL Dounce homogenizer (2–3 strokes with a large-clearance pestle and 8–12 strokes with small-clearance pestle) until the suspension appears homogeneous.
5. Transfer the homogenate to a fresh microcentrifuge tube and spin at 1000 ×*g* for 8 min at 4 °C.
6. Discard supernatant and gently resuspend the pellet in 700 μL of 25% iodixanol solution.
7. Prefill microcentrifuge tubes with 700 μL of 29% iodixanol solution.
8. Gently layer the resuspension onto the 29% iodixanol solution, taking care not to mix the solutions, such that there is a colorless, clear bottom phase (containing the 29% iodixanol) and an opaque, milky top phase (containing the tissue resuspended in 25% iodixanol).
9. Centrifuge the tube at 10,300 ×*g* for 20 min at 4 °C, and carefully remove the supernatant, leaving 25–50 μL in the bottom to avoid disturbing the pellet.
10. Place the pellet on ice, and immediately proceed to the next step (*see* **Note 7** for storage options for nuclear suspensions).

3.1.2 Immunodetection of Inclusions or Cell-Type Markers (See Note 8)

1. Immediately after isolation, resuspend the pellet in 500 μL of pre-chilled 10% goat serum in PBS, and incubate at 4 °C for 30 min, with gentle shaking, to block non-specific binding sites.
2. Add primary antibody (to Sox6, Olig2, or αSyn) to the nuclear suspension, at the optimal concentration (*see* **Note 8**), and incubate for 1 h at 4 °C with gentle shaking.
3. Spin the cell suspension at 800 ×*g* at 4 °C for 10 min.
4. Remove the supernatant, taking care not to disturb the pellet, and gently resuspend it in 500 μL pre-chilled PBS, to wash off excess primary antibody.

5. Spin the cell suspension again at 800 ×*g*, at 4 °C for 10 min.
6. Discard the supernatant and resuspend the pellet in 500 μL pre-chilled PBS.
7. Add 2 μg/mL goat anti-mouse and/or goat anti-rabbit secondary antibody and 1 μg/mL DAPI.
8. Incubate the cell/antibody suspension for 1 h at 4 °C.
9. Re-pellet the suspension by spinning again at 4 °C for 10 min, and then gently remove the supernatant, and resuspend the pellet in 200 μL fresh PBS.
10. Use a hemocytometer to estimate the number of nuclei per ml in the final suspension; in a "good" nuclear fraction, most of the nuclei should be seen as singles, with few clumps and little cellular debris (*see* **Note 9**).

3.1.3 Manual Isolation of Single Nuclei Using a Device Fixed to an Inverted Microscope (See **Note 2**)

1. Remove the *QIAscout* or *CellRaft* array from the sealed pouch and wash with 2 mL of PCR-grade dH_2O.
2. After 3 min, remove the water and repeat with another 2 mL of dH_2O for 3 min.
3. Remove the water and apply 1 mL of Cell-Tak (or equivalent) following array manufacturer's instructions.
4. After the recommended coating time, wash 3 times (as in **Steps 1** and **2**) with sterile PBS.
5. Dilute 4000–6000 nuclei (according to the nuclear concentrations obtained in Subheading 3.1.2, **Step 10**) in 2 mL of PBS, add them to the array, and allow them to settle for 2–8 h at 4 °C (*see* **Note 10** and Fig. 1). Seal the array with parafilm to prevent desiccation, and store it at 4 °C for up to 2 weeks.
6. Using a microscope equipped with a 10× objective and appropriate fluorescence filters, select rafts containing a single nucleus with the desired characteristics (e.g., positive for Sox6 and with a clear nucleolus) (*see* Fig. 1).
7. Individually capture arrays using the magnetic wand, and transfer them to a microcentrifuge tube containing 4 μL PBS (for MDA) or 5 μL TE (for *PicoPLEX*), according to manufacturer's instructions (*see* **Note 11** and Fig. 1).
8. As controls, use at least one of each of the following, per experiment: (a) a tube with no raft, into which the magnetic wand has been dipped after it was dipped into the array covering fluid; ideally, one such control should be included after every 5–10 rafts with a nucleus; (b) a raft with no nucleus; and (c) 10 pg of genomic DNA (*see* **Note 12**).

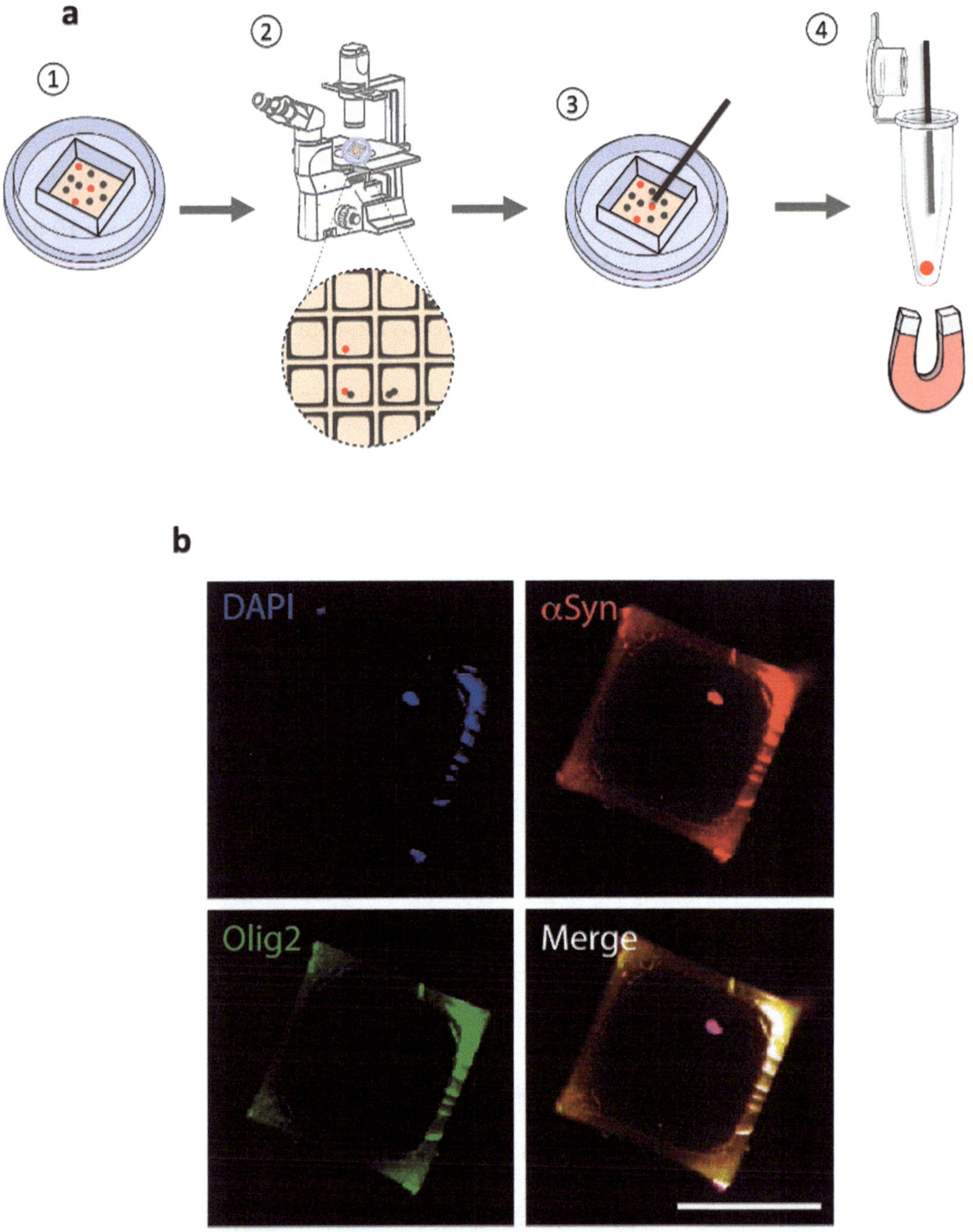

Fig. 1 (**a**) Schematic representation of single-nuclei capturing using *QIAscout* or *CellRaft* devices. Nuclei are plated at an optimal concentration (**1**), the arrays are visually inspected under the microscope (**2**), and these arrays containing single nuclei (**3**) are captured using a magnetic wand and transferred to a microcentrifuge tube (**4**). (**b**) Cell raft containing a single nucleus positive for αSyn and negative for Olig2, hence likely representing a neuronal nuclear inclusion. The cell raft was isolated and released over a glass slide before imaging on an epifluorescence microscope. *Scale bar* = 150 μm

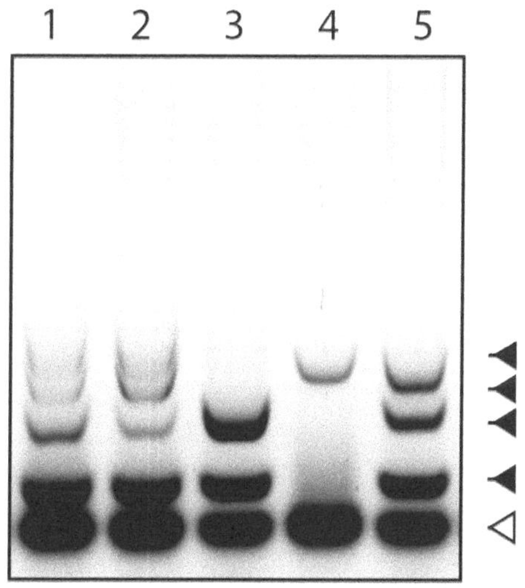

Fig. 2 Multi-locus PCR of single nuclei amplified with *REPLI-g*. Bold arrows indicate the expected bands and clear arrows primer leftovers. Samples 3 and 4 were discarded for downstream library preparation due to poor genome coverage

3.2 Single-Cell Whole-Genome Amplification, Library Preparation, and Sequencing (See Note 4)

3.2.1 Multiple Displacement Amplification (MDA)

1. Allow the rafts containing single nuclei to thaw at 4 °C if not freshly used.
2. Lyse single nuclei for 10 min at 65 °C (or 3 min for the genomic DNA control), following the manufacturer's protocol.
3. Proceed immediately to isothermal DNA amplification, for 2 h at 30 °C, per the manufacturer's protocol. At this point, samples can be stored at −20 °C.
4. Determine the DNA concentration in the amplified samples with Qubit Broad-range reagents. Optimally, the amplification reactions should yield 10–40 μg of DNA; any samples containing <10 μg should be discarded.
5. Optional: Perform a multi-locus PCR to confirm broad genome amplification (*see* **Note 13** and Fig. 2).
6. Prepare Illumina-compatible libraries with QIASeq FX single-cell DNA library kit: Use 1:10 dilutions of the amplified samples (containing around 500 ng of DNA) as input, and follow the manufacturer's protocol. Using a fragmentation time of 10 min should yield a fragment size of ~500 bp.
7. Clean up the libraries using 0.8× Agencourt AMPure XP beads, following the manufacturer's protocol, and elute the samples in 52.2 μL TE.
8. Perform a second cleanup on 50 μL of the eluted samples, using 1× Agencourt AMPure XP beads, and then elute the samples in 25 μL of TE buffer.
9. Store libraries at −20 °C until sequencing.

3.2.2 Illumina-Compatible Library Preparation Using SMARTer PicoPLEX Gold Single-Cell DNA-Seq Kit

1. Allow isolated rafts containing each nucleus to thaw at 4 °C, if not freshly used.
2. Perform nuclear lysis as indicated in the manufacturer's protocol.
3. Immediately after lysis, perform the preamplification reaction (multiple cycles of linear amplification of the released DNA) following the manufacturer's instructions. If necessary, the protocol can be stopped at this point and the samples stored at −20 °C.
4. Remove the excess primers from the preamplification reaction using 1× Agencourt AMPure XP beads, per the manufacturer's instructions, eluting the preamplified DNA in 20 μL TE.
5. Perform amplification/indexing reactions as indicated in the *PicoPLEX* manufacturer's protocol, using the SMARTer DNA HT Dual Index kit and the following additional specifications:
   (a) Replace the water in the Amplification Master Mix with the fluorescent dye stock solution.
   (b) Use a real-time PCR system.
   (c) Cycling conditions: 95 °C for 3 min, 4 cycles of 95 °C for 30 s, 63 °C for 25 s, 68 °C for 1 min, and 12 cycles of; 95 °C for 30 s and 68 °C for 1 min.
6. Use amplification plots to determine the quality of each individual library and ensure that there is no overamplification (*see* **Note 3** and Fig. 3). Amplified libraries can be stored at −20 °C.
7. Clean up libraries using 1× AMPure XP beads, and elute the libraries in 20 μL TE (but *see* **Step 8** if performing high-multiplexing sequencing).
8. For high-multiplexing sequencing reactions, pool equal volumes of the individual libraries (*see* **Note 14**), and clean 40 μL of the final pool. Elute the pooled libraries with 20 μL TE.
9. Store libraries at −20 °C until sequencing.

3.2.3 Single-Cell Whole-Genome Sequencing (scWGS) for CNV Detection (See Note 15)

1. Analyze the size and concentration of each individual library (low multiplexing) or pool of libraries (high multiplexing), per manufacturers' instructions (*see* **Note 14**).
2. Calculate the molarity for each library or pool of libraries using the formula:

$$\text{Picomoles/L} = \text{DNA concentration in g/mL}/(0.66 \times \text{DNA size in bp})$$

3. Pool the different libraries or library pools into a single tube and use the formula:

$$V_i = (V_f \times C_f)/(\# \times C_i)$$

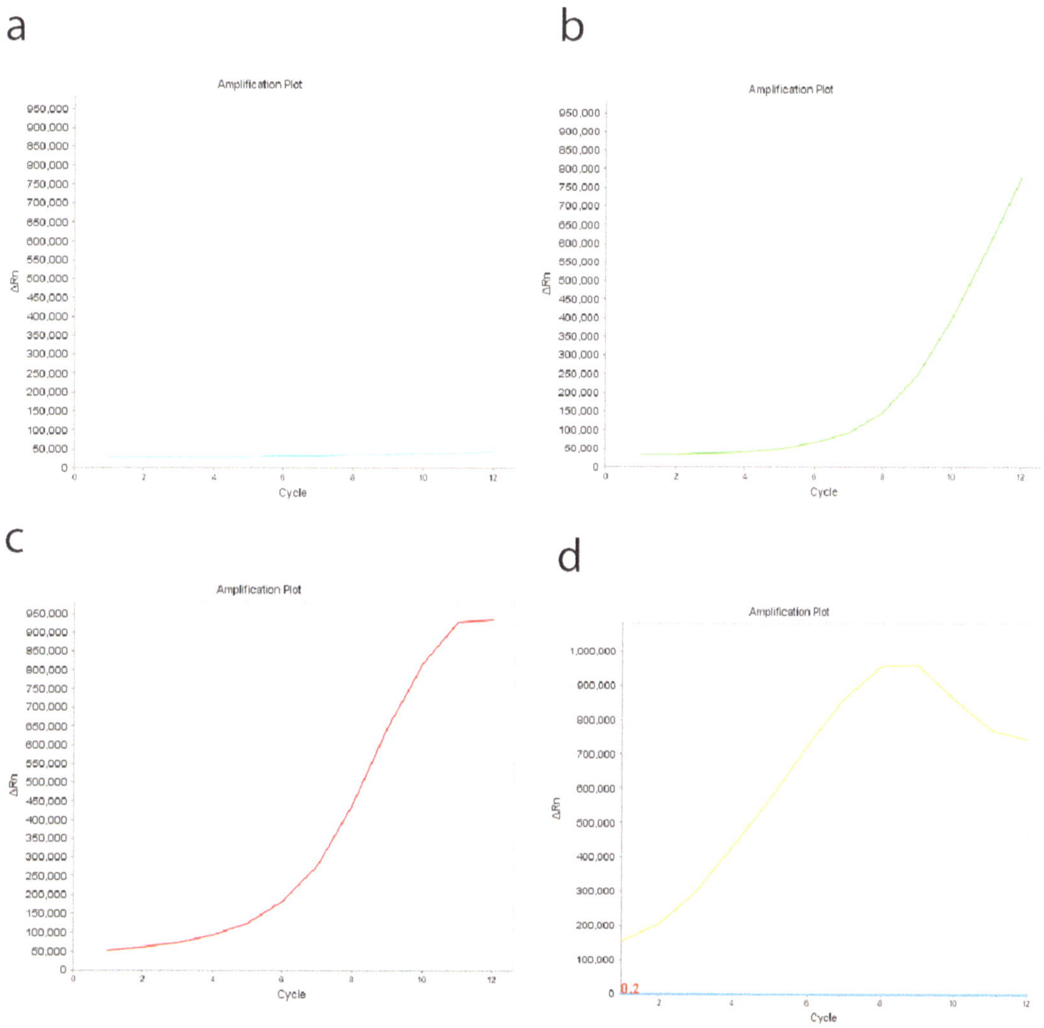

Fig. 3 Amplification plots (**a**–**d**) and *Bioanalyser* traces (**e**–**f**) of *PicoPLEX* libraries. (**a**) Empty raft. (**b**) Single nucleus. (**c**) Single nucleus reaching plateau phase. (**d**) 50 pg gDNA showing clear overamplification. If this were seen in a single nuclear reaction, it would indicate that the DNA template amount was far higher than 6.6 pg, due to large-scale contamination (*see* **Note 2**). (**e**) *Bioanalyser* profile for library shown in (**b**). (**f**) *Bioanalyser* profile for library shown in (**c**). *Arrowhead* indicates "shoulder effect" due to large DNA. This is due to the reaching of the plateau phase. Once libraries reach plateau, primers and other components get used up. In this situation, denaturing and re-annealing cycles can cause library molecules to anneal at their adapter sites, forming partially double-stranded molecules (*see manufacturer's instructions*)

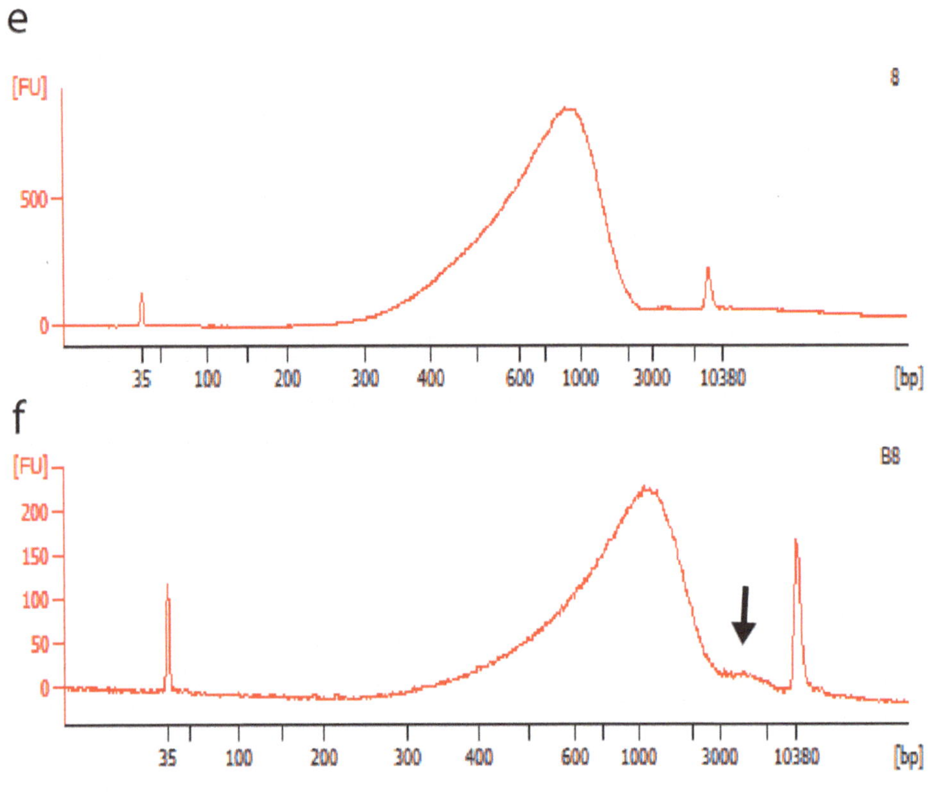

Fig. 3 (continued)

where V_i is the volume of sample to add to the final pool, V_f is the final volume of the pool, C_f is the final concentration of the library, # is the number of samples to pool, and C_i is the library concentration.

4. On the day of sequencing, determine again the concentration (in nM) of the final pool of libraries using the Bioanalyser for DNA size determination and the Qubit high-sensitivity reagents for DNA concentration. If the concentration estimate done via the Bioanalyser is different, utilize the Qubit values.
5. Run the sequencing reaction, following Illumina guidelines (*see* **Notes 15** and **16**).

3.3 Bioinformatic Analysis (See Note 17)

The bioinformatic pipeline described below is illustrated in Fig. 4 and Table 1.

1. Trim reads if appropriate (*see* **Note 18**). For *Picoplex*, remove the 14-bp adapter, using *Trimmomatic* [36].
2. Check data with *FastQC* (optional, but highly recommended for all raw data).

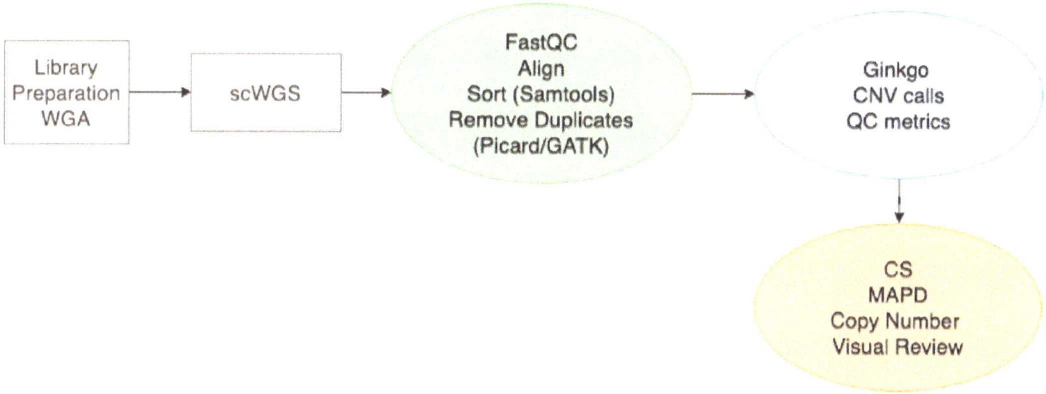

Fig. 4 Workflow and analysis pipeline

Table 1
Example commands used in bioinformatic pipeline

Define sample name	sample_name="Sample_A"
Trim adapter (for *Picoplex*)	java -jar /path/to/Trimmomatic-0.39/trimmomatic-0.39.jar PE ${sample_name}.R1_001.fastq.gz ${sample_name}.R2_001.fastq.gz ${sample_name}.1P.fastq.gz ${sample_name}.1U.fastq.gz ${sample_name}.2P.fastq.gz ${sample_name}.2U.fastq.gz HEADCROP:14
Align to human genome	bowtie2 -x /path/to/reference -1 ${sample_name}.1P.fastq.gz -2 ${sample_name}.2P.fastq.gz -S ${sample_name}.bt2.sam
Convert sam to bam, sort, index	samtools view -Sb ${sample_name}.bt2.sam > ${sample_name}.bam && samtools sort -o ${sample_name}.sorted.bam ${sample_name}.bam && samtools index ${sample_name}.sorted.bam
Filter for mapping quality (optional)	samtools view ${sample_name}.sorted_dup_removed.bam -q 1 -o ${sample_name}.sorted_dup_removed_filtered.bam
Remove duplicates	java -jar /path/to/picard.jar MarkDuplicates I=${sample_name}.sorted.bam O=${sample_name}.sorted_dup_removed.bam M=${sample_name}_marked_dup_metrics.txt REMOVE_DUPLICATES=true
Convert bam to bed	bamToBed -i ${sample_name}.sorted_dup_removed_filtered.bam > ${sample_name}.sorted_dup_removed_filtered.bed

3. Align to the human genome using *Bowtie2* (*see* **Note 19**) [37] (use hg19 for current *Ginkgo* compatibility or hg38 if different tools will be used for downstream analysis).
4. Use *samtools* [38] to sort bam files, merge from different runs if required, index, and filter for mapping quality, if required (*see* **Note 20**).
5. Remove duplicates using *MarkDuplicates* (*see* **Note 21**) or *GATK* [39].

6. Remove cells with too few (<800,000) unique reads from further analysis (*see* **Note 22**).
7. Visualize final bam files using *IGV* (this is optional, as there will be very few reads in any given window, and it cannot be used to assist or confirm variant calling at this coverage).
8. Generate bed files from bam files using *BEDtools* [40].
9. Obtain QC metrics and CNV calls by analyzing data using *Ginkgo* [41] (*see* **Note 23**).
10. Review quality to exclude any samples not suitable for analysis, and compare the results of different methods if needed (*see* **Note 24**).
11. Optional: Filter further to retain only the highest-quality cells, calculating the confidence score [42] using *R* (*see* www.github.com/proukakis and **Note 25**).
12. Obtain CNV calls from *Ginkgo*, and consider post-processing to retain only the highest-confidence CNVs (*see* **Note 26**).
13. Perform final curation by visual review, removing CNVs that are (a) smaller than 3 bins (although a minimum of 5 has also been used [12]); (b) in a region where the sequencing coverage is wavy, rather than showing a clear increase or decrease; and (c) called with borderline values in multiple cells, usually around centromeres, where mapping quality is often poor.

4 Notes

1. Selection of markers may be helped by reviewing publicly available RNA sequencing data, although most of it is from mouse. We expect more such data sources to become available soon. Current options with open access through a web interface include *mousebrain.org* [43], *dropviz.org* [44], and http://zylkalab.org/datamousecortex for mouse cortex [45]. For human neurons, including SN DA, data are available at www.humanbraincode.org [46].
2. For manual isolation of single nuclei on an inverted microscope, two very similar devices can be used: the *QIAscout* (Qiagen) with 12,000 microraft arrays and the *CellRaft* accessory device (Cell Microsystems), with 10,000 microrafts within each single reservoir *CytoSort*™ array, although only the latter is commercially available at the time of writing. We found them identical for practical purposes, and we use them interchangeably. The device is mounted on an inverted microscope with a motorized stage. It is recommended to consult the isolation device manufacturer for compatibility with a specific model of microscope. For curation purposes, pictures of isolated nuclei were acquired with a CCD camera.

The core of each device is an array formed of microrafts, which can be released and recovered with a magnetic wand. Despite not being suitable for high-throughput experiments, it presents three main advantages with respect to FANS:

(a) There is no need for gating adjustments (typically FANS requires around 50,000 events for this purpose), enabling the use of as few as 5000 nuclei per experiment.

(b) Visual inspection of each nucleus allows selection based on morphological criteria and immunofluorescence, while discarding doublets, nuclear clumps, or visibly damaged nuclei.

(c) It enables image acquisition of each nucleus, to correlate with the quality of WGA and CNV calls.

3. We consider contamination by DNA derived from any lysed cells of the sample, present in the solution covering the array, a major potential hazard. To avoid this, after retrieving each raft, perform 3 sequential washes of the wand with: (1) 200 U/mL DNase solution, (2) 70% EtOH, and (3) sterile PBS. If there is a large-scale contamination leading to starting-material quantities well over 6.6 pg, this may be visible in the WGA plots, revealed as "dipping" after the plateau is reached (an example is shown in Fig. 3d). If there is any dipping after the plateau is reached, do not sequence this product. Please note, however, that this may not be enough to detect small-scale contamination, and therefore, the controls outlined in Subheading 3.1.3 (*see* **Step 8**) are crucial.

4. In order to prepare sequencing libraries from a single cell, which contains ~6.6 pg DNA, the starting material must be amplified. WGA is usually carried out by one of the three commercially available methods: isothermal MDA, PCR methods such as degenerate-oligonucleotide-PCR (DOP-PCR), or hybrid methods (e.g., *PicoPLEX*). Each method presents advantages and disadvantages, extensively reviewed in ref. 7. One important point to consider is whether the WGA reaction should proceed straight to library preparation or whether this should be a separate process. In the former case, there are obvious time savings, enabling well-optimized, high-throughput protocols. On the other hand, separated WGA and library preparation enable the user to perform careful quality control over the WGA products (*see* Subheading 3.2.1, **Step 5**). The *Qiagen MDA* kit used here requires separate library preparation, whereas the *PicoPLEX Gold* kit yields libraries ready for Illumina sequencing, although a *PicoPLEX*-WGA-only kit is also available. WGA products that have not been processed into libraries may be more suitable for targeted enrichment (although we have not tested this) and also allow: (a) volume

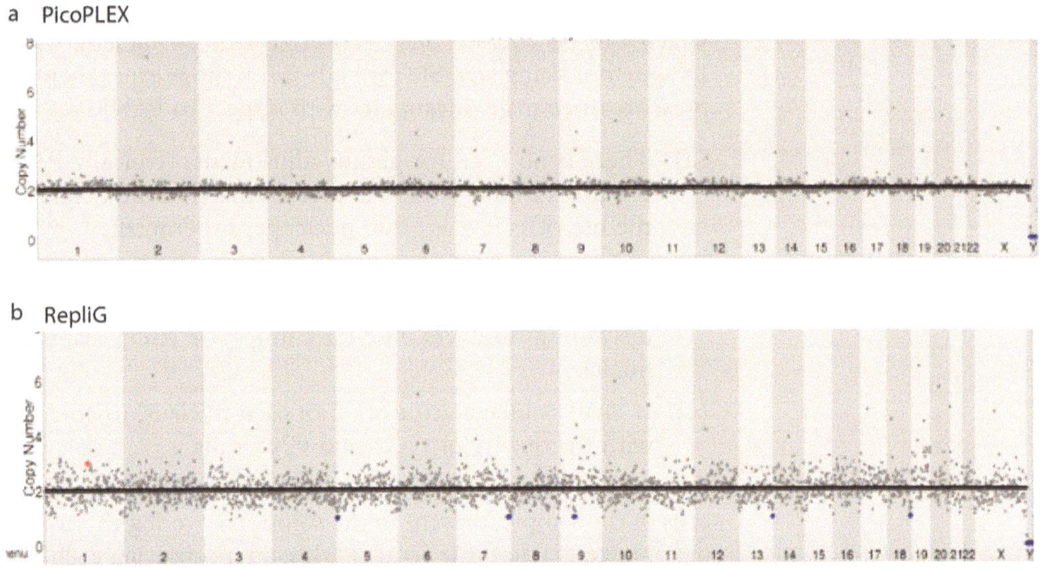

Fig. 5 Single-cell whole-genome sequencing (scWGS) profile of one typical cell amplified with each whole-genome amplification (WGA) method. (**a**) *PicoPLEX*, (**b**) *RepliG*. Note the increased deviation from the baseline copy-number 2 in most genomic segments in (**b**), which results in higher median absolute pairwise deviation (MAPD) values

reduction during library preparation, if suitable equipment is available, which reduces costs considerably, and (b) alternative library construction, such as for use in long-read sequencing. During optimization of our workflow, we performed an unbiased comparison of *PicoPLEX* and *REPLI-g*, using the SN of a 92-year-old female with no neurological illness documented during her life and whose autopsy only showed signs of cerebrovascular disease and pathological aging. Nuclei were selected based on their large size, the presence of a nucleolus, and Sox6 positivity and so are likely to represent DA neurons (*see* Subheading 1). We present our conclusions here, using **steps 1–10** of the bioinformatic pipeline in Subheading 3.3, Fig. 5, which shows scWGS profiles for a representative cell, and Fig. 6 illustrates the median absolute pairwise deviation (MAPD; *see* **Note 24**) scores for groups of cells amplified by *PicoPLEX* ($n = 12$) and *REPLI-g* ($n = 6$). The profiles of the *PicoPLEX* are much less noisy, with most genomic regions very close to the copy-number 2 baseline, and accordingly have a significantly better MAPD. These results were obtained using bin size = 500 kb, which influences the size of CNVs that are detectable. We found that using smaller bins led to slight deterioration of MAPD (*see* Fig. 7), although these values would still be potentially acceptable if a less stringent cutoff were used. Our findings are in line with several published studies of older versions of the same kits, which have shown

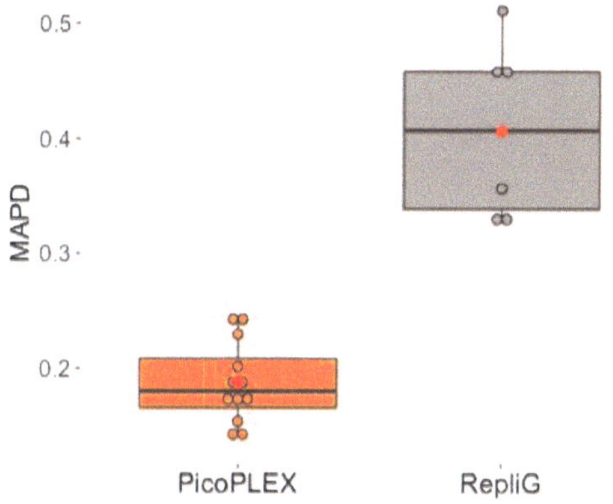

Fig. 6 Boxplots of median absolute pairwise deviation (MAPD; *see* **Note 24**) values in *PicoPLEX* and *RepliG* cells. *Red rhombus* in each boxplot represents the mean value of cells for each category ($p = 0.0001$ using Mann–Whitney *U* test)

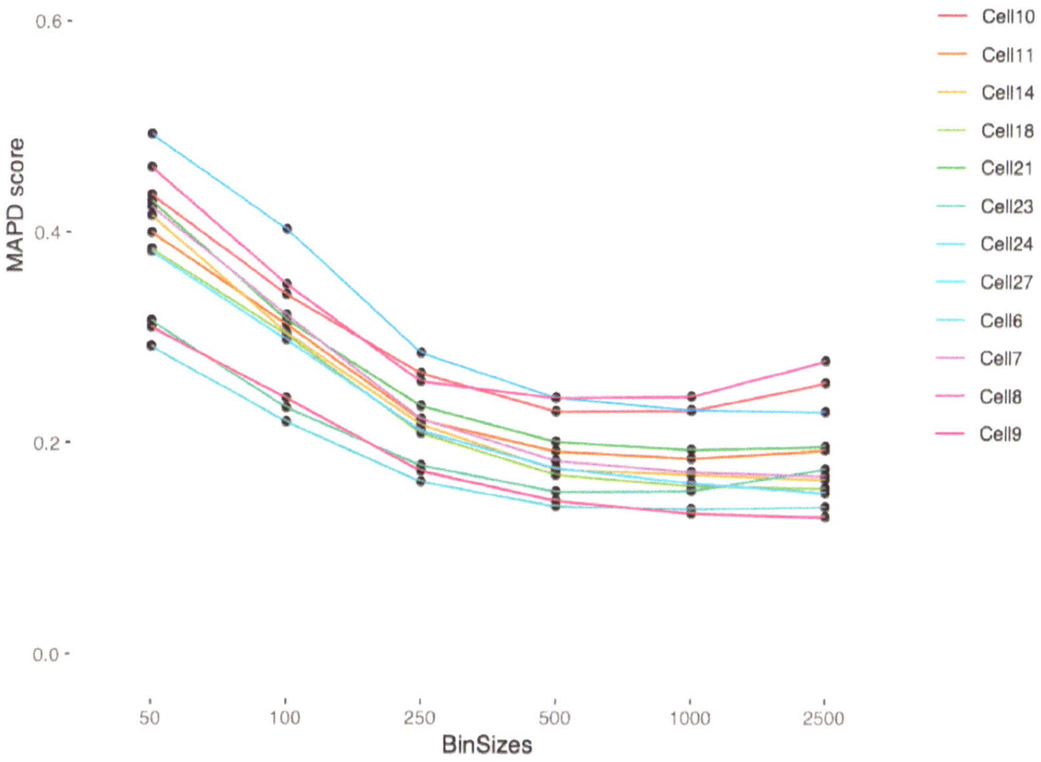

Fig. 7 The effect of bin size used in analysis on the median absolute pairwise deviation (MAPD) values in *PicoPLEX* cells

that *PicoPLEX* and similar chemistries such as *MALBAC* are more suitable than MDA for CNV analysis, although MDA may be more suitable for single-nucleotide variant detection [47] (*see* **Note 27**). We have also performed preliminary work with laser-capture microdissection (*see* **Note 28**).

5. Our protocol for nuclear isolation is based on a discontinuous iodixanol gradient. Despite the principle being similar to sucrose gradients for organelle sedimentation, the lower viscosity of the iodixanol allows a faster sedimentation (20 min at 10,000 ×*g*) of the nuclei. Moreover, iodixanol creates an isosmotic environment that prevents the disruption of macromolecular structures [48].
6. Mincing the tissue into small fragments on a glass plate (on ice), using a scalpel, significantly decreases the time needed for homogenization and facilitates the trituration with the pipette tips.
7. We have chosen an indirect protocol for immunolabeling, using an unconjugated primary antibody and a fluorophore-conjugated secondary antibody, to improve the signal of low-abundance antigens. This avoids the need for on-site primary antibody labeling when no commercial options are available, thus increasing flexibility in the epitopes that can be detected. However, it presents two main drawbacks with respect to the use of a labeled primary antibody (direct protocol): (1) it increases the length of time and the number of centrifugation steps and subsequently decreases the final yield and the integrity of the nuclei and (2) it requires a more thorough optimization of the antibody protocols, including for the secondary-only controls. Direct labeling has been successfully used by others; for a detailed protocol *see* ref. 49. Before beginning, determine the optimal working concentration for each antibody to be used, via titration assays.
8. We have used Sox6 (0.5 μg/mL) for dopaminergic neurons, Olig2 (1 μg/mL) for oligodendrocytes, and αSyn (1 μg/mL) to detect inclusions in the nucleus, although cytoplasmic ones may also be retained, if contiguous with the nuclear membrane. We have also combined two antibodies, provided they are raised in different species to allow appropriate secondary antibody recognition. The optimal working concentration for each antibody must be determined in advance by titration assays.
9. We found, quite frequently, several clumped nuclei, or even larger tissue debris, such as blood vessels. If this is the case, filter the nuclear suspension with 70 μm cell strainers and re-count the number of nuclei. Additionally, cytoplasmic inclusions and other components may be retained during the isolation, if they are contiguous with the nuclear membrane. We expected to detect only nuclear inclusions, which can be found

in MSA, but we detected some inclusions that appeared extra-nuclear, but which co-isolated with the nuclei [19]. The presence of cytoplasmic membranous components has previously been noted in neuronal nuclear fractions [50]. This may depend on the stringency of the nuclear isolation process, but we have not tried to vary this, and we do not know what proportion of cytoplasmic inclusions in MSA or PD are retained in this protocol.

10. The number of nuclei seeded on the array is a key factor in success, as too many nuclei will lead to having few microrafts with a single nucleus, whereas too few will lead to most of the microrafts being empty, following a Poisson distribution. Thus, seeding the array with a nucleus:microraft ratio of 1:2 or 1:3 increases the proportion of wells capturing a single nucleus.

11. When performing *QiaScout* or *CellRaft* isolation for the first time, it is highly recommended to ensure the rafts are properly released at the bottom of the tube and not in the tube wall, as they will desiccate. To do so, visually inspect every tube, after collection, under a stereoscopic microscope; rafts will appear as small orange squares.

12. Although a single-cell genome is ~6.6 pg, the same quantity randomly taken from a sample extracted from "bulk" tissue may have some DNA regions absent and others over-represented, so 10 pg genomic DNA is usually used as a positive control in single-cell WGA. Please note that some *RepliG* negative controls could yield to up to 20 μg of DNA, which is due to random extension of primer dimers, but they will not provide human genome sequencing data.

13. Multi-locus PCR: WGA may not amplify the entire genome of a given cell. In order to only sequence the best WGA products, the user optionally can confirm broad genome amplification by performing a multi-locus PCR. We amplify four loci on different chromosomes across the genome in a single multiplexed reaction for convenience, using the primers below (product size in brackets) [51]:

Chr5 (140 bp)	F: GGAGTCATCCTCCAGGTTATTGTTACCATC
	R: CCTTGGAAGAGGGAGAAATTCCTTGGTTA
Chr10 (523 bp)	F: CTTTCCGCCTAACTAGAATGCAGACCA
	R: CGCTCGTGTTGGGAAGAAGACTCC
Chr15 (423 bp)	F: TGCTGGAGCAATACTCAGAACTGTTGC
	R: GCTAATCCCTGCAGTAATTTCAAATGGCT
Chr20 (291 bp)	F: CTGGACCAAGTGGCTTCTTCGACTAG
	R: GCGTGCCGAAGTCTAGGTCTTTATATCTAG

We perform reactions in a 25 μL volume, which contains 0.5 μL of each primer at a stock concentration of 10 μM, plus 0.25 μL of 20 mM dNTP, made up with the enzyme, buffer, and ddH_20. The PCR conditions are optimized for the Hot Start DNA polymerase (NEB #0174). Any enzyme should be suitable but will require adjustment according to the manufacturer's protocol.

(i) 94 °C, 30 s.

(ii) Thirty cycles of 94 °C, 15 s/55 °C, 45 s/68 °C, 1 min.

(iii) 68 °C, 5 min.

We recommend sequencing only WGA samples showing 3–4 bands, as these are the ones where it is most likely that a large portion of the genome was amplified.

14. We visually inspected each library amplification plot (*see* Fig. 3e–f) and pooled together libraries with similar amplification profiles, as differences between libraries are indicative of different library concentrations and/or average sizes and may lead to unequal reads from each library. If the sequencing protocol is likely to lead to several million reads per library, then this may not be necessary.

15. The concentration and size of the libraries must be determined before sequencing. The gold standards for size analysis are electrophoretic methods such as the Agilent Bioanalyser or TapeStation. However, DNA concentration can also be determined by RT-qPCR, electrophoretic, and fluorometric methods. This needs to be carefully optimized in every lab and for every library chemistry used. In our lab, we perform size analysis with the Agilent 2100 Bioanalyser and determine DNA concentration via RT-qPCR, using the *QIAseq Library Quant Assay* kit for *REPLIg*-generated libraries, and *Qubit* high-sensitivity reagents for *PicoPLEX* ones. Different Illumina sequencing platforms (*MiSeq, NextSeq, NovaSeq*) and different reagent cartridges are available, based on the number of samples and the desired genome coverage. We strongly recommend asking for support from the supplier for final loading concentrations and optimal cluster density, although the user may need to perform final optimization (therefore, one should budget for additional runs). It is crucial to determine the conditions leading to an optimal clustering, including PhiX for diversity, since under- or over-clustering seriously compromises the success of the sequencing run. We used *MiSeq* (loading 12 pM) and *NextSeq* (loading 1.2 pM), with 10% PhiX (for *REPLI-g*) and 15% PhiX (for *PicoPLEX*), following the manufacturer's suggestions.

16. For low-coverage sequencing, we use *MiSeq* and *NextSeq*, although *NovaSeq* can be used for larger-scale multiplexing. In general, single-end 75-bp sequencing is adequate, and we have not found any advantage in paired-end reads. These would obviously be crucial if structural variant calling were to be attempted on higher coverage (30× or more), but the chimeric artifacts of WGA would require careful identification, although proof-of-principle has been provided [52], and the cost would be orders-of-magnitude higher.
17. Considerable effort has been invested in developing pipelines, relying on read depth, to detect CNVs in low-coverage, single-cell, whole-genome sequencing (scWGS) data. Orthogonal validation is, by definition, not possible, and in most algorithms, only megabase-scale CNVs can be confidently detected. Steps common in all tools used for CNV calling from scWGS are: binning (partition of the genome), GC correction, mappability correction, removal of outlier bins, and removal of outlier cells [53]. Analysis tools are generally based on circular binary segmentation (CBS), originally developed for array comparative genomic hybridization (CGH) data [54] (such as *Ginkgo*, *SCNV* [43]), or Hidden Markov Models (such as *HMMcopy* [55] and *AneuFinder* [56]). Although there is no accepted gold standard, *HMMcopy* and *Ginkgo* may be the most robust, with *HMMcopy* having a higher precision, but with *Ginkgo* inferring CNVs more accurately when the ploidy of the cell is unknown [57]. We have mainly used *Ginkgo* but have not performed any extensive comparisons. We note other promising tools, one based on machine learning trained on lymphocyte V(D)J recombination [15], which can detect CNVs <1 Mb, and another one using Bayesian statistics [12]. *PASD-QC*, a dedicated tool for QC, is worth considering if deeper sequencing is planned [36], although it is optimal at coverage >1, which is not achieved with the strategy we adopted.
18. *Ginkgo* is optimized for read lengths of 48, 76, and 101. Most of our data had a read length of 76 but included a 14-bp adapter, which requires trimming in the case of *PicoPLEX*. We, therefore, performed *Ginkgo* analysis after all reads were trimmed to 48. We did not see any difference if we trimmed *PicoPLEX* reads to 62 bp instead, by removing only the adapter, but we did not perform extensive comparisons.
19. To align sequencing data, we use *Bowtie2*, although *BWA mem* and other aligners can be used. Benchmarking suggests that there is no "perfect" aligner, as each aligner performs better in different regions and with different data [58]. For this reason, multiple aligners could be tested, or simulations performed, e.g., with *TEASER* (https://github.com/Cibiv/Teaser [59]).

In limited comparisons, we found that *Bowtie2* was faster and produced slightly better MAPD scores than BWA, but no obvious differences in CNV calling.

20. To minimize errors due to misalignment, we filter bam files for a mapping quality of 30. Although filtering is not common practice, it should be considered, at least for q1 (uniquely mapping reads).
21. Note that there is an option for marking duplicates and retaining them, but we prefer to remove them completely.
22. We accepted 800,000 reads (or read-pairs) as the minimum acceptable number per single-cell library, although lower values have also been used, e.g., 600,000 [12]. We performed limited comparisons to determine whether more reads would be beneficial. In data from 40 MSA cells sequenced with 0.8–10 million reads [19], the MAPD did indeed improve with more reads (Spearman $r = -0.5$, $p < 0.001$), but detailed work would be required to determine whether this improves CNV detection. On balance, a strategy aiming for ~1 million reads per cell is an adequate compromise between coverage and cost.
23. The user can optimize several parameters. We use bin size = 500 kb, variable bin size, masking bad bins, independent analysis of each sample, clustering using ward linkage, Euclidean distance, and no ploidy specified. *Ginkgo* was run on a Mac with OS X El Capitan and 32 Gb RAM, using *R*, with additional data visualization on the *Ginkgo* server (http://qb.cshl.edu/ginkgo).
24. A key QC metric is the median absolute pairwise deviation (MAPD) value, which reflects technical noise due, for example, to uneven genome amplification, and is not unduly influenced by rare CNV break points (note that *Ginkgo* can be fully run on the web server, but this does not at present provide the MAPD per individual sample). We exclude cells with MAPD >0.3, but higher values have been considered acceptable, e.g., 0.4 [15].
25. Single cells should have integer copy numbers in all regions (= 2 if normal, = 0 or 1 for a loss, or ≥3 for a gain). Cells with frequent intermediate copy-number states are difficult to interpret and could signify uneven amplification.

    The confidence score is calculated as the median of the absolute distances between the predicted copy-number values and the nearest integers for all genomic segments [15]. The formula is shown below, where C_i is the predicted copy number of a particular genomic interval i:

$$\mathrm{CS} = 1 - 2\mathrm{Med}\sum\nolimits_{i=0}^{n}([C_i] - C_i, C_i - [C_i])$$

We retain cells with a confidence score ≥ 0.8, as used in one study [15], although we initially used 0.85 [14]. Another promising quality metric for CNV segmentation scoring, which we have not tested, is the Bayesian Information Criterion (BIC) [12]. BIC rewards segmentations that allow high variance within the segments and avoids creating too many segments. The best-quality cells will have low BIC scores.

26. *Ginkgo* provides only integer-value copy-number states and, as discussed above, intermediate copy-number states are difficult to interpret. To determine the apparent copy number of each CNV called by *Ginkgo*, calculate the mean of the copy-number ratio of all bins after segmentation and GC correction within the CNV (output step of the *Ginkgo* pipeline), multiplied by the reported ploidy of that cell. CNV calls that deviate too much from integer values can then be removed. Rather than a fixed threshold, we use *Mixtools* [42] to fit a three-cluster Gaussian mixture model of the copy number of all called segments around 1, 2, and 3, as previously suggested [12]. Determine the precise copy-number value from a cumulative two-tailed probability, which we set at 1%, using the centered Gaussian distribution near 2, and only retain calls with copy numbers outside these values. In our MSA data, these numbers were <1.45 for losses and >2.55 for gains [20]. We note that different thresholds can be used for lenient and stringent calls [12], and the user can test different parameters.

27. We have not tested other kits, notably *Ampli*1™ (Silicon Biosystems), which uses specific primers for PCR WGA and has shown even better CNV detection potential in a study using array CGH [60], or the very recently developed *primary template-directed amplification* [61], which appears to match *PicoPLEX* and *Ampli*1 for evenness of amplification, and MDA for base-level accuracy [8].

28. The ideal method for single-cell isolation may be laser-capture microdissection, as this will provide spatial information and enable detailed cell (rather than nucleus) characterization, although it is low-throughput and laborious. To our knowledge, this only has been used successfully for single-cell CNV detection in breast cancer [62]. We have not optimized this, but DA neurons are easily recognizable by their cytoplasmic neuromelanin pigment, and we did achieve successful amplification using the original versions of both *PicoPLEX* and *REPLI-g*. The best results were obtained using *PicoPLEX* preceded by overnight incubation at 55 °C with a lysis buffer [63] (1 mM EDTA, 20 mM Tris–HCl, pH 8.0, 0.2 mg/mL Proteinase K, and 0.001% Triton-X), followed by heat inactivation at 75 °C for 10 min and then at 95 °C for 4 min. We did not succeed with another MDA kit (*TruePrime*, *Expedeon*

Cygnis), which we have not pursued further, but it was felt to be unsuitable for CNV calling in another study [64]. Section thickness is a crucial parameter that requires optimization. We only tested 10 μm, which may result in partial-thickness nuclei, at least for neurons. Gentle fixation and staining are required before identifying the cells of interest, and these could be combined with rapid immunohistochemistry, e.g., to identify alpha-synuclein aggregates, but these steps will require careful optimization for each brain region and cell type.

Acknowledgments

We are grateful to the Michael J. Fox Foundation for Parkinson's Research for funding this work, to the Queen Square Brain Bank for supplying brain samples, and to all the individuals who kindly donated their brains for research. Katya Mokretar and Tammaryn Lashley assisted with laser-capture microdissection.

References

1. D'Gama AM, Walsh CA (2018) Somatic mosaicism and neurodevelopmental disease. Nat Neurosci 21(11):1504–1514
2. McConnell MJ, Moran JV, Abyzov A et al (2017) Intersection of diverse neuronal genomes and neuropsychiatric disease: the Brain Somatic Mosaicism Network. Science 356(6336):1641
3. Leija-Salazar M, Piette C, Proukakis C (2018) Review: somatic mutations in neurodegeneration. Neuropathol Appl Neurobiol 44(3): 267–285
4. Lodato MA, Walsh CA (2019) Genome aging: somatic mutation in the brain links age-related decline with disease and nominates pathogenic mechanisms. Hum Mol Genet 28(R2): R197–R206
5. Proukakis C (2020) Somatic mutations in neurodegeneration: an update. Neurobiol Dis 144: 105021. https://doi.org/10.1016/j.nbd.2020.105021
6. Rohrback S, Siddoway B, Liu CS, Chun J (2018) Genomic mosaicism in the developing and adult brain. Dev Neurobiol 78(11): 1026–1048. https://doi.org/10.1002/dneu.22626
7. Gawad C, Koh W, Quake SR (2016) Single-cell genome sequencing: current state of the science. Nat Rev Genet 17(3):175–188
8. Gonzalez V, Natarajan S, Xia Y et al (2020) Accurate genomic variant detection in single cells with primary template-directed amplification. bioRxiv. https://doi.org/10.1101/2020.11.20.391961
9. Sanders A, Meiers S, Ghareghani M et al (2019) Single-cell tri-channel-processing reveals structural variation landscapes and complex rearrangement processes. bioRxiv. https://doi.org/10.1101/849604
10. Lähnumann D, Köster J, Szczurek E et al (2020) Eleven grand challenges in single-cell data science. Genome Biol 21:31
11. Cai X, Evrony GD, Lehmann HS, Elhosary PC, Mehta BK, Poduri A, Walsh CA (2014) Single-cell, genome-wide sequencing identifies clonal somatic copy-number variation in the human brain. Cell Rep 8(5):1280–1289
12. Chronister WD, Burbulis IE, Wierman MB et al (2019) Neurons with complex karyotypes are rare in aged human neocortex. Cell Rep 26(4):825–835
13. Gole J, Gore A, Richards A et al (2013) Massively parallel polymerase cloning and genome sequencing of single cells using nanoliter microwells. Nat Biotechnol 31(12): 1126–1132
14. McConnell MJ, Lindberg MR, Brennand KJ et al (2013) Mosaic copy number variation in human neurons. Science 342(6158):632–637

15. Rohrback S, April C, Kaper F, Rivera RR et al (2018) Submegabase copy number variations arise during cerebral cortical neurogenesis as revealed by single-cell whole-genome sequencing. Proc Natl Acad Sci 115(42):10804–11080
16. Sekar S, Tomasini L, Proukakis C et al (2020) Complex mosaic structural variations in human fetal brains. Genome Res 30(12):1695–1704
17. Lee MH, Siddoway B et al (2018) Somatic APP gene recombination in Alzheimer's disease and normal neurons. Nature 563(7733):639–645
18. Mokretar K, Pease D, Taanman JW et al (2018) Somatic copy number gains of α-synuclein (SNCA) in Parkinson's disease and multiple system atrophy brains. Brain 141(8): 2419–2431
19. Perez-Rodriguez D, Kalyva M, Leija-Salazar M et al (2019) Investigation of somatic CNVs in brains of synucleinopathy cases using targeted SNCA analysis and single-cell sequencing. Acta Neuropathol Commun 7(1):219
20. Lake BB, Ai R, Kaeser GE et al (2016) Neuronal subtypes and diversity revealed by single-nucleus RNA sequencing of the human brain. Science 352(6293):1586–1590
21. Westra JW, Rivera RR, Bushman DM et al (2010) Neuronal DNA content variation (DCV) with regional and individual differences in the human brain. J Comp Neurol 518(19): 3981–4000
22. Yang Y, Shepherd C, Halliday G (2015) Aneuploidy in Lewy body diseases. Neurobiol Aging 36(3):1253–1260
23. Yang Y, Shepherd CE, Halliday G (2017) Increased aneuploidy is not a universal feature across α-synucleinopathies. Mov Disord 32(3): 475–476
24. Sukhorukova EG, Grigoriev IP, Kirik OV et al (2013) Intranuclear localization of iron in neurons of mammalian brain. J Evol Biochem Phys 49:370–372
25. Cannon JR, Greenamyre JT (2009) NeuN is not a reliable marker of dopamine neurons in rat substantia nigra. Neurosci Lett 464(1): 14–17
26. Kumar SS, Buckmaster PS (2007) Neuron-specific nuclear antigen NeuN is not detectable in gerbil substantia nigra pars reticulata. Brain Res 1142:54–60
27. Gasparotto J, Ribeiro CT, Bortolin RC et al (2017) Targeted inhibition of RAGE in substantia nigra of rats blocks 6-OHDA-induced dopaminergic denervation. Sci Rep 7(1):8795
28. La Manno G, Gyllborg D, Codeluppi S et al (2016) Molecular diversity of midbrain development in mouse, human, and stem cells. Cell 167(2):566–580
29. Nichterwitz S, Chen G, Aguila Benitez J et al (2016) Laser capture microscopy coupled with Smart-seq2 for precise spatial transcriptomic profiling. Nat Commun 7:12139
30. Poulin JF, Zou J, Drouin-Ouellet J et al (2014) Defining midbrain dopaminergic neuron diversity by single-cell gene expression profiling. Cell Rep 9(3):930–943
31. Panman L, Papathanou M, Laguna A et al (2014) Sox6 and Otx2 control the specification of substantia nigra and ventral tegmental area dopamine neurons. Cell Rep 8(4):1018–1025
32. Nott A, Schlachetzki JCM, Fixsen BR, Glass CK (2021) Nuclei isolation of multiple brain cell types for omics interrogation. Nat Protoc 16(3):1629–1646. https://doi.org/10.1038/s41596-020-00472-3
33. Zeisel A, Hochgerner H, Lönnerberg P et al (2018) Molecular architecture of the mouse nervous system. Cell 174(4):999–1014
34. Ahmed Z, Asi YT, Lees AJ et al (2013) Identification and quantification of oligodendrocyte precursor cells in multiple system atrophy, progressive supranuclear palsy and Parkinson's disease. Brain Pathol 23(3):263–273
35. Yeung MSY, Djelloul M, Steiner E et al (2019) Dynamics of oligodendrocyte generation in multiple sclerosis. Nature 566(7745):538–542
36. Bolger AM, Lohse M, Usadel B (2014) Trimmomatic: a flexible trimmer for Illumina sequence data. Bioinformatics 30(15): 2114–2120
37. Langmead B, Salzberg SL (2012) Fast gapped-read alignment with Bowtie 2. Nat Methods 9(4):357–359
38. Li H, Durbin R (2009) Fast and accurate short read alignment with Burrows-Wheeler transform. Bioinformatics 25(14):1754–1760
39. McKenna A, Hanna M, Banks E et al (2010) The genome analysis toolkit: a mapreduce framework for analyzing next-generation DNA sequencing data. Genome Res 20(9): 1297–1303
40. Quinlan AR (2014) BEDTools: the Swiss-Army tool for genome feature analysis. Curr Protoc Bioinformatics 47:11.12.1–11.1234
41. Garvin T, Aboukhalil R, Kendall J et al (2015) Interactive analysis and assessment of single-cell copy-number variations. Nat Methods 12(11):1058–1060
42. Benaglia T, Chauveau D, Hunter D et al (2009) Mixtools: an R package for analyzing mixture models. J Stat Soft 32(6):1–29
43. Wang X, Chen H, Zhang NR (2017) DNA copy number profiling using single-cell sequencing. Brief Bioinforma 19(5):731–736

44. Sanders A, Macosko EZ, Wysoker A et al (2018) Molecular diversity and specializations among the cells of the adult mouse brain. Cell 174(4):1015–1030
45. Loo L, Simon JM, Xing L, McCoy ES et al (2019) Single-cell transcriptomic analysis of mouse neocortical development. Nat Commun 10(1):134
46. Dong X, Liao Z, Gritsch D et al (2018) Enhancers active in dopamine neurons are a primary link between genetic variation and neuropsychiatric disease. Nat Neurosci 21(10): 1482–1492
47. Deleye L, Tilleman L, Vander Plaetsen AS et al (2017) Performance of four modern whole-genome amplification methods for copy number variant detection in single cells. Sci Rep 7(1):3422
48. Graham JM (2002) Rapid purification of nuclei from animal and plant tissues and cultured cells. Sci World J 2:1551–1554
49. Wierman MB, Burbulis IE, Chronister WD et al (2017) Single-cell CNV detection in human neuronal nuclei. In: Frade J, Gage F (eds) Genomic mosaicism in neurons and other cell types. Neuromethods, vol 131. Humana Press, New York, NY
50. Xu X, Stoyanova EI, Lemiesz AE et al (2018) Species and cell-type properties of classically defined human and rodent neurons and glia. elife 7:e37551
51. Evrony GD, Cai X, Lee E et al (2012) Single-neuron sequencing analysis of L1 retrotransposition and somatic mutation in the human brain. Cell 151(3):483–496
52. Voet T, Kumar P, Van Loo P et al (2013) Single-cell paired-end genome sequencing reveals structural variation per cell cycle. Nucleic Acids Res 41(12):6119–6138
53. Mallory XF, Edrisi M, Navin N et al (2020) Methods for copy number aberration detection from single-cell DNA-sequencing data. Genome Biol 21:208
54. Olshen AB, Venkatraman ES, Lucito R, Wigler M (2004) Circular binary segmentation for the analysis of array-based DNA copy number data. Biostatistics 5(4):557–572
55. Shah SP, Xuan X, DeLeeuw RJ, Khojasteh M et al (2006) Integrating copy number polymorphisms into array CGH analysis using a robust HMM. Bioinformatics 22(14): 431–439
56. Bakker B, Taudt A, Belderbos ME, Porubsky D et al (2016) Single-cell sequencing reveals karyotype heterogeneity in murine and human malignancies. Genome Biol 17(1):115
57. Mallory XF, Edrisi M, Navin N, Nakhleh L (2020) Assessing the performance of methods for copy number aberration detection from single-cell DNA sequencing data. PLoS Comput Biol 16(7):1008012
58. Hatem A, Bozdağ D, Toland AE, Çatalyürek ÜV (2013) Benchmarking short sequence mapping tools. BMC Bioinformatics 14:184
59. Sherman MA, Barton AR, Lodato MA et al (2018) PaSD-qc: quality control for single-cell whole-genome sequencing data using power spectral density estimation. Nucleic Acids Res 46(4):e20
60. Normand E, Qdaisat S, Bi W et al (2016) Comparison of three whole-genome amplification methods for detection of genomic aberrations in single cells. Prenat Diagn 36(9): 823–830
61. Gonzalez-Pena V, Natarajan S, Xia Y et al (2021) Accurate genomic variant detection in single cells with primary template-directed amplification. Proc Natl Acad Sci U S A 118(24):2024176118
62. Casasent AK, Schalck A, Gao R et al (2018) Multiclonal invasion in breast tumors identified by topographic single-cell sequencing. Cell 172(1–2):205–217
63. Keinath MC, Timoshevskiy VA, Timoshevskaya NY et al (2015) Initial characterization of the large genome of the salamander Ambystoma mexicanum using shotgun and laser capture chromosome sequencing. Sci Rep 5: 16413
64. Deleye L, De Coninck D, Dheedene A et al (2016) Performance of a TthPrimPol-based whole-genome amplification kit for copy number alteration detection using massively parallel sequencing. Sci Rep 6:31825

Part IV

Lipids Analyses of Human Brain

Chapter 12

Mass Spectrometry Analysis of the Human Brain Sphingolipidome

Xin Ying Chua, Ryan Huang, Deron Herr, Mitchell K. P. Lai, Markus R. Wenk, and Federico Torta

Abstract

In recent decades, mass spectrometry-based lipidomics has provided a fertile environment for scientific investigations of biochemical and mechanistic processes in biological systems. Notably, this approach has been used to characterize physiological and pathological processes relevant to the central nervous system by identifying changes in the sphingolipid content in the brain, cerebral spinal fluid, and blood plasma. However, despite a preponderance of studies identifying correlations between specific lipids and disease progression, this powerful resource has not yet substantively translated into clinically useful diagnostic assays. Part of this gap may be explained by insufficient depth of the lipidomic profiles in many studies, by lab-to-lab inconsistencies in methodology, and a lack of absolute quantification. These issues limit the identification of specific molecular species and the harmonization of results across independent studies. In this chapter, we contextualize these issues with recent reports identifying correlations between brain lipids and neurological diseases, and we describe the workflow our group has optimized for analysis of the blood plasma sphingolipidome, adapted to the characterization of the human brain tissue.

Key words Alzheimer's disease, Brain, Clinical correlation, Lipidomics, LC-MS/MS methodology, Mass spectrometry, Neurological disease, Sphingolipidomics, Sphingolipids, Sphingosine 1-Phosphate, Plasma

1 Introduction

Blood plasma is an accessible body fluid that is highly amenable to lipidomic characterization. As a result, many studies have been performed in an effort to associate alterations in plasma sphingolipids (SPs) with neurological disease. Findings from these studies have identified significant changes that provide potential biomarkers and suggest metabolic perturbations that may contribute to disease pathology. For example, specific species of saturated ceramides were shown to be significantly lower in the plasma of patients with prodromal Alzheimer's disease (AD), compared to cognitively normal controls [1]. Coordinately, in a study of women aged

Jerold Chun (ed.), *Alzheimer's Disease: Methods and Protocols*,
Methods in Molecular Biology, vol. 2561, https://doi.org/10.1007/978-1-0716-2655-9_12,

70–79 without dementia, plasma content of specific saturated ceramides (d18:1/16:0, d18:1/22:0, d18:1/24:0) was positively correlated with subsequent incident AD [2]. Conversely, plasma sphingomyelins were shown to be reduced in AD patients, at the expense of elevated ceramides [3]. In a large-scale lipidomic study that evaluated 1912 subjects across two independent cohorts, significant and complex derangements in SPs were found in the plasma of AD patients. These included increases in GM3 gangliosides and decreases in sphingomyelins and dihydroceramides [4]. Furthermore, work from our labs evaluated the plasma content of "noncanonical" sphingosine 1-phospate (S1P) species and identified a reduction in d16:1 S1P specifically in patients with vascular dementia (VaD) but not AD [5].

Despite the above findings, it is unclear how faithfully plasma content reflects SP levels in the brain. SPs may not efficiently cross the blood–brain barrier, raising the possibility that reported associations between plasma SPs and neurological disease may be the result of purely peripheral processes associated with disease pathology. Considering the fundamental differences in CNS versus plasma sphingolipidoses, it is unlikely that specific changes in the brain parenchyma are detectable in the blood. For example, while both plasma and brain contain similar complements of ceramides and sphingomyelins with regard to their N-acyl fatty acid content, one notable difference between these two tissues relates to the sphingoid backbone. Specifically, whereas approximately 10% of plasma SPs contain a d16:1 backbone, these noncanonical species are essentially absent from the brain [6], likely due to the low expression of *SPTLC3* in this tissue [7]. Studies that perform reciprocal evaluation of both brain and plasma are needed to understand how neurological disease results in alterations of the plasma sphingolipidome.

Collection of cerebral spinal fluid (CSF) via lumbar puncture offers a potential avenue to interrogate SP content of the brain. CSF provides a good representation of brain SP content and can be processed using essentially the same protocols used for blood plasma. Indeed, electrospray ionization MS (ESI/MS) analysis was used to identify a reduction in total sulfatides (SHexCers) in the CSF of patients with incipient dementia (very mild impairment) relative to cognitively normal, age-matched controls [8]. Another study, using a similar technique, identified an increase in total CSF sphingomyelins in a similar cohort [9]. Interestingly, in the CSF of multiple sclerosis patients, ceramides showed both increases and decreases depending on the species. Notably, d18:1/16:0 and d18:1/24:0 ceramides were elevated in the CSF of patients and exhibited neurodegenerative effects in vitro, consistent with the disease pathology [10]. However, there are several shortcomings to a reliance on CSF: (1) Due to the risk of trauma or infection during collection, CSF cannot be routinely or repeatedly accessed in healthy subjects. (2) CSF cannot be used to identify SP alterations that have regional heterogeneity within the brain. (3) Intracellular alterations in SPs may not be reflected in the CSF. (4) It is often not possible to obtain CSF from experimental animals (i.e., mice) in sufficient quantities for analysis. (5) Lipid content in the CSF is approximately 500-fold lower compared to the plasma [11], so many biologically relevant SP species may fall below the threshold of reliable quantification. Therefore, there remains a need to characterize the brain sphingolipidome from postmortem samples and animal disease models.

Despite the technical challenges in obtaining accurate, in-depth sphingolipidomes from CNS tissue, several studies have reported successful development of liquid chromatography MS (LC-MS) protocols that facilitated the identification of significant correlations between neuropathological processes and the content of specific SPs in postmortem brains [12]. An early study identified a marked reduction in total sulfatides in the gray matter and white matter of AD patients at all stages, from mild to severe dementia [13]. More recently, LC-MS/MS profiling corroborated this by demonstrating a decrease in hexosylceramides (HexCer) in the neocortex of patients with mild AD but found increased HexCer in advanced AD [14]. Interestingly, this study also showed a similar trend for sphingomyelins and reported a consistent increase in ceramides for both mild and severe AD. A cross-sectional study of postmortem hippocampal tissue from cognitively normal subjects >65 years of age was performed to identify correlations between SPs and age, sex, and *APOE* genotype [15]. Although no associations were identified with *APOE* genotype, several SPs, such as ceramides, sphingomyelins, and sulfatides, were positively correlated with age.

In addition to dementia, postmortem brains have been interrogated in the context of other neurodegenerative diseases. One study evaluated postmortem tissue from patients with multiple sclerosis to compare active white matter lesions or inactive lesions with "normal" brain regions. This characterization identified several distinct changes; notably, specific dihydroceramides (d18:0) were lower in inactive lesions but elevated in active plaques [16]. By contrast, brains of patients with multiple system atrophy (MSA) demonstrated a very different SP profile, whereby most species of sphingomyelins and sulfatides were decreased specifically in the motor cortex [17]. This suggests that SP alterations in neurodegenerative diseases have disease-specific etiologies and characteristic profiles.

In experimental animal models, LC-MS/MS was used to identify an increase in saturated ceramide species in the striatum and cortex of Huntington's disease mice [18]. Similarly, a shotgun lipidomic method was used to characterize alterations in *ApoE* knockout and transgenic mice, revealing derangements of sphingomyelins and HexCers [19]. Another study characterized differences in brain SPs in a mouse model for autism spectrum disorder, showing a decrease in most sphingomyelins [20]. This illustrates the value of lipidomic analyses for non-degenerative neurological diseases. The importance of addressing anatomical variability was recently emphasized via a detailed characterization of SP profiles from different brain regions in wild-type mice [21] that revealed marked differences among regions in SP composition, within and among SP classes.

The recent clinical and experimental observations provided by LC-MS/MS-mediated SP quantification and the preponderance of emerging high-quality sphingolipidomic data demonstrate the extent to which the field has advanced. However, reflecting the existence of CNS-specific methodologies in different laboratories, standardization among protocols and analytical approaches is improving. Here, we provide details of the LC-MS/MS technique optimized by our groups to provide reproducible, in-depth SP quantification from CNS tissues.

2 Materials

2.1 Processing of Brain Tissues

1. Ultra-Turrax homogenizer (or comparable mechanical homogenizer).
2. *Tris-HCl buffer* (1 mL per 50 mg tissue): 50 mM Tris–HCl, 120 mM sodium chloride (NaCl), 5 mM potassium chloride (KCl), adjusted to pH 7.4 with 1 N sodium hydroxide (NaOH). Store at 4 °C for up to 1 month.
3. *Homogenization solution*: Add one tablet of cOmplete™ protease inhibitor cocktail (or equivalent) and one tablet of PhosSTOP™ phosphatase inhibitor (or equivalent) to 10 mL of Tris-HCl buffer. Store at 4 °C for up to 1 month.
4. Vortex.
5. Forceps.
6. Scalpel.

2.2 Lipid Extraction

1. Eppendorf Safe-Lock 1.5 mL polypropylene tubes (or equivalent).
2. Eppendorf ThermoMixer (or equivalent).
3. Microcentrifuge.
4. SpeedVac™ vacuum concentrator (or equivalent).
5. Chloroform ($CHCl_3$), HPLC grade.
6. Methanol (MeOH), HPLC grade.
7. Milli-Q water.
8. Micropipettor and corresponding pipette tips.

2.3 LC-MS/MS

1. UHPLC system: Agilent 1290 Infinity LC (or equivalent).
2. Triple-quadrupole mass spectrometer: Agilent 6490 or 6495 QQQ (or equivalent).
3. Chemically inert autosampler vials (glass or polypropylene) with airtight polytetrafluoroethylene (PTFE)-sealed screw septum caps.
4. Reversed-phase Agilent ZORBAX RRHD Eclipse Plus (or equivalent) C18 column (2.1 × 50 mm, 1.8 μm).
5. Schott Duran glass bottles (or equivalent) for preparation and storage of internal standards (ISTD) and mobile phases.
6. *Resuspension solution*: chloroform/methanol ($CHCl_3$/MeOH) solution, 1:1 (v/v), chilled.
7. Chloroform/methanol ($CHCl_3$/MeOH) solution, 1:2 (v/v), chilled.
8. Internal standard solution (ISTD), prepared in $CHCl_3$/MeOH (1:2; v/v) (360 μL per sample, quality control [QC], or blank), prepared in fume hood and stored at −20 ° C.

Internal standards	Chemical formula	Concentration (ng/mL)
SM d18:1/12:0250	$C_{35}H_{71}N_2O_6P$	200
GluCer d18:1/12:0	$C_{36}H_{69}NO_8$	100
GalCer d18:1/12:0	$C_{36}H_{69}NO_8$	86.5
Cer d18:1/12:0	$C_{30}H_{59}NO_3$	100
Sphingosine-d7	$C_{18}H_{30}D_7NO_2$	25
Sphinganine-d7	$C_{18}H_{32}D_7NO_2$	25

9. *Mobile phase A*: 400 mL acetonitrile, 600 mL ultrapure water, 10 mM ammonium formate. Mix well and sonicate for 10 min.
10. *Mobile phase B*: 100 mL acetonitrile, 900 mL 2-propanol, 10 mM ammonium formate. Mix well and sonicate for 10 min.

3 Methods

3.1 Processing of Brain Tissues (See Notes 1–3)

1. Thaw brain tissues (0.5–1.0 g) on ice.
2. Dissect the tissues with disposable forceps and surgical scalpels in a petri dish, on ice, until they are free of white matter and meninges.

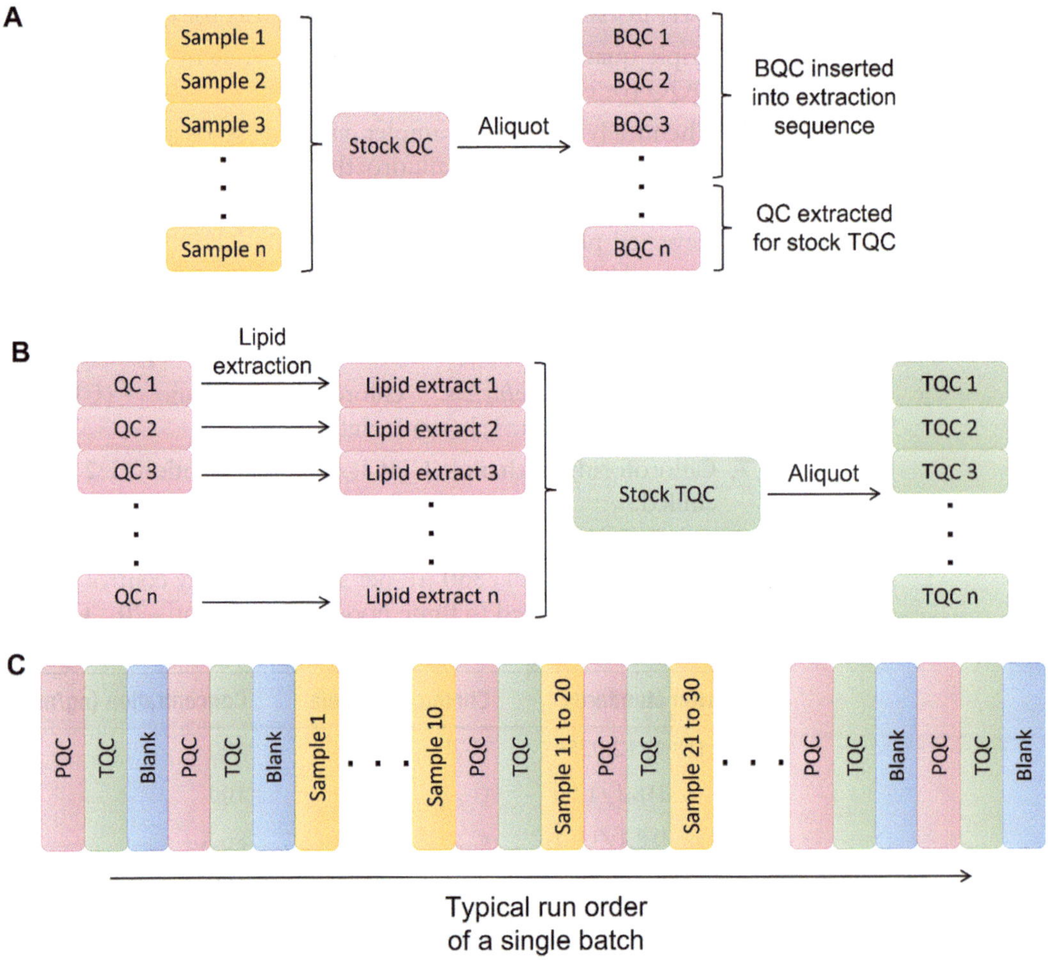

Fig. 1 Schematic diagram of extraction procedure and injection sequence that integrate quality controls (QCs). Illustration of steps for (**a**) generation of batch quality controls (BQCs), (**b**) generation of technical quality controls (TQCs), and (**c**) injection sequence

3. Homogenize the dissected tissue with cold homogenization solution maintaining ratio of 50 mg tissue (wet weight)/mL solution.
4. Store at −80 ° C until use.

3.2 Preparation of Quality Control (QC) Samples and Blanks (See Note 4)

Batch QC (BQC) samples (refer to Fig. 1a):

1. Thaw brain homogenates on ice.
2. Vortex the samples.
3. Estimate the volume of (pooled) BQCs required (*see* **Note 4**).

4. Pool equal aliquots of each sample to generate the target volume of pooled BQC. Vortex to mix.
5. Aliquot 20 μL of pooled BQC to labeled Eppendorf Safe-Lock 1.5-mL polypropylene tubes.
6. Store at −80 ° C until use.

Technical QC (TQC) samples (refer to Fig. 1b):

1. Estimate the volume of (pooled) TQCs required (*see* **Note 4**).
2. Pool equal aliquots of each lipid extract to generate the target volume of pooled TQC. Vortex to mix.
3. Aliquot 100 μL of pooled TQC into labeled vials.
4. Store at −80 ° C until use.

Blanks:

1. Aliquot 100 μL of Milli-Q water into labeled tubes.

3.3 Two-Phase Extraction

1. Add 360 μL of chilled ISTD solutions [in $CHCl_3$/MeOH, 1:2 (v/v)] to 20 μL of homogenized sample, quality controls (QCs), or blanks.
2. Incubate on ThermoMixer at 300 rpm, 4 °C, for 1 h, in the dark.
3. Add 120 μL of chilled $CHCl_3$ and 100 μL of chilled Milli-Q water to sample.
4. Vortex for 30 s.
5. Centrifuge at 9000 ×*g*, for 2 min, to separate the organic phase from the aqueous phase (*see* **Note 5**).
6. Using a micropipettor, transfer the lower organic phase into a clean microcentrifuge tube (*first organic extract*), taking care not to take up the aqueous phase as the pipette tip passes through the upper phase.
7. Add 500 μL of chilled $CHCl_3$ to the remaining aqueous phase.
8. Vortex for 30 s.
9. Centrifuge at 9000 ×*g*, for 2 min, to separate the organic and aqueous phases.
10. Transfer the lower organic phase to the previously collected first organic extract.
11. Dry the pooled organic extract completely, in a vacuum concentrator.
12. Resuspend the lipid extract with 100 μL of chilled $CHCl_3$/MeOH (1:1; v/v).
13. Transfer the lipid extract into chemically inert autosampler vials.
14. Store at −80 ° C until analysis.

3.4 LC-MS/MS Analysis

This protocol is adapted from our previously published method [22]; see this reference for a list of lipid multiple-reaction monitoring (MRM) transitions, as optimized for brain tissue.

1. Perform lipid separations using a UHPLC Agilent 1290 chromatographic system (or equivalent), with a reversed-phase Agilent ZORBAX Agilent Zorbax RRHD Eclipse Plus C18 column (2.1 mm × 50 mm, 1.8 μm) connected to an Agilent QqQ 6495 LC-MS system.
2. Before the injection of study samples, condition the system by injecting two blanks and several TQCs until the retention times and signals are stable.
3. Use Mobile Phases A and B, applying the following gradient: 0–2 min, 20–60% B; 2–7 min, 60–100% B; 7–9 min, 100% B; 9.01–10.8 min, 20% B. Keep the oven temperature at 40 °C, the flow rate at 0.4 mL/min, and the sample injection volume at 2 μL. Set the positive ionization spray voltage and nozzle voltage to 3500 V and 500 V, respectively. Set the drying gas and sheath gas temperatures to 200 °C and 250 °C, respectively, and the drying gas and sheath gas flow rates to 12 L/min. Set the iFunnel high- and low-pressure radio frequency (RF) to 80 V and 40 V, respectively. Use the Dynamic MRM positive ion mode for data acquisition.
4. Injection sequence (refer to Fig. 1c):
   (a) Inject samples, BQCs, and blanks in the same sequence used for extraction.
   (b) Inject TQCs between the different batches to monitor the instrument performance.

3.5 Data Analysis

1. Use Agilent MassHunter software to analyze the data (https://www.agilent.com/en/product/software-informatics/mass-spectrometry-software/data-analysis).
2. Calculate the signal-to-noise (S/N) ratios using the raw peak areas of each lipid in the study samples and blanks. Exclude any lipids from the analysis that have an S/N ratio <10 and a coefficient of variation (CV) >20%, in the QC samples, and do not show a linear behavior (coefficient of determination [R^2] < 0.8) in the dilution curves (which can be obtained by injecting different volumes of the TQC sample and then plotting the raw peak areas against the volume).
3. Use the ISTDs to normalize the raw peak areas in the corresponding lipid class, and then further normalize the concentrations to either the tissue weight of the original sample or to the protein concentration of the homogenate (*see* **Note 6**).

4 Notes

1. Samples should be collected quickly, rinsed in buffer, and stored at $-80\ ^\circ C$ until used, to minimize degradation. Ensure tissue quality by obtaining samples with low postmortem delays (PMI < 18 h) and high RNA integrity numbers (RIN $\geq$ 6).
2. Human brain tissues should be handled as though they may contain infectious agents (e.g., prions). Use appropriate personal protective equipment, such as surgical gloves, protective eyewear, and masks. Dispose of sharp items in proper containers after use. Ensure that the work area and equipment are thoroughly cleansed and decontaminated after use.
3. In large studies, samples may be divided into batches for extractions. Each batch should be representative of the overall study population by using a stratified randomization procedure.
4. Especially for large studies, quality control (QC) samples should be included for quality assurance and to correct for signal-to-noise differences within and between batches. Lipid extraction from BQC and TQC samples should be performed together with the study samples and following the injection sequence. BQCs monitor the variability of the extraction process, and both BQCs and TQCs monitor that of the analytical system.
5. At least two BQC and two TQC samples should be added to the start and end of each extraction batch. In addition, one BQC and one TQC should be added after every third, fifth, or tenth study sample, depending on the total number of study samples to be analyzed. For example, for a run with 100 samples, with QCs added every fifth sample, there will be 24 BQCs and 24 TQCs required in total; hence, 480 μL of BQC and 480 μL of TQC are required. The speed and time can be increased if the two phases do not separate well, but note that increasing speed and time may cause the protein layer to mix with the organic phase.
6. Endogenous species are usually quantified using one standard per lipid class; thus, this method can only deliver relative quantitation for lipid species, except for the few species corresponding to the respective stable-isotope standards.

References

1. Mielke MM, Haughey NJ, Bandaru VVR et al (2010) Plasma ceramides are altered in MCI and predict cognitive decline and hippocampal volume loss. Alzheimers Dement 6:378–385. https://doi.org/10.1016/j.jalz.2010.03.014
2. Mielke MM, Bandaru VVR, Haughey NJ et al (2012) Serum ceramides increase the risk of Alzheimer disease: the Women's Health and Aging Study II. Neurology 79:633–641. https://doi.org/10.1212/WNL.0b013e318264e380
3. Han X, Rozen S, Boyle SH et al (2011) Metabolomics in early Alzheimer's disease: identification of altered plasma sphingolipidome using shotgun lipidomics. PLoS One 6:e21643. https://doi.org/10.1371/journal.pone.0021643
4. Huynh K, Lim WLF, Giles C et al (2020) Concordant peripheral lipidome signatures in two large clinical studies of Alzheimer's disease. Nat Commun 11:5698. https://doi.org/10.1038/s41467-020-19473-7
5. Chua XY, Chai YL, Chew WS et al (2020) Immunomodulatory sphingosine-1-phosphates as plasma biomarkers of Alzheimer's disease and vascular cognitive impairment. Alzheimers Res Ther 12:122. https://doi.org/10.1186/s13195-020-00694-3
6. Narayanaswamy P, Shinde S, Sulc R et al (2014) Lipidomic "deep profiling": an enhanced workflow to reveal new molecular species of signaling lipids. Anal Chem 86:3043–3047. https://doi.org/10.1021/ac4039652
7. Lam BWS, Yam TYA, Chen CP et al (2021) The noncanonical chronicles: emerging roles of sphingolipid structural variants. Cell Signal 79:109890. https://doi.org/10.1016/j.cellsig.2020.109890
8. Han X, Fagan AM, Cheng H et al (2003) Cerebrospinal fluid sulfatide is decreased in subjects with incipient dementia. Ann Neurol 54:115–119. https://doi.org/10.1002/ana.10618
9. Kosicek M, Zetterberg H, Andreasen N et al (2012) Elevated cerebrospinal fluid sphingomyelin levels in prodromal Alzheimer's disease. Neurosci Lett 516:302–305. https://doi.org/10.1016/j.neulet.2012.04.019
10. Vidaurre OG, Haines JD, Katz Sand I et al (2014) Cerebrospinal fluid ceramides from patients with multiple sclerosis impair neuronal bioenergetics. Brain 137:2271–2286. https://doi.org/10.1093/brain/awu139
11. Sol J, Jové M, Povedano M et al (2021) Lipidomic traits of plasma and cerebrospinal fluid in amyotrophic lateral sclerosis correlate with disease progression. Brain Commun 3:fcab143. https://doi.org/10.1093/braincomms/fcab143
12. Hejazi L, Wong JWH, Cheng D et al (2011) Mass and relative elution time profiling: two-dimensional analysis of sphingolipids in Alzheimer's disease brains. Biochem J 438:165–175. https://doi.org/10.1042/BJ20110566
13. Han X, Holtzman DM, McKeel DW et al (2002) Substantial sulfatide deficiency and ceramide elevation in very early Alzheimer's disease: potential role in disease pathogenesis. J Neurochem 82:809–818. https://doi.org/10.1046/j.1471-4159.2002.00997.x
14. Akyol S, Ugur Z, Yilmaz A et al (2021) Lipid profiling of Alzheimer's disease brain highlights enrichment in glycerol(phospho)lipid, and sphingolipid metabolism. Cell 10:2591. https://doi.org/10.3390/cells10102591
15. Couttas TA, Rustam YH, Song H et al (2020) A novel function of sphingosine kinase 2 in the metabolism of Sphinga-4,14-diene lipids. Meta 10:236. https://doi.org/10.3390/metabo10060236
16. Podbielska M, Szulc ZM, Ariga T et al (2020) Distinctive sphingolipid patterns in chronic multiple sclerosis lesions. J Lipid Res 61:1464–1479. https://doi.org/10.1194/jlr.RA120001022
17. Don AS, Hsiao J-HT, Bleasel JM et al (2014) Altered lipid levels provide evidence for myelin dysfunction in multiple system atrophy. Acta Neuropathol Commun 2:150. https://doi.org/10.1186/s40478-014-0150-6
18. Di Pardo A, Basit A, Armirotti A et al (2017) De novo synthesis of sphingolipids is defective in experimental models of Huntington's disease. Front Neurosci 11:698. https://doi.org/10.3389/fnins.2017.00698
19. Fitzner D, Bader JM, Penkert H et al (2020) Cell-type- and brain-region-resolved mouse brain Lipidome. Cell Rep 32:108132. https://doi.org/10.1016/j.celrep.2020.108132

20. Ju J, Yang X, Jiang J et al (2021) Structural and lipidomic alterations of striatal myelin in 16p11.2 deletion mouse model of autism spectrum disorder. Front Cell Neurosci 15: 718720. https://doi.org/10.3389/fncel.2021.718720
21. Wang C, Palavicini JP, Han X (2021) A lipidomics atlas of selected sphingolipids in multiple mouse nervous system regions. Int J Mol Sci 22:11358. https://doi.org/10.3390/ijms222111358
22. Burla B, Muralidharan S, Wenk MR, Torta F (2018) Sphingolipid analysis in clinical research. Methods Mol Biol 1730:135–162. https://doi.org/10.1007/978-1-4939-7592-1_11

Chapter 13

Mass Spectrometry for the Advancement of Lipid Analysis in Alzheimer's Research

Jonatan Martínez-Gardeazabal, Marta Moreno-Rodríguez, Estíbaliz González de San Román, Beatriz Abad, Iván Manuel, and Rafael Rodríguez-Puertas

Abstract

Recent technical advances in mass spectrometry, as applied to the analytical chemistry of lipid molecules, enable the simultaneous detection of the multiplicity of lipid complex species present in the human brain. This, in combination with quantitative studies carried out in plasma samples, helps to identify disease biomarkers including for Alzheimer's disease (AD). Mass spectrometry imaging (MSI) is particularly powerful for the anatomical localization of lipids in brain slices, identifying lipid modifications in postmortem frozen samples from AD patients.

Human brain tissues are sectioned in a cryostat and then covered with a chemical matrix, such as mercaptobenzothiazole (MBT) or α-cyano-4-hydroxycinnamic acid (CHCA), to ionize the lipid molecules either by sublimation or by spraying. We describe the use of matrix-assisted laser desorption ionization (MALDI) in an LTQ–Orbitrap–XL mass spectrometer to scan brain tissue slices with high spatial resolution, analyzing 50 μm cell layers. The lipid spectra obtained for each pixel are transformed to color-coded intensity maps of hundreds of lipid species included those within a single tissue slice.

Key words Lipid, MALDI-MSI, UHPLC-MS, Biomarkers, Alzheimer's disease, Neurolipid, Mass spectroscopy

1 Introduction

Lipidomics—the systematic study of lipids and their functions in biological systems—has become an area of great interest in biomedical research during the last few years. The technological advances in mass-spectroscopy imaging (MSI) have not only enabled a faster and more accurate identification of myriad lipid species but have also shed light on their anatomical localizations. Indeed, MSI has achieved a milestone in neuroscience research, allowing, for the first time, the simultaneous identification of multiple lipid species in tiny brain region nuclei where lipids are highly specialized [1]. These

Jerold Chun (ed.), *Alzheimer's Disease: Methods and Protocols*,
Methods in Molecular Biology, vol. 2561, https://doi.org/10.1007/978-1-0716-2655-9_13,

lipids are well known for their functions as structural molecules and in cell energy metabolism, but MSI has revealed that they are also involved in signaling in complex biological pathways, even behaving as neurotransmitters (neurolipids). Although lipid analysis has traditionally been performed using gas-chromatography mass spectrometry (GC-MS), development of matrix-assisted laser desorption/ionization (MALDI) method, together with the use of ultrahigh-performance liquid chromatography tandem mass spectrometry (UHPLC-MS), has significantly increased both the lipid sensitivity and the breadth of lipids that can be detected simultaneously. These techniques have also been used to search for biomarkers for the early detection of lipid-related diseases, which may also include Alzheimer's disease (AD) [2].

MALDI-mass spectrometry imaging (MALDI-MSI) has an advantage over conventional MS techniques because, in addition to identifying a multiplicity of lipid molecules, it offers unbiased visualization of the spatial distributions of lipids in tissues. This feature allows one to infer the physiological functions of lipids in the brain under both healthy and pathological conditions [3, 4].

Mass spectrometry techniques are also being used to identify specific biomarkers for different pathologies in human plasma samples. For example, certain phospholipids (PLs) have been proposed as potential serum biomarkers that could predict mild cognitive impairment or AD, 2 or 3 years before the onset of clinical symptoms [2, 5]. Furthermore, MALDI-MSI has been used to find lipid alterations in AD postmortem brain samples [4, 6]. Thus, these lipidomic approaches are contributing to our understanding of the complex relationships among the lipids in cell structures, the roles of lipids in energy metabolism, and neurolipid signaling, in the development of pathological processes that underly different neurodegenerative diseases.

2 Materials

2.1 MALDI-MSI

1. Cryostat.
2. Gelatin-coated slides (recommendation: 26 × 76 mm).
3. Transparent plastic tape.
4. 2-Mercaptobenzothiazole (MBT), 300 mg.
5. Glass sublimator (recommendation: ACE glass 8023, 152 × 250 mm), or equivalent.
6. Vaseline for vacuum.
7. Hot plate that measures 100 °C with an error of ±0.5 °C.
8. Vacuum pump that reaches 1 bar.

9. Deep, glass Petri dish (approximately 50 × 100 mm), with the top covered in aluminum foil.
10. Methanol (MeOH) gradient, liquid chromatography grade.
11. *Matrix solution*: α-cyano-4-hydroxycinnamic acid (CHCA), 5 mg/mL in MeOH/dd$H_2$0, 70:30 (v/v) (*see* **Note 1**).
12. Tissue MALDI sprayer (TM-sprayer, HTX Imaging).
13. LTQ-Orbitrap XL mass spectrometer (*see* Fig. 1).

2.2 Plasma Lipid Extraction for UHPLC-MS Detection

1. Internal standard: Splash™ LipidoMix™, or equivalent.
2. Ceramide/sphingoid internal standard mixture I (Avanti Polar Lipids), or equivalent.
3. Microtube, 1.5 mL.
4. Optima LC/MS-grade water (H_2O), or equivalent.
5. Optima LC/MS-grade acetonitrile (ACN), or equivalent.
6. Optima LC/MS-grade 2-propanol (IPA), or equivalent.
7. Optima LC/MS-grade formic acid, or equivalent.
8. Ammonium formate.
9. *Dry-sample reconstitution solution*: ACN:IPA (50:50, v/v).
10. IPA, pre-cooled to −20 °C.
11. Vortex mixer.
12. Maximum-recovery chromatography vials.
13. LTQ Velos ESI positive-ion calibration solution, or equivalent.
14. LTQ Velos ESI negative-ion calibration solution, or equivalent.
15. Refrigerated Hettich Mikro 200R centrifuge (maximum speed, 15,000 RPM; maximum RCF, 21,382 ×*g*; 85-mm radius; 4 °C minimum temperature; dimensions [H × W × D], 260 mm × 281 mm × 553 mm), or equivalent.
16. SAVANT SC250EXP SpeedVac concentrator, or equivalent.
17. Reverse-phase column: Acquity UPLC C_{18} CSH™ 2.1 mm × 100 mm, 1.7 μm (Waters), or equivalent.
18. Pre-column: Acquity UPLC C_{18} CSH™ 2.1 mm × 5 mm, 1.7 μm (Waters), or equivalent.
19. *Mobile phase A solution*: ACN:H_2O, 40:60 (v/v); ammonium formate, 10 mM; formic acid, 0.1%. This solution should be prepared just before use.
20. *Mobile phase B solution*: ACN:IPA, 10:90 (v/v); ammonium formate, 10 mM; formic acid, 0.1%. This solution should be prepared just before use.
21. *Blank sample*: 50:50 (v/v) ACN:IPA solution.

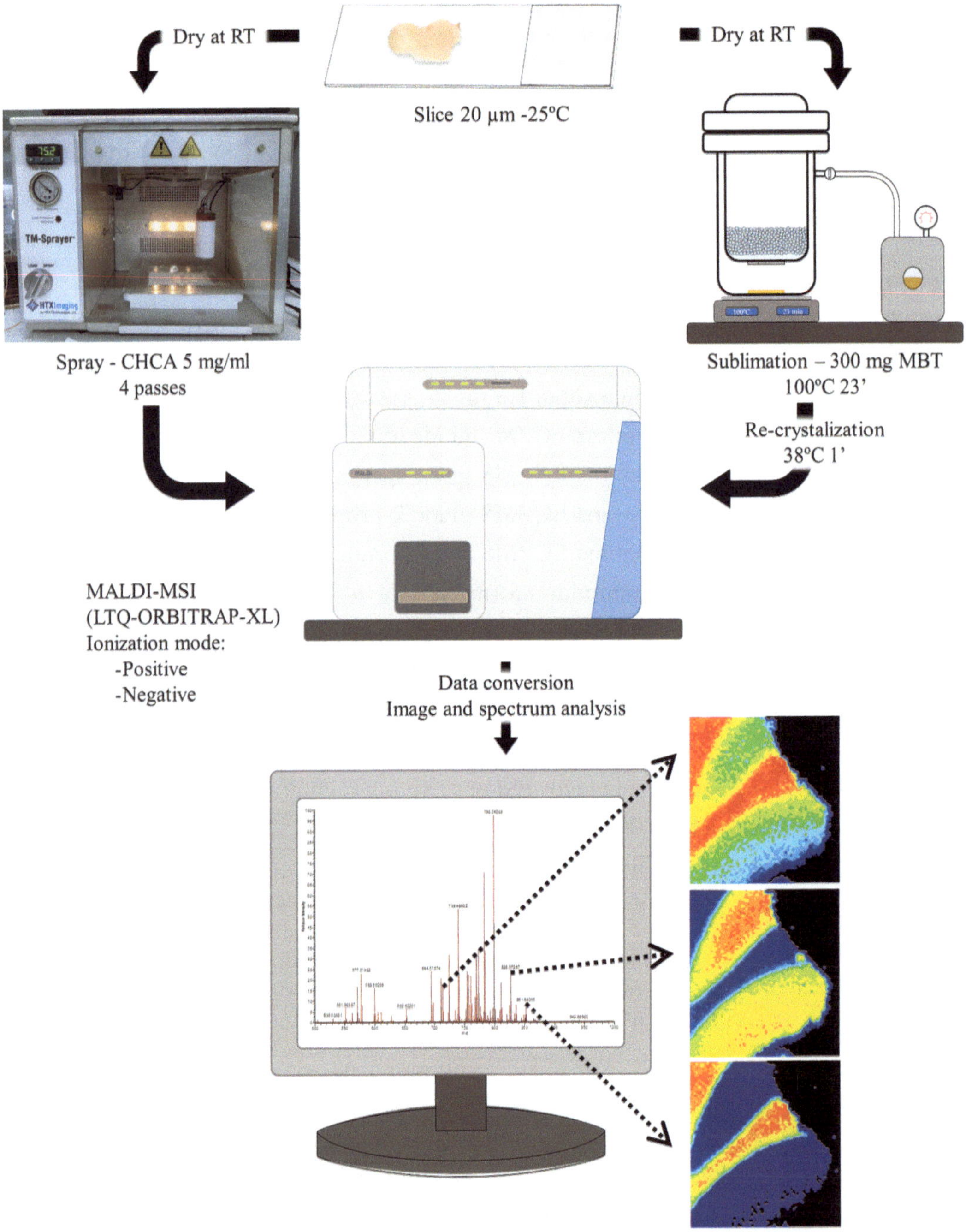

Fig. 1 Matrix-assisted laser desorption ionization-mass spectrometry imaging (MALDI-MSI) workflow. Mount 20 μm thick slices of frozen tissue onto gelatin-coated slides and keep at −25 °C until used. Thaw the samples at room temperature (RT) in a desiccator for 15 min before spraying (*left pathway*) or sublimating (*right pathway*). Analyze the samples with the mass spectrometer to obtain the data, and then transform them into color-coded images

22. *Diluted internal standard solution (ISS)*: A solution of Splash™ LipidoMix™ standards and ceramide/sphingoid internal standard mixture I (1:1).
23. *Quality control (QC) sample*: prepared by combining equal aliquots from each of the samples (in each group) of the study.
24. *QC system (QCsys) sample*: These samples are prepared as for the QC sample but are only used to condition the apparatus.
25. *QC dilution (QCdilution) sample*: QC diluted (1:10) with ACN–IPA (50:50, v/v).
26. Vanquish Ultimate 3000 UHPLC system coupled to electrospray ionization (EIS) system.
27. Q Exactive HF-X hybrid quadrupole-Orbitrap mass spectrometer, or equivalent.
28. *Xcalibur* software, v 4.1 or later.
29. *LipidSearch* software, v 4.2.2 or later.

3 Methods

3.1 Tissue-Sample Preparation for MALDI-MSI

1. Cut frozen brain tissue into 20 μm thick slices using a cryostat (*see* **Note 2**).
2. Mount tissue sections onto gelatin-coated slides, and keep them at −25 °C until use. Use commercially coated slides, or dip slides in 2% gelatin solution at 37 °C.
3. Thaw the slides at room temperature (RT) in a desiccator for 15 min before use (*see* **Note 3**).
4. Coat the tissue with the matrix, either using matrix sublimation with MBT (*see* Subheading 3.2) or matrix spraying with CHCA (*see* Subheading 3.3).

3.2 Matrix Sublimation for MALDI-MSI

1. Preheat the hot plate to 100 °C.
2. Place 300 mg of MBT at the bottom of the outer body of the sublimator (*see* Fig. 2 and **Note 4**).
3. Using tape, affix the slide with the tissue samples to the bottom of the *inner unit* of the sublimator (facing downward), and then place the inner unit into the outer body of the sublimator, sealing the junction with Vaseline to enable a vacuum (*see* Fig. 2 and **Note 5**).
4. Fill the inner unit of the sublimator with ice water, to keep the tissues cool during sublimation (*see* Fig. 2 and **Note 6**).
5. Once the hot plate reaches 100 °C, connect the vacuum pump to the sublimator, and set it to 1 bar. Immediately place the sublimator onto the hot plate, and proceed with the matrix deposition for 23 min at 100 °C (*see* **Note 7**).

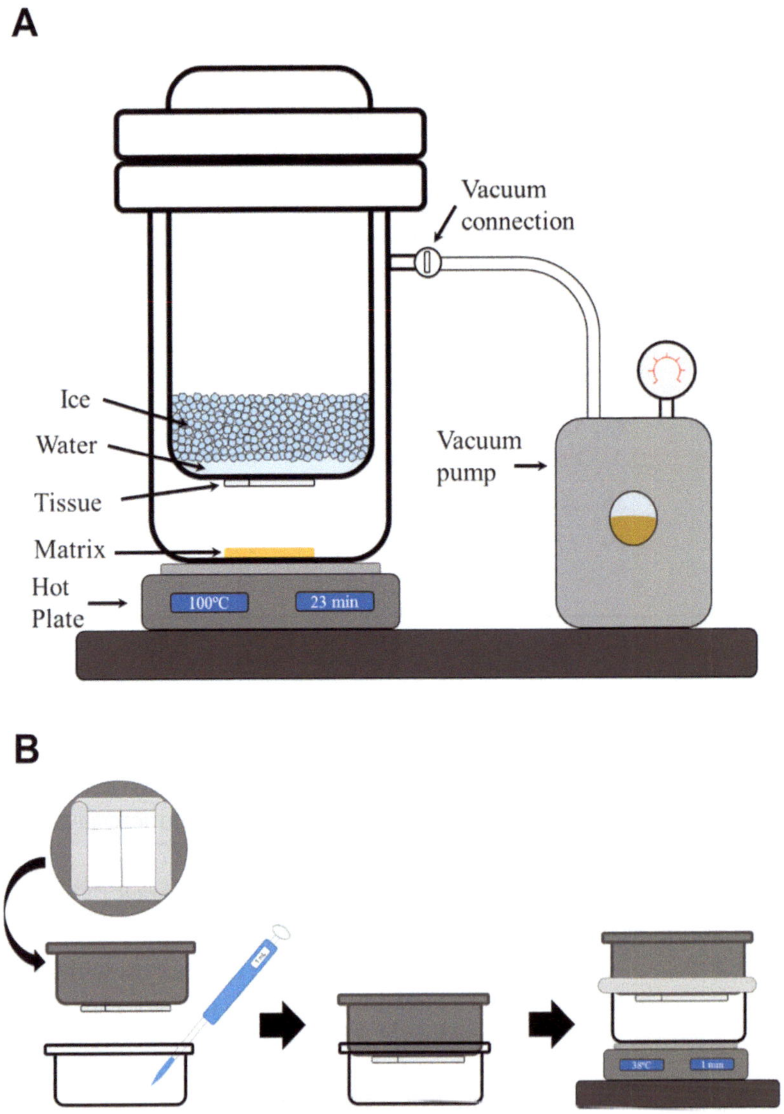

Fig. 2 Diagram of sublimation. (**a**) Sublimation process. Place the matrix (*yellow*) at the bottom of the (outer chamber of the) sublimator, and use tape to affix the sample slide (facing downward) on the bottom of the inner chamber. Place the inner chamber inside the outer one, using Vaseline to seal the joint. Fill the top of the inner chamber with ice water, connect to vacuum pump, and place the entire apparatus on a hot plate at 100 °C for 23 min. (**b**) Re-crystallization process. Cover the lid of a deep Petri dish with aluminum foil (*dark gray*), and then use tape to affix the slide to the outside of the lid, with the samples facing outward. Add 1 mL of methanol (MeOH) to the inside of the bottom of the Petri dish (*left panel*), then invert the top, and place it on top of the bottom, so the sample on the slide is facing downward (*middle panel*). Finally, set the Petri dish assembly on a 38 °C hot plate (*right panel*) for 1 min

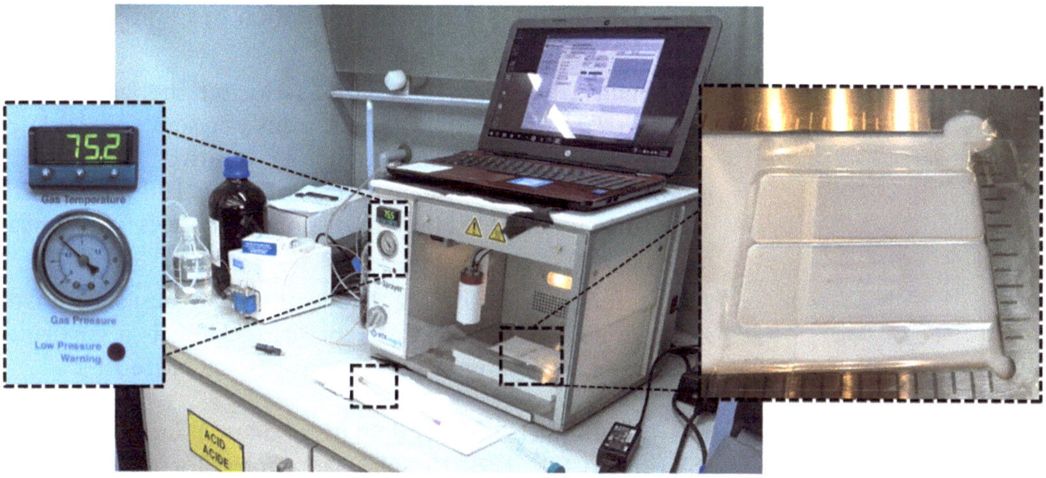

Fig. 3 Diagram of spraying apparatus. *Left*: Set the nozzle temperature to 75 °C and the nitrogen pressure to 10 psi. *Center*: Connect a syringe (*dashed box*) to the loop of the sprayer, and fill the loop with 5 mL of the matrix solution. *Right*: Place the samples on the plate, and begin the spraying process (*see* **Note 9**)

6. To re-crystallize after matrix deposition, preheat the hot plate to 38 °C, then remove the slide from the sublimation apparatus, and use tape to affix it to the inside of the top of a deep Petri dish (which was previously covered with aluminum foil). Add 1 mL of MeOH to the bottom of the Petri dish, and then cover the bottom with the top. Set the Petri dish assembly on the hot plate (at 38 °C) for 1 min (*see* Fig. 2 and **Note 8**).

3.3 Matrix Spraying for MALDI-MSI

1. For this (alternative) method of matrix deposition, turn on the sprayer, and fill the loop of the sprayer with 5 mL of the matrix solution. For optimal spraying performance, set up the device with the following settings, for six consecutive passes: nozzle temperature, 75 °C; separation of the nozzle from the glass slide, 40 mm; track spacing, 3 mm; change the spray pattern by a 90 °C angle between each pass; spray velocity, 1200 mm/min; peristaltic pump flow rate, 0.70 mL/min; and nitrogen gas pressure, 10 psi (*see* Fig. 3).
2. Place the sample on the plate and start the process (*see* **Note 9**).
3. When complete, place the slide in a closed desiccator for 15 min to ensure complete drying of the coated material, before loading the sample into the mass spectrometer.

3.4 MALDI-MSI

1. To obtain a reference image for analyzing the samples, scan the samples that have been previously coated (either by sublimation or by spraying) together with the slide holder.
2. Load the samples into the MALDI-MSI device (e.g., an LTQ-Orbitrap XL mass spectrometer), and use its control software to configure the region of interest (ROI) to be analyzed and the measurement parameters.

3. Choose the appropriate spatial resolution to scan the ROI (*see* **Note 10**).
4. To obtain a good signal-to-noise ratio, set the laser energy between 10 and 15 μJ, with 2-μ scans per step and 10 laser pulses. Set the m/z range to either positive- or negative-ionization mode, and the mass resolution to 100 k full-width at half-maximum (FWHM), to distinguish species with similar m/z ranges.
5. After scanning, visualize the data with ImageQuest software or an equivalent, assigning different mass channels a particular color code, according to their intensities (*see* **Note 11**).

3.5 Plasma Lipid Extraction for UHPLC-MS Detection

1. In a 1.5 mL microtube, add 20 μL of diluted internal standard solution to 80 μL of the plasma samples, for a final sample volume of 100 μL.
2. Add 4 volumes of IPA, pre-cooled to −20 °C, to the plasma samples prepared in the previous step.
3. Vortex for 1 min, and then incubate at RT for 10 min.
4. Centrifuge the microtubes at 13,000 ×*g* for 20 min, at 4 °C.
5. Transfer the supernatant to a new tube and dry using a Speed-Vac concentrator (at RT, for 60 min). Store the dried samples at −80 °C until the assays can be performed (*see* **Note 12**).

3.6 Plasma Lipid Extraction for UHPLC-MS

An overview of this protocol can be found in Fig. 4.

1. Reconstitute the dried samples with 120 μL of dry-sample reconstitution solution, and transfer to maximum-recovery chromatography vials for subsequent injection into the UHPLC-Q-ORBITRAP-HF-X. Maintain all sample vials (study samples, blank, diluted internal standard solution, QC, and QCsys) at 10 °C on the injector.
2. Prepare mobile-phase solutions (*mobile phase A solution* and *mobile phase B solution*), and transfer them to the mobile-phase reservoirs.
3. Place the reverse-phase column and the pre-column in the column oven of the UHPLC, and edit the UHPLC operating conditions (*see* **Note 13**).
4. Set up the mass spectrometer parameters (*see* **Note 14**), and load the UHPLC-MS/MS protocol.
5. Design the analysis sequence as follows (*see* Fig. 5 and **Note 15**):
   (a) Inject a blank sample and a diluted internal standard solution sample, at the beginning and at the end of the run.

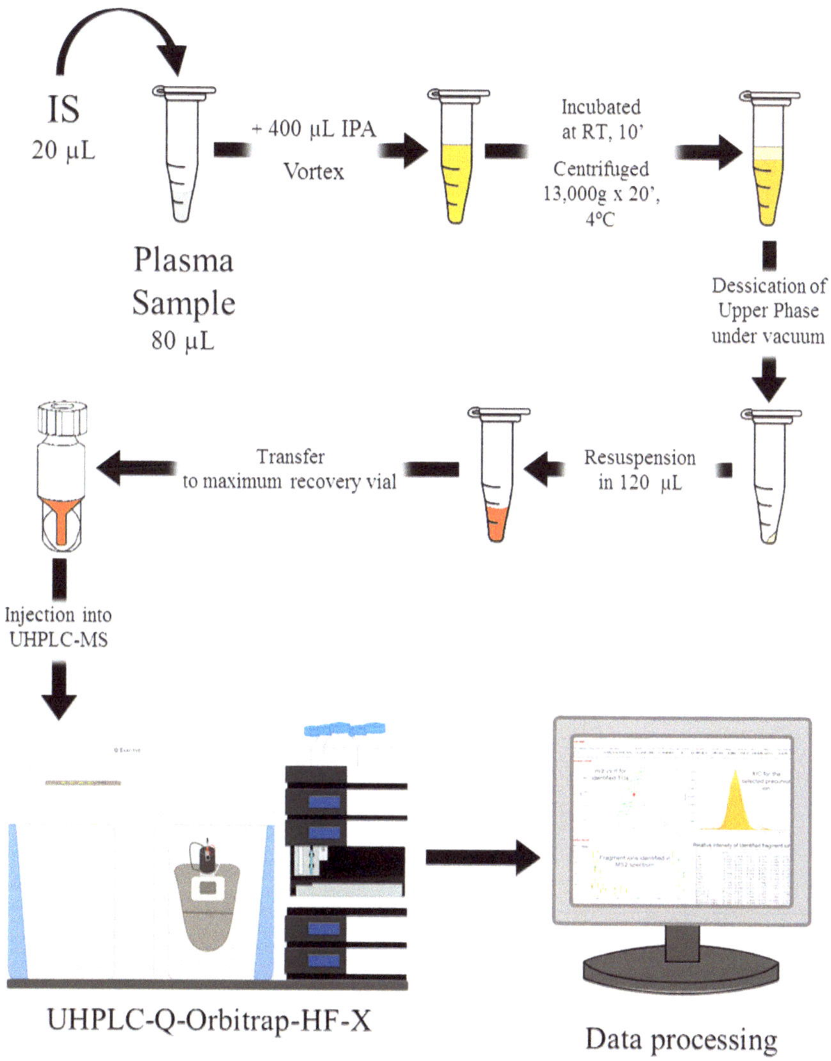

Fig. 4 Ultrahigh-performance liquid chromatography tandem mass spectrometry (UHPLC-MS) workflow. A known amount (20 μL) of diluted internal standard solution (*IS*) is added to 80 μL of each plasma samples. For the lipid extraction, add 4 volumes of IPA (2-propanol, pre-cooled to −20 °C) to the mixtures, and vortex for 1 min. After incubation at room temperature (RT) for 10 min, centrifuge at 13,000 × *g* for 20 min, at 4 °C. Once completed, dry the supernatant in a vacuum centrifuge, and store at −80 °C until the analysis can be performed. Just before the analysis, reconstitute the samples in 120 μL of dry-sample reconstitution solution, and transfer the solution to a maximum-recovery chromatography vial for subsequent injection into the UHPLC-Q-ORBITRAP-HF-X. Process the data using LipidSearch software (v4.2.2 or later)

(b) Inject the QCsys 12 times at the beginning of the sequence, to confirm that the analytical system is stabilized before the sample batch is analyzed. Use the same procedure for the QCsys sample as for study samples.

(c) Inject a QC sample that should be injected regularly (e.g., every 5 injections).

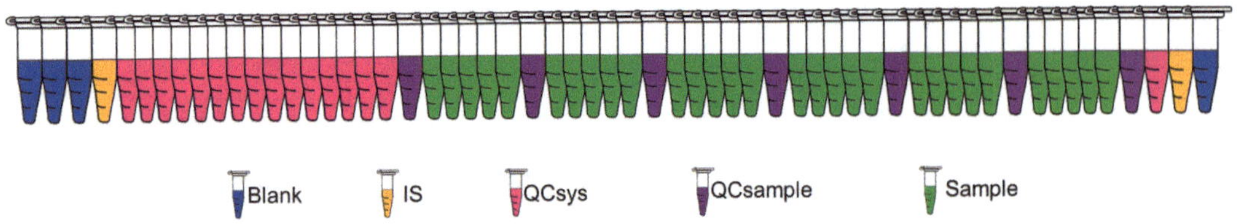

Fig. 5 Outline of injection sequence. (1) Three blanks (2-propanol:acetonitrile [IPA:CAN], 50:50); (2) diluted internal standard solution (ISS); (3) 15 quality controls for system conditioning (QCsys); (4) one QC sample, run after every 5 injections, to check for sensitivity and stability; (5) five randomized experimental samples; and then (6) a quality control (QC) sample followed by an ISS and, finally, a blank

(d) When injecting the experimental samples, randomize their order of injection, to minimize the effect of the instrumental drift that can arise from column degradation or contamination of the MS source.

(e) At the end of the sequence, inject (diluted) 1, 2, 5, 7, and 10 μL of QC dilution sample. These injections are used to calculate the coefficients of correlation between dilution factors of QC samples and areas of chromatographic peaks (above 0.8), to validate the automatic integration performed by LipidSearch and eliminate false positives.

6. Use the software package, Xcalibur (v4.1 or later), to acquire and process the MS data, and LipidSearch (v4.2.2 or later) to align the chromatograms and identify (*see* **Note 16**) and quantify (*see* **Note 17**) the peaks.

4 Notes

1. First, dissolve CHCA in methanol, ensure proper dissolution of the matrix, and then add the water.
2. Frozen tissues must be kept at −80 °C during storage, but the tissue samples to be cut (0.2–2.0 cm in thickness) must equilibrate at −20°/−25 °C for at least 2 h before cutting in the −20 °C cryostat. It is highly recommended to avoid freeze–thaw cycles. The postmortem interval (PMI; the time elapsed between death and the initial freezing of the tissues [at −80 °C]) should be as short as possible and should definitely be less than 72 h. In addition to matching for age, sex, or other population characteristics, control samples should be matched for PMI and storage duration.
3. Prior to sublimation or spraying, the samples should be completely dried by placing in a closed desiccator for 15 min, to remove moisture generated during the thawing process.

4. For successful sublimation, it is important to distribute the matrix evenly, in a thin layer at the bottom of the sublimator, so all of the matrix attains the proper sublimation temperature (100 °C).
5. Fix the slide in place using transparent plastic tape, taking care not to cover the tissue slice.
6. Seal the sublimator carefully with Vaseline to ensure sufficient vacuum. Ice blocks should not be in direct contact with the tissue; it is better to add some water or crushed ice to get uniform matrix deposition.
7. The vacuum should be released very gradually to prevent the lifting of the non-sublimated matrix when air returns to the sublimator, to avoid sample damage or artifacts.
8. In the re-crystallization process, MeOH vapor must be generated, for which it is necessary to seal the Petri dish with adhesive tape before placing it on the hot plate. The vapor enhances the signal, thereby increasing the intensity of the detected peaks [7].
9. During the spraying process, the matrix is applied via a stream of nitrogen gas; therefore, the slides must be held in place with tape to prevent movement of the slides as the gas stream passes through (*see* Fig. 3).
10. If using a very high resolution, the sample scanning time increases proportionally; therefore, one must compromise between the length of the scan and the spatial resolution. For large samples (>1 cm^2) of the human brain tissue, good spatial resolution can be obtained at 50–150 μm.
11. The intensities of the different molecules can be compared using MSiReader software or an equivalent [8], allowing one to export a ROI and make statistical calculations. MSiReader is a powerful and open-source option that enables a large number of processing tasks to be performed on MS data. However, it does not support the ".RAW" file format, which is generated with Thermo Fisher Scientific (TFS) spectrometers, so the files must be converted to the ".imzML" file format, using TFS's latest ImageQuest software, or open-source .RAW-to-.imzML converters. This analysis can detect a large number of molecules with very similar masses. Therefore, the m/z values should be assigned using different reference databases, such as Lipid Maps [9] or Human Metabolome Database [10], or be based on previously published results [1, 4]. All assignments may be made using a 5-ppm mass accuracy as the tolerance window.
12. Samples can be stored under an inert atmosphere, such as nitrogen, for long-term preservation.

13. The UHPLC-Q-ORBITRAP-HF-X should be equipped with a binary solvent-delivery pump for the mobile phases, a thermostat-controlled autosampler, and a column oven, for separation of individual lipids. For this purpose, a reverse-phase column and a pre-column are used at 65 °C. The mobile phases consist of ACN and H_20 (40:60, v/v) with 10 mM ammonium formate and 0.1% formic acid (*mobile phase A solution*) or ACN plus IPA (10:90, v/v) with 10 mM ammonium formate and 0.1% formic acid (*mobile phase B solution*). A better chromatographic separation can be achieved using the following gradient: 0–2 min, 40–43% *phase B*; 2–2.1 min, 43–50% *phase B*; 2.1–12 min, 50–54% *phase B*; 12–12.1 min, 54–70% *phase B*; then 12.1–18.0 min, 70–100% *phase B*. Then, equilibrate the system for 5 min prior to the next injection. The flow rate should be 500 μL/min and the injection volume 2 μL.
14. Settings: MS experiments can be performed in positive- and negative-ion modes using an electrospray ionization (ESI) source. The tune parameters must be optimized using the Splash™ LipidoMix™ internal standard mixture (or equivalent). The flow rates of sheath gas, sweep gas, and auxiliary gas for both polarities should be adjusted to 35, 0, and 10, respectively (arbitrary units). For both ionization modes, maintain the capillary temperature at 285 °C and the heater temperature at 370 °C, and set the spray voltages to 3.90 KV for positive and 3.20 kV for negative ions. Set the S-lens RF level to 40. Operate the Orbitrap mass analyzer at a resolving power of 120 k FWHM at m/z 200 in full-scan mode (scan range, 250–2000 m/z; automatic gain control target, $1e^6$) and 7.5 k FWHM at m/z 200 in Top15 data-dependent MS^2 mode (HCD fragmentation with stepped normalized collision energy, 25 and 30 in positive-ion mode and 20, 30, and 40 in negative-ion mode; injection time, 11 ms; isolation window, 1 m/z; automatic gain control target, $1e^5$) with a dynamic exclusion setting of 6.0 s.
15. The sequence of sample injections plays an important role in preventing bias in the interpretation of the data (*see* Fig. 5). First, blanks and diluted internal standards must be injected at the beginning and at the end of the run, in order to test for possible contamination of the samples or carry-over effects on the sensitivity or the shape of the chromatographic peaks. Moreover, the QCsys could be injected before the analyses, for system conditioning. In addition, quality control (QC) samples can be prepared by homogenizing together equal aliquots (20 μL) from each of the samples (in each group) of the study. These QC samples should be injected regularly (e.g., every 5 injections) to check sensitivity and

stability. Finally, the experimental samples must be injected randomly, to reduce bias introduced by column degradation or possible contamination of the MS source.

16. LipidSearch software is used for lipid identification and quantification. The lipid identification workflow is as follows: first, the individual data files are searched for product ion MS/MS spectra of lipid precursor ions. MS/MS fragment ions are predicted for all precursor adduct ions measured within ±5 ppm. The product ions that match the predicted fragment ions within the ±5 ppm mass tolerance are used to calculate a match score, and those candidates providing the highest-quality match are identified. Next, the search results from the individual positive- and negative-ion files from each sample group are aligned within a retention window (±0.1 min), and the data are merged for each annotated lipid. The annotated lipids are then filtered to reduce false positives according to the following criteria:

    (a) Main ion: In ESI^+, the main ion $[M + H]^+$ is observed for lysophosphatidylcholines (LPC), phosphatidylcholines (PC), lysophosphatidylethanolamines (LPE), phosphatidylethanolamines (PE), phosphatidylserines (PS), sphingomyelins (SM), ceramides (Cer), hexosylceramides (Hex1Cer), dihexosylceramides (Hex2Cer), sphingosines (SPH), sulfatides (ST), cardiolipins (CL), and acylcarnitines (AcCA); $[M + NH_4]^+$ is observed for lysophosphatidylinositols (LPI), phosphatidylinositols (PI), phosphatidylglycerols (PG), triacylglycerols (TG), cholesterol esters (ChE), and diacylglycerols (DG); $[M + Na]^+$ is observed for DG.

        In ESI^-, the main ion $[M - H]^-$ is observed for LPE, PE, PS, PG, LPI, PI, fatty acid (FA), and CL; and $[M + HCOO]^-$ is observed for LPC, PC, SM, Cer, Hex1Cer, and Hex2Cer.

    (b) Identification grade (ID quality filter): (A) lipid class and FA are completely identified, (B) lipid class and some FA are identified, (C) lipid class or FA is identified; and (D) lipid ID obtained via other fragment ions (H_2O loss, etc.).

    (c) The coefficients of variation of the areas of the chromatographic peaks that are related to lipid detected in QC samples should be <30%.

    (d) The coefficients of correlation between dilution factors of QC samples and the areas of chromatographic peaks should be >0.8.

    (e) The ratio of the areas of the chromatographic peaks of biological vs. blank samples should be >1.5.

17. Regarding the quantification: the concentrations of Cer, SM, PC, and PE can be determined by using the corresponding standards—Cer(d18:1/12:0), SM(d36:2D_9), PC(18:1D_7/15:0), and PE(18:1D_7/15:0), respectively, in both positive- and negative-ion detection modes. Moreover, LPE, DG, and TG concentrations can be determined with LPE(18:1D_7), DG(15:0/18:1D_7), and TG(15:0/18:1D_7/15:0), respectively, in positive-ion detection mode. The quantification should be conducted via normalization of the intensity of the monoisotopic peak of each lipid species to the intensity of the monoisotopic peak of the internal standard for each lipid class.

Acknowledgments

This work was supported by the Department of Education of the Basque Government (Grant #IT1454-22 to the "Neurochemistry and Neurodegeneration" Consolidated Research group) and the ISCIII Spanish Ministry for Health (#PI20/00153) co-funded by ERDF "A way to make Europe". Technical support and personnel, provided by General Research Services (SGiker), University of the Basque Country (UPV/EHU), the Ministry of Economy and Competitiveness (MINECO), the Basque Government, the European Regional Development Fund (ERDF), and the European Social Fund (ESF) are gratefully acknowledged. J.M.-G. is the recipient of Margarita Salas fellowship funded by the European Union-Next GenerationEU.

References

1. Martínez-Gardeazabal J, de San Román EG, Román E, Moreno-Rodríguez M, Llorente-Ovejero A, Manuel I, Rodríguez-Puertas R (2017) Lipid mapping of the rat brain for models of disease. Biochim Biophys Acta Biomembr 1859:1548–1557
2. Mapstone M, Cheema AK, Fiandaca MS, Zhong X, Mhyre TR, MacArthur LH, Hall WJ, Fisher SG, Peterson DR, Haley JM, Nazar MD, Rich SA, Berlau DJ, Peltz CB, Tan MT, Kawas CH, Federoff HJ (2014) Plasma phospholipids identify antecedent memory impairment in older adults. Nat Med 20:415–418
3. Veloso A, Astigarraga E, Barreda-Gómez G, Manuel I, Ferrer I, Giralt MT, Ochoa B, Fresnedo O, Rodríguez-Puertas R, Fernández JA (2011) Anatomical distribution of lipids in human brain cortex by imaging mass spectrometry. J Am Soc Mass Spectrom 22:329–338
4. de San Román EG, Román E, Manuel I, Giralt MT, Ferrer I, Rodríguez-Puertas R (2017) Imaging mass spectrometry (IMS) of cortical lipids from preclinical to severe stages of Alzheimer's disease. Biochim Biophys Acta Biomembr 1859:1604–1614
5. González-Domínguez R, Rupérez FJ, García-Barrera T, Barbas C, Gómez-Ariza JL (2016) Metabolomic-driven elucidation of serum disturbances associated with Alzheimer's disease and mild cognitive impairment. Curr Alzheimer Res 13:641–653
6. Yuki D, Sugiura Y, Zaima N, Akatsu H, Takei S, Yao I, Maesako M, Kinoshita A, Yamamoto T, Kon R, Sugiyama K, Setou M (2014) DHA-PC and PSD-95 decrease after loss of synaptophysin and before neuronal loss in patients with Alzheimer's disease. Sci Rep 4:7130

7. Yang J, Caprioli RM (2011) Matrix sublimation/recrystallization for imaging proteins by mass spectrometry at high spatial resolution. Anal Chem 83:5728–5734
8. Robichaud G, Garrard KP, Barry JA, Muddiman DC (2013) MSiReader: an open-source interface to view and analyze high resolving power MS imaging files on Matlab platform. J Am Soc Mass Spectrom 24:718–721
9. Sud M, Fahy E, Cotter D, Brown A, Dennis EA, Glass CK, Merrill AHJ, Murphy RC, Raetz CRH, Russell DW, Subramaniam S (2007) LMSD: LIPID MAPS structure database. Nucleic Acids Res 35:D527–D532
10. Wishart DS, Feunang YD, Marcu A, Guo AC, Liang K, Vázquez-Fresno R, Sajed T, Johnson D, Li C, Karu N, Sayeeda Z, Lo E, Assempour N, Berjanskii M, Singhal S, Arndt D, Liang Y, Badran H, Grant J, Serra-Cayuela A, Liu Y, Mandal R, Neveu V, Pon A, Knox C, Wilson M, Manach C, Scalbert A (2018) HMDB 4.0: the human metabolome database for 2018. Nucleic Acids Res 46: D608–D617

Part V

Protein Analyses of Human Brain

Chapter 14

Use of Affinity Purification–Mass Spectrometry to Identify Phosphorylated Tau Interactors in Alzheimer's Disease

Geoffrey Pires, Beatrix Ueberheide, Thomas Wisniewski, and Eleanor Drummond

Abstract

Phosphorylated tau is the main protein present in neurofibrillary tangles, the presence of which is a key neuropathological hallmark of Alzheimer's disease (AD). The toxic effects of phosphorylated tau are likely mediated by interacting proteins; however, methods to identify these interacting proteins comprehensively in human brain tissue are limited. Here, we describe a method that enables the efficient identification of hundreds of proteins that interact with phosphorylated tau (pTau), using affinity purification–mass spectrometry (AP-MS) on human, fresh-frozen brain tissue from donors with AD. Tissue is homogenized using a gentle technique that preserves protein–protein interactions, and co-immunoprecipitation of pTau and its interacting proteins is performed using the PHF1 antibody. The resulting protein interactors are then identified using label-free quantitative liquid chromatography-mass spectrometry (LC-MS)/MS. The Significance Analysis of INTeractome (SAINT) algorithm is used to determine which proteins significantly interact with pTau. This approach enables the detection of an abundance of all 6 isoforms of tau, 23 phosphorylated residues on tau, and 125 significant pTau protein interactors, in human AD brain tissue.

Key words Phosphorylated tau, pTau, Neurofibrillary tangles, Alzheimer's disease, Tauopathy, Proteomics, Affinity purification–mass spectrometry, AP-MS, Immunoprecipitation, Human brain, Tau interaction

1 Introduction

Neurofibrillary tangles are a key pathological hallmark of Alzheimer's disease (AD) that consist primarily of aggregated, hyperphosphorylated tau (pTau). Pathological aggregation of pTau is hypothesized to mediate synaptic impairment, neurodegeneration, and dementia in AD [1, 2]. However, the disease mechanisms involved are unknown. Determining the identity of proteins that interact with pTau is an excellent way to uncover the specific mechanisms that underlie pTau toxicity; indeed, some pTau protein interactors have already been identified in isolated, targeted studies [3–13]. The study of pTau protein interactions can enhance our

Jerold Chun (ed.), *Alzheimer's Disease: Methods and Protocols*,
Methods in Molecular Biology, vol. 2561, https://doi.org/10.1007/978-1-0716-2655-9_14,

understanding of AD pathogenesis, providing new insight into how tau mediates toxicity in AD and leading to the identification of new biomarkers and, potentially, new therapeutic avenues. However, our current understanding of pTau protein interactions in the human brain is still limited, in part because all prior studies examining them have focused on *targeted* proteins of interest. The approach described here—examining pTau interactions using an unbiased, mass spectrometry–based approach—enables the simultaneous quantification of hundreds of protein interactions at once. Although recent studies have used this approach to identify proteins that all tau species interact with in the human brain [14–16], these studies do not discriminate among interactions with physiological and pathological tau species.

In this chapter, we describe an affinity purification–mass spectrometry (AP-MS) method that enables simultaneous identification of hundreds of proteins that interact with a neurodegenerative disease–specific pTau species (for a schematic diagram of the overall methodology, *see* Fig. 1). The species that we have targeted is phosphorylated at S396 and S404, sites recognized by the widely used PHF1 antibody. This pathological pTau species is abundant in AD and in other primary tauopathies, such as corticobasal degeneration, progressive supranuclear palsy, and Pick's disease. Our approach uses a gentle homogenization method designed to maintain protein–protein interactions, using a low-salt homogenization buffer [17]. Next, affinity purification of pTau and its interactors is performed using the Dynabeads™ Protein G immunoprecipitation kit, but we made an important modification in binding the antigen (pTau) to the antibody (PHF1) *before* using the Dynabeads™ to immunoprecipitate the resulting antigen–antibody complex. Second, we extended the recommended incubation times from two 10-min incubations to two overnight incubations, to enhance the binding process. Additionally, to control for differences in non-specific binding between donors, we performed control immunoprecipitations using a mouse IgG isotype control, for each patient.

Our approach of incubating the antibody with brain homogenate prior to co-immunoprecipitation with Dynabeads™ means that cross-linking the antibody to the beads is not an option. To overcome the limited dynamic range caused by the excess of IgG in the eluate, the eluate is separated by sodium dodecyl sulfate polyacrylamide gel electrophoresis (SDS PAGE), and the antibody bands excised, digested, and analyzed by liquid chromatography–mass spectrometry (LC-MS), keeping them separated from the rest of the immunoprecipitated proteins. LC-MS analysis is performed using an EASY nLC coupled to an Orbitrap Fusion Lumos mass spectrometer. Proteins are identified using SEQUEST within Proteome Discoverer, and potential pTau interactors are ranked by significance using the Significance Analysis of INTeractome

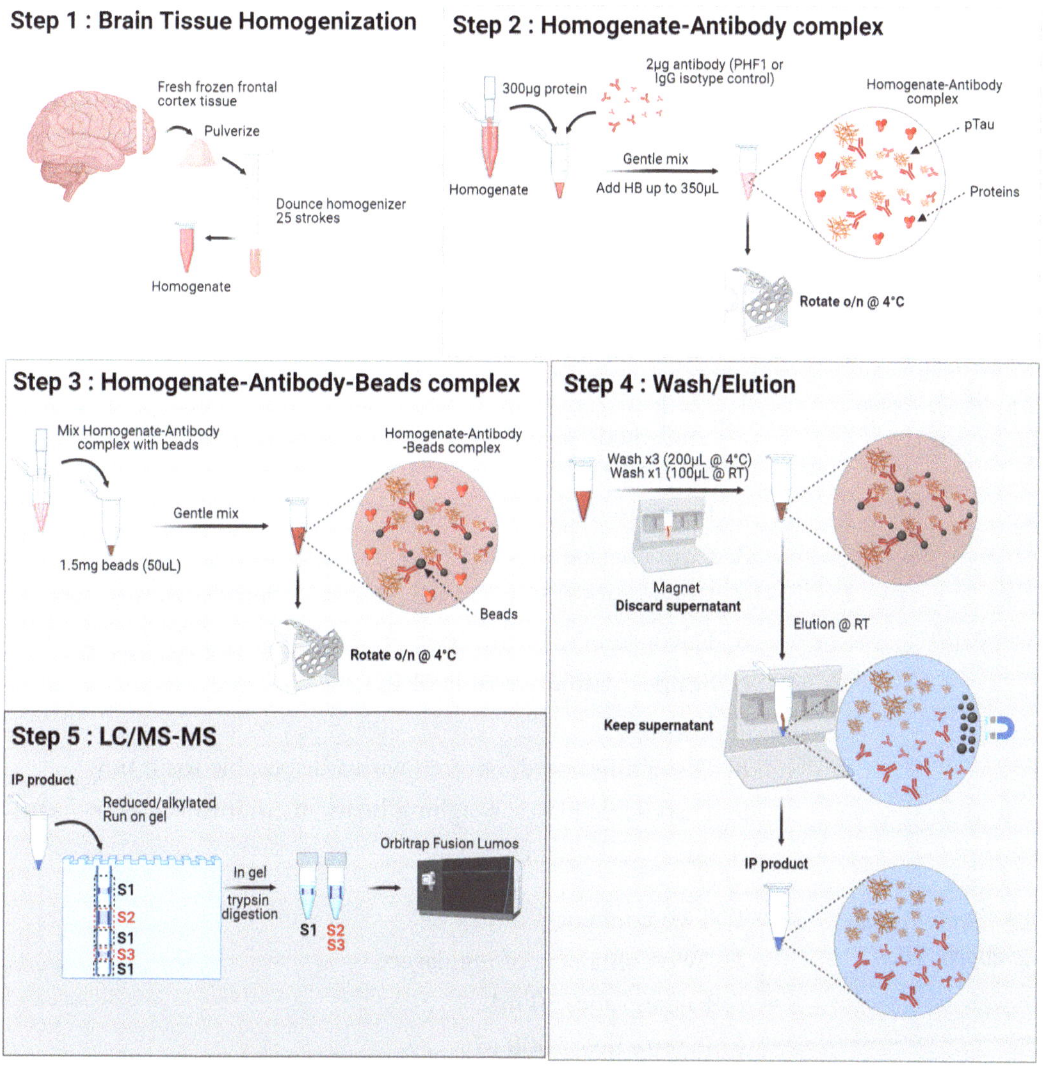

Fig. 1 Schematic of the affinity purification–mass spectrometry workflow. This approach includes three major procedures: brain homogenization (***Step 1***), co-immunoprecipitation of pTau and binding partners (***Steps 2–4***), and mass spectrometry analysis (***Step 5***). *See* text for details

(SAINT) algorithm [18]. Our successful immunoprecipitation of pTau is evidenced by pTau being the most abundant protein detected; we found all 6 isoforms of tau (0N3R, 1N3R, 2N3R, 0N4R, 1N4R, and 2N4R) and 23 phosphorylated residues on tau. Importantly, the most abundant pTau species we enriched was phosphorylated on serine at residues 396 and 404, consistent with the epitope (GAEIVYKpSPVVSGDTpSPRHLSNV) of PHF1. In total, we identified 1164 proteins that co-immunoprecipitated with pTau in human AD brain tissue. Of those, we categorized 125 proteins as probable pTau interactors [19].

Given the important role of pTau in many neurodegenerative diseases, this method has great potential to provide insight into the specific pTau-mediated disease mechanisms in different types of neurodegenerative disease [20]. In addition, the use of our technique with antibodies that target different pTau species could provide new insight into whether specific species are particularly toxic in neurodegenerative disease and whether this toxicity is mediated by interactions with specific proteins.

2 Materials

2.1 Brain Tissue Dissection

1. Fresh-frozen frontal cortex tissue from neuropathologically confirmed severe AD cases fulfilling the (A3, B3, C3) criteria [21].
2. Bench underpads.
3. Container of dry ice.
4. Falcon tubes, 50 mL (one per case, labeled with the case ID).
5. Falcon tubes, 15 mL (one per case, labeled with the case ID).
6. Scalpel, with disposable blades.
7. Forceps.
8. Dissecting board, covered with a disposable foam tray.
9. Disposable plastic weighing boats (medium sized, 7 × 7 cm).
10. Analytical balance.

2.2 Brain Tissue Homogenization

1. Two containers of dry ice.
2. Two containers of regular ice.
3. Falcon tube, 50 mL.
4. Pipette tips, 1000 μL.
5. *Homogenization buffer*: HEPES-NaOH, 50 mM, pH 7.0; sucrose, 250 mM; EDTA, 1 mM (this can be prepared in advance in 50 mL tubes and stored at 4 °C for future use).
6. Protease inhibitor cocktail, 10× (we use cOmplete™ ULTRA tablets, mini, EDTA-free), or equivalent.
7. Phosphatase inhibitor cocktail, 50× (we use PhosphoSTOP™ EASYpack), or equivalent.
8. Aluminum foil.
9. Liquid nitrogen.
10. Hammer.
11. Metal heating block, or any other metal flat surface.
12. Spatula.
13. Vortex.

14. Dounce tissue homogenizer (7 mL), composed of a stainless steel plunger with a polytetrafluoroethylene (PTFE) head and a size-matched glass vessel.
15. Glass Pasteur pipettes, with pipette bulb.
16. Polypropylene tubes, 2 mL (1 per case, labeled with case ID).
17. Polypropylene tubes, 1.5 mL (2 per case, labeled with case ID).
18. Beaker, 1 L, containing 500 mL of 40% bleach (to use as a waste container).
19. Ethanol, 200 proof.
20. Double-distilled water (ddH_2O).
21. Bicinchoninic acid (BCA) protein assay kit, or equivalent.

2.3 Co-immunoprecipitation

1. Dynabeads™ Protein G immunoprecipitation kit.
2. Homogenization buffer.
3. PHF1 antibody (*see* **Note 1**).
4. Mouse IgG isotype control antibody.
5. Low-binding tubes, 1.5 mL (we use Axygen® 1.5 mL Snaplock Microcentrifuge Tubes, catalog # MCT-150-C-S), or equivalent, 10 tubes per case; label them with the appropriate antibody combination, and number them from 1 through 5 for duplicates (i.e., IgG #1, IgG #2, …, IgG #5, PHF1 #1, … PHF1 #5).
6. Micro-pipettors, P1000 and P200.
7. Micro-pipette tips, 1000 μL and 200 μL.
8. Revolver lab rotator.
9. Vortex.
10. Magnetic rack capable of holding 16 × 1.5 mL tubes.
11. Large tray containing wet ice (at least 38 × 25 × 8 cm).
12. Pipette tips for loading gels, 200 μL.

2.4 LC-MS: Sample Preparation

1. Dithiothreitol (DTT) (≥99.5% pure).
2. Iodoacetamide (≥99.5% pure).
3. Thermomixer R, or equivalent.
4. NuPAGE™ gel, 4–12% Bis-Tris, 1.0-mm thick, or equivalent.
5. GelCode Blue stain reagent, or equivalent.
6. Scalpel blade.
7. Methanol.
8. Ammonium bicarbonate (≥99.5% pure).
9. Acetonitrile, LC-MS grade (≥99.5% pure).
10. SpeedVac concentrator, or equivalent.

11. Trypsin, sequencing grade.
12. R2 POROS™ beads, 50 μM, or equivalent solid-phase extraction system for peptides.
13. Formic acid, LC-MS grade.
14. Trifluoroacetic acid (TFA), LC-MS grade.
15. Shaker (or equivalent) for light agitation of Eppendorf tubes.
16. C18 ZipTips®, or equivalent, for peptide desalting (one per sample).
17. Microcentrifuge, or equivalent, with a g-force of ~100–170 × *g*. The g-force should not be increased above 190 × *g*.
18. Acetic acid, LC-MS grade.
19. Formic acid (LC-MS grade), 5%, and TFA (LC-MS grade), 0.2% in water (LC-MS grade).
20. Acetonitrile, 40% in 0.5% acetic acid in water (all LC-MS grade).
21. Acetonitrile, 80% in 0.5% acetic acid in water (all LC-MS grade).

2.5 LC-MS/MS Analysis

1. Trap column (Acclaim® PepMap 100 pre-column, 75 μm × 2 cm, C18, 3 μm, 100 Å) connected to an analytical column (EASY-Spray column, 50 m × 75 μm ID, PepMap RSLC C18, 2 μm, 100 Å), or equivalent nano-LC columns. Alternatively, the sample could be loaded directly onto an analytical column.
2. EASY-nLC 1000 HPLC, or equivalent.
3. HPLC Solvent A: acetonitrile, 2% in 0.5% acetic acid in water (all LC-MS grade).
4. HPLC Solvent B: acetonitrile, 95% in 0.5% acetic acid in water (all LC-MS grade).
5. Orbitrap Fusion Lumos mass spectrometer, or equivalent high-end mass spectrometer.

3 Methods

3.1 Brain Tissue Dissection

1. Cover all bench areas with absorbent underpads.
2. Obtain fresh-frozen postmortem prefrontal cortex brain tissue from pathologically confirmed AD cases (*see* Fig. 1, **Step 1**).
3. Using containers of dry ice, a dissection board covered with a disposable foam tray, forceps, and a scalpel, excise several small (~5–10-mm^2 block face) portions of gray matter from each tissue sample, to obtain a stock of gray matter from each case. *See* **Note 2**.

4. For each case, store tissue in a labeled, 50 mL Falcon tube on dry ice (storage tube).
5. Fill one weighing boat with dry ice (WB#1).
6. Place a second, empty weighing boat on dry ice (WB#2).
7. Place a labeled, 15 mL Falcon tube on dry ice (holding tube).
8. Transfer the brain tissue into the cap of the 50 mL storage tube and place the cap onto the dry ice of WB#1.
9. Place the pre-chilled WB#2 on the weighing pan of an analytical balance.
10. Obtain ~250-mg portions of grey matter by pooling several of the small portions obtained in **Step 3**, keeping the tissue frozen at all times, using pre-chilled forceps for transfer. Record the exact weight of the piece to be homogenized. *See* **Note 3**.
11. Transfer the tissue into the 15 mL holding tube. Keep in a container with dry ice until Subheading 3.2, **Step** 7.
12. Transfer the remaining tissue back into the storage tube and store at −80 °C until further use.

3.2 Brain Tissue Homogenization

1. Pre-chill the following on dry ice: a 50 mL tube of homogenization buffer, the Dounce tissue homogenizer, 2 mL polypropylene tubes (1 per sample), 1.5 mL polypropylene tubes (2 per sample), the hammer, the spatula, and the metal heating block (with the reverse side on top; described in **Note 4**).
2. Calculate the volume of homogenization buffer required to homogenize each tissue sample (5 mL buffer/g of tissue).
3. Calculate the *total* volume of homogenization buffer required to homogenize *all* tissue samples, and put it in a Falcon tube (15 mL or 50 mL, depending on how many cases are to be homogenized). Store any remaining homogenization buffer at 4 °C for future use.
4. Add enough protease (10×) and phosphatase (50×) inhibitors to the above tube, such that the final concentration of each is 1×. *See* **Note 5**.
5. Keep the mixture on ice for 10 min, vortexing every 2 min to allow the tablets to dissolve.
6. For each tissue sample, add 5 mL/g of the above mixture to the pre-chilled Dounce homogenizer tube, and keep on ice.
7. To prepare for homogenization, pulverize the tissue sample following the method in **Note 6**, working as quickly as possible so that the tissue does not thaw (*see* Fig. 1, **Step 1**).
8. Use the pre-chilled spatula to transfer the pulverized tissue into the homogenizer tube (containing 5 mL of the homogenization mixture of **Step 5**), working very quickly to prevent the

powdered material from thawing. Take care when transferring the tissue to the very bottom of the homogenizer, and avoid letting the tissue touch the tube walls. From this point forward, discard any pipette tip, tube, etc. that has come into contact with the brain tissue or homogenate into a waste container containing 40% bleach.

9. Homogenize the brain tissue on ice, using a total of 25 strokes of the Dounce homogenizer pestle. Stop after 12 strokes, and push down all pieces sticking to the wall of the homogenizer, using a P1000 pipette, and then proceed with the last 13 strokes (*see* Fig. 1, **Step 1**).
10. Using a glass Pasteur pipette and a pipette bulb, transfer all the homogenate into a 2 mL polypropylene tube, and keep on ice. Discard glass Pasteur pipette after use in the appropriate sharps container. *See* **Note** 7.
11. Aliquot homogenate into two 1.5 mL polypropylene tubes. Set aside one small aliquot of homogenate in regular ice to determine protein concentration. Flash-freeze the 1.5 mL tubes immediately, by placing on dry ice (*see* Fig. 1, **Step 1**).
12. Repeat **Steps 6–11**, disinfecting three times with 200-proof ethanol and rinsing the homogenizer three times with ddH_2O, prior to each new homogenization.
13. Store all homogenates at −80 °C until use.
14. Determine protein concentrations using a BCA assay kit or an equivalent system. *See* **Note 8**.

3.3 AP-MS of Phosphorylated Tau and Interacting Partners

For each sample, ten individual co-immunoprecipitations are performed: $n = 5$ replicates using the antibody PHF1, to affinity-enrich pTau and its binding partners, and $n = 5$ replicates using the mouse IgG isotype control antibody, to control for non-specific binding. *See* **Note 9**.

3.3.1 Antigen–Antibody Binding (All Steps to Be Done on Ice)

1. Transfer the volume of brain homogenate needed for 300 μg of protein into 10 × 1.5 mL low-binding tubes per case, and dilute to a final volume of 300 μL (a final concentration of 1 μg/μL), with homogenization buffer (*see* Fig. 1, **Step 2**).
2. Add 2 μg of either PHF1 or mouse IgG isotype control antibody to each tube. Gently triturate using a P1000 pipette. *See* **Note 10** and Fig. 1, **Step 2**.
3. To normalize, add sufficient homogenization buffer to reach a final volume of 350 μL, and triturate again. *See* **Note 11** and Fig. 1, **Step 2**).
4. Incubate overnight at 4 °C, with rotation, to allow antigen–antibody binding. Check that the liquid is moving within the tube when rotating (*see* Fig. 1, **Step 2**).

3.3.2 Antigen–Antibody Binding to Beads (All Steps on Ice, Unless Specified)

1. The next day, completely resuspend Dynabeads® by gently vortexing for 30 s.
2. For each sample, add 50 μL (1.5 mg) of Dynabeads® slurry to a new, labeled 1.5 mL low-binding tube, and pre-wash the beads as detailed below (this step is performed at room temperature [RT]) (*see* Fig. 1, **Step 3**).
3. Place the tube on the magnet, to separate the beads from the solution, and discard the supernatant.
4. Remove the tube from the magnet.
5. Add 200 μL of the washing and binding buffer (included with the kit) to each tube. Mix gently 10–12 times by trituration, avoiding bubbles.
6. Place the tube against the magnet and discard the supernatant.
7. Remove the tube from the magnet.
8. Optional: aliquot 20 μL of homogenate–antibody solution into a microtube (labeling it "Sup0"), and store at −20 °C for later use, to confirm successful antibody binding to Dynabeads®.
9. Transfer the homogenate–antibody complex to the tube containing the beads, and use a P1000 pipette to gently triturate the solution, 4–5 times, avoiding introducing bubbles (*see* Fig. 1, **Step 3**).
10. Incubate overnight at 4 °C, with rotation, to allow antigen–antibody binding to Dynabeads®. Check that the liquid is moving appropriately when rotating (*see* Fig. 1, **Step 3**).

3.3.3 Washing (All Steps on Ice, Unless Specified)

1. Optional: Before washing, place tube on the magnet, and aliquot 20 μL of solution into a microtube (labeling it "Sup1"); this is the unbound portion. Store at −20 °C for later use, to compare it to Sup0, to confirm successful antibody binding to Dynabeads®. *See* **Note 12**.
2. The next day, wash the Dynabeads®-antibody-antigen complex 3 times, using 200 μL washing buffer for each wash. Mix by gently triturating 10–12 times, using a P200 pipette, avoiding making bubbles and ensure that the beads are well-suspended. Discard supernatant after each washing step, using the magnet (Fig. 1, **Step 4**).
3. Resuspend the Dynabeads®–antibody–antigen complex in 100 μL washing buffer, and transfer the bead suspension to a clean, labeled 1.5 mL low-binding tube (*see* **Note 13**). This step is performed at RT, to prepare for the elution step (*see* Fig. 1, **Step 4**).

3.3.4 Elution (Non-denaturing; All Steps Done at RT)

1. Place the tube against the magnet, as before, and discard the supernatant.
2. Gently resuspend the Dynabeads®–antibody–antigen complex in 20 μL of the elution buffer (which comes with the kit). Make sure all the beads are immersed in the elution buffer by thoroughly triturating with a P10 pipette. Incubate for 5 min at RT (*see* Fig. 1, **Step 4**).
3. Place tube against the magnet and transfer the supernatant (labeled, "IP product") into a clean, 1.5 mL low-binding tube, using a 200 μL gel-loading pipette tip (*see* Fig. 1, **Step 4**).
4. Pool all replicates of a given IP product into one 1.5 mL low-binding tube, using a P200 pipette.
5. Freeze at −20 °C until use.
6. Optional: To confirm pTau immunoprecipitation prior to LC-MS, perform a regular Western blot, using 5% (5 μL) of the pooled IP products. *See* **Note 14**.

3.4 AP-MS: Sample Preparation

1. Reduce proteins in the eluate by incubating with 2 μL of a 200 mM DTT solution in SDS loading buffer, at 57 °C for 1 h.
2. Alkylate samples with 2 μL of a 500 mM iodoacetamide solution in SDS loading buffer at RT, in the dark, for 45 min.
3. Separate samples on a NuPAGE™ 4–12% Bis-Tris, 1.0-mm-thick gel, or equivalent.
4. Stain gel with GelCode Blue stain reagent (or equivalent).
5. Excise the IgG bands using a scalpel blade, cut them into 1-mm^3 pieces, combine them in a microtube, and process them separately from the rest of the lane (as shown in Fig. 1, **Step 5**, "S2" and "S3").
6. Excise the remainder of the lane, and cut the excised gel into 1-mm^3 pieces (*see* Fig. 1, **Step 5**, "S1"). Place in a 1.5 mL Eppendorf tube.
7. Destain the gel pieces in a freshly made, 1:1 (v/v) methanol/100 mM ammonium bicarbonate solution, under gentle agitation, for 10–30 min. Replace the destaining solution at least three times.
8. Dehydrate the samples: First, add enough acetonitrile to cover the gel pieces, and gently pipet the solution up and down three times, and then discard the solution. Next, to further dehydrate the samples, place them in a SpeedVac concentrator until the gel pieces are dry (approximately 15 min).
9. Rehydrate the dehydrated gel pieces with a freshly made solution of 200-ng trypsin in 100 mM ammonium bicarbonate, using enough volume to cover the gel pieces (note the volume required, for the next step). Allow the digest to proceed overnight at RT, with gentle agitation.

10. Add a slurry of R2 50-μM POROS™ beads in 5% formic acid and 0.2% TFA to samples, using double the volume of the liquid required for **Step 9**.
11. Place the samples on a rotary shaker for 1–3 h at 4 °C.
12. Load the POROS bead suspension onto equilibrated C18 ZipTips®, and use a microcentrifuge or equivalent for 30 s at ~100–170 × *g* to force the liquid through the tips (*see* **Note 15**).
13. Wash the POROS beads three times with 50 μL 0.1% TFA, removing the wash solution each time by microcentrifuging (or an equivalent method) for 30 s at ~100–170 × *g* (as in **Step 12**).
14. Wash the POROS beads three more times with 50 μL of 0.5% acetic acid, as described in the previous step.
15. Elute peptides off the beads by adding 50 μL of 40% acetonitrile in 0.5% acetic acid, followed by adding 50 μL of 80% acetonitrile in 0.5% acetic acid (collecting the eluents in the same vial).
16. Remove the organic solvent using a SpeedVac concentrator (usually ~20 min), and reconstitute the samples in 0.5% acetic acid.
17. Store the desalted peptide mixture at −80 °C until further analysis.

3.5 LC-MS/MS Analysis

1. Load an aliquot of each peptide mixture (adjusted so that each mixture contains the same number of μg/μL) either onto a trap column connected to an analytical column or directly onto an analytical column, using the autosampler of an EASY-nLC 1200 HPLC or equivalent nano-LC.
2. The peptides will be gradient eluted directly into an Orbitrap Fusion Lumos mass spectrometer (or equivalent). The choice of gradient is flexible. We use the following: 5 min at 5% Solvent B and then to 25% Solvent B over 80 min, followed by 15 min to 45% Solvent B and in 10 min to 100% Solvent B. *See* **Note 16**.
3. For high-resolution full mass spectra (MS), use a resolution of 240,000 (@200 m/z), an automatic gain control (AGC) target of 1e6, a maximum injection time of 50 ms, and a scan range of 400–1500 m/z. However, the exact settings will depend on the instrument type and desired speed.
4. Following each full MS analysis, acquire higher-energy C-trap dissociation (HCD) MS/MS spectra in the ion trap (rapid-scan mode) with a TopN methodology (3 s), using an automatic gain control (AGC; for the ion population) target of 2e4, a maximum injection time of 18 ms, one microscan, a 0.7-m/z

isolation window, a fixed first mass of 110 m/z, and a normalized collision energy (NCE) of 30 (30%). However, the exact settings will depend on the instrument type and available fragmentation methods.

5. The MS/MS spectra can be searched against the Uniprot human reference proteome database using SEQUEST within Proteome Discoverer. The recommended search parameters are for the instrument settings in **Steps 3** and **4**: precursor mass tolerance ±10 ppm; fragment mass tolerance ±0.4 Da; trypsin cleavage with two missed cleavages allowed; variable modification of oxidation of methionine; phosphorylation on serine, threonine, and tyrosine; deamidation of glutamine and asparagine; and fixed modification of carbamidomethylation of cysteine. Peptides and proteins should be filtered to a lower-than-1% false-discovery rate (FDR) using a target–decoy database strategy; proteins require at least two unique peptides to be reported. However, there are other equivalent search engines, and the search parameters must be adjusted to the mass accuracy of the respective instrument.
6. To obtain a probabilistic score (SAINT score) that a protein is a bona fide pTau interactor, the data can be analyzed using the SAINTexpress algorithm [18] (https://reprint-apms.org). Proteins that have a SAINT score ≥0.65 are considered, in this study, to be probable pTau interactors and should be further evaluated.

4 Notes

1. The PHF1 antibody was formerly available from the late Dr. Peter Davies. However, Dr. Davies' team at the Litwin–Zucker Research Center will be continuing to provide PHF1 antibodies to the academic community.
2. Work quickly to make sure the tissue does not thaw at any time. Avoid white matter and any large blood vessels, bloody regions, meninges, or arachnoid matter. Start by cutting along the white-grey matter border to obtain a large and straight piece of cortex, prior to cutting into several smaller pieces (~5–10-mm^2 face each). We usually obtain a stock of around 20 small pieces of tissue/case so that we have enough tissue for several experiments.
3. If the portions excised in **Step 2** are too large and the required weight cannot be achieved in **Step 3**, repeat **Step 2** with the excised portions. Keep the tissue frozen during the cutting process by working quickly.

4. For the pulverization process in Subheading 3.2, **Step** 7, we place the tissue wrapped in aluminum foil on top of the reverse side of a dry-bath metal block, using it as a flat surface. However, any flat metal surface should be suitable.
5. The volume of homogenization buffer required to homogenize ~250 mg of tissue is approximately 1.25 mL. For $n = 5$ cases, we aliquoted 10 mL of homogenization buffer and supplemented it with one tablet of protease inhibitor cocktail (10×) and one-fifth of a tablet of phosphatase inhibitors (50×) and used ~6.25 mL for the whole process.
6. Get a 15 × 15 cm sheet of aluminum foil, fold it in half to make it thicker, and fold up the edges. Place the frozen tissue onto the sheet of aluminum foil; tightly fold foil in half to enclose the tissue, and then fold up the edges again. Place the foil packet on top of the reverse side of the pre-chilled dry-bath metal block. Pour a small amount (~10 mL) of liquid nitrogen on top of the aluminum foil. Pulverize the tissue with ~10 hammer blows, until it is turned into powder. The liquid nitrogen used here completely freezes the tissue, making tissue pulverization faster and more effective.
7. Here, the use of the Pasteur pipette is important for two reasons: (1) to check that there are no big chunks of tissue remaining and (2) to reach easily the bottom of the glass tube of the homogenizer.
8. The average protein concentration of homogenates prepared at 5 mL/g of tissue should be between 10 and 15 mg/mL, according to previous studies [17].
9. According to the manufacturer's instructions, the required amount of Dynabeads® to use for downstream proteomics applications should be five times that of a regular immunoprecipitation. This is the reason we are performing replicates for each antibody combination. Doing one large immunoprecipitation per condition, rather than in five replicates, should be feasible, but we have not tested it yet.
10. Regarding the PHF1 antibody, this step is batch dependent, as it is an unpurified antibody, which varies significantly in concentration between batches. The concentration of the PHF1 antibody should be determined in the laboratory before use; ours was 200 μg/mL.
11. This volume was tested to ensure free rotation of liquid inside the tube. This is important for optimal antigen–antibody binding to Dynabeads®.
12. To do this, perform a western blot on Sup0 and Sup1, incubating only with an anti-mouse secondary antibody, which will determine whether there is a decrease in antibody in the solution (which confirms binding to the Dynabeads®).

13. Using a new tube at this step is recommended, to avoid co-elution of proteins bound to the tube wall.
14. Perform the western blot in denaturing, but non-reducing conditions by incubating samples for 20 min at RT after denaturation with loading buffer. Avoid using reducing agents or boiling, to preserve the aggregated structure of paired helical filaments of pTau. Load samples and run a standard western blot from this point onward.
15. There are several protocols available for in-gel digestion and tryptic peptide desalting. We describe here the one that we used for this study, but other published protocols should be equally able to generate desalted tryptic peptides suitable for LC-MS analysis. Alternatively, the sample could be prepared either by direct digestion "on-beads" or by eluting the bound proteins from the beads using low pH and then digesting the eluted proteins "in-solution." Here, we decided to use the additional step of 1D separation via SDS-PAGE to increase the dynamic range of our analysis. Removing the antibody fraction and analyzing it separately facilitate enhanced detection of binding partners.
16. The peptide mixtures can be analyzed by any modern mass spectrometry instrument that affords sufficient sensitivity. For this specific project, we used the fastest scan speed available on the instrument, as the excess antibody reduces the dynamic range of the analysis, especially in the 50- and 25-kDa fraction. The low-resolution MS/MS scan rate is significantly faster and will help facilitate the detection of binding partners of lower abundance.

Acknowledgments

This study was supported by funding from the Bluesand Foundation and Dementia Australia to E.D., the Philippe Chatrier Foundation to G.P., and National Institutes of Health grants P30AG066512, P01AG060882, and RF1AG058267 to T.W. The proteomics work was in part supported by the NYU School of Medicine and an NIH shared instrumentation grant (1S10OD010582-01A1) for the purchase of an Orbitrap Fusion Lumos.

References

1. Nelson PT et al (2012) Correlation of Alzheimer disease neuropathologic changes with cognitive status: a review of the literature. J Neuropathol Exp Neurol 71(5):362–381
2. Scheltens P et al (2021) Alzheimer's disease. Lancet 397(10284):1577–1590
3. Maziuk BF et al (2018) RNA binding proteins co-localize with small tau inclusions in tauopathy. Acta Neuropathol Commun 6(1):71
4. Wang P et al (2017) Tau interactome mapping based identification of Otub1 as Tau deubiquitinase involved in accumulation of pathological Tau forms in vitro and in vivo. Acta Neuropathol 133(5):731–749
5. Wang X et al (2019) Tau interactome analyses in CRISPR-Cas9 engineered neuronal cells reveal ATPase-dependent binding of wild-type but not P301L Tau to non-muscle myosins. Sci Rep 9(1):16238
6. Babu JR, Geetha T, Wooten MW (2005) Sequestosome 1/p62 shuttles polyubiquitinated tau for proteasomal degradation. J Neurochem 94(1):192–203
7. Baumann K et al (1993) Abnormal Alzheimer-like phosphorylation of tau-protein by cyclin-dependent kinases cdk2 and cdk5. FEBS Lett 336(3):417–424
8. Chen Y et al (2019) 14-3-3/Tau interaction and Tau amyloidogenesis. J Mol Neurosci 68(4):620–630
9. Lee MJ, Lee JH, Rubinsztein DC (2013) Tau degradation: the ubiquitin-proteasome system versus the autophagy-lysosome system. Prog Neurobiol 105:49–59
10. Litersky JM et al (1996) Tau protein is phosphorylated by cyclic AMP-dependent protein kinase and calcium/calmodulin-dependent protein kinase II within its microtubule-binding domains at Ser-262 and Ser-356. Biochem J 316(Pt 2):655–660
11. Lovestone S et al (1996) Phosphorylation of tau by glycogen synthase kinase-3 beta in intact mammalian cells: the effects on the organization and stability of microtubules. Neuroscience 73(4):1145–1157
12. Sluchanko NN, Gusev NB (2011) Probable participation of 14-3-3 in tau protein oligomerization and aggregation. J Alzheimers Dis 27(3):467–476
13. Pires G et al (2019) Secernin-1 is a novel phosphorylated tau binding protein that accumulates in Alzheimer's disease and not in other tauopathies. Acta Neuropathol Commun 7(1):195
14. Ayyadevara S et al (2016) Proteins that mediate protein aggregation and cytotoxicity distinguish Alzheimer's hippocampus from normal controls. Aging Cell 15(5):924–939
15. Hsieh YC et al (2019) Tau-mediated disruption of the spliceosome triggers cryptic RNA splicing and neurodegeneration in Alzheimer's disease. Cell Rep 29(2):301–316.e10
16. Meier S et al (2015) Identification of novel Tau interactions with endoplasmic reticulum proteins in Alzheimer's disease brain. J Alzheimers Dis 48(3):687–702
17. Diner I, Nguyen T, Seyfried NT (2017) Enrichment of detergent-insoluble protein aggregates from human postmortem brain. J Vis Exp 128:55835
18. Choi H et al (2011) SAINT: probabilistic scoring of affinity purification-mass spectrometry data. Nat Methods 8(1):70–73
19. Drummond E et al (2020) Phosphorylated tau interactome in the human Alzheimer's disease brain. Brain 143(9):2803–2817
20. Pires G, Drummond E (2021) It takes more than tau to tangle: using proteomics to determine how phosphorylated tau mediates toxicity in neurodegenerative diseases. Neural Regen Res 16(11):2211–2212
21. Montine TJ et al (2012) National Institute on Aging-Alzheimer's Association guidelines for the neuropathologic assessment of Alzheimer's disease: a practical approach. Acta Neuropathol 123(1):1–11

Chapter 15

In Vitro Amplification of Pathogenic Tau Seeds from Neurodegenerative Disease Patient Brains

Hong Xu and Virginia M. -Y. Lee

Abstract

Aggregated microtubule-associated protein tau (tau) is the hallmark lesion of a group of neurodegenerative diseases, termed tauopathies. Normal endogenous tau is highly soluble and intrinsically disordered when it is not bound to microtubules. Pathological tau proteins are aggregates of bioactive filaments capable of inducing their normal counterparts into pathological conformations that are human tauopathy dependent. Taking advantage of this feature, we established an in vitro seeding reaction to amplify faithfully human-derived tau strains. This approach allows us to expand the quantity and improve the quality of pathogenic tau strains derived from human patient postmortem brains and to further understand tau pathogenesis in models of tauopathy. Here, we describe the approach to generate human pathogenic tau using human-derived tau seeds and recombinant human tau in vitro.

Key words In vitro seeding, Tau strains, Tau spreading, Tauopathy, Amplification, Cell free

1 Introduction

Tauopathies are neurodegenerative diseases characterized by the abnormal aggregation of tau protein in patient brains [1]. The strong association of aggregated tau with the progress of neurodegeneration and cognitive decline makes tau protein a primary pathological and clinical diagnostic biomarker for tauopathy and a therapeutic target for these diseases [2]. Pathological tau aggregates stereotypically spread between interconnected regions of patient brains that correlate with disease progression [3]. Emerging evidence supports that these aggregates play a central role in this spreading.

To mimic the spatiotemporal development of tauopathy in the human brain, we used human patient-derived tau filaments (tau seeds) to generate spreading cell and animal models of pathological tau [4]. These pathogenic tau seeds are potent inducers of tau aggregation in wild-type animals expressing tau at the physiological

Jerold Chun (ed.), *Alzheimer's Disease: Methods and Protocols*,
Methods in Molecular Biology, vol. 2561, https://doi.org/10.1007/978-1-0716-2655-9_15,

level. Compared with traditional transgenic models of tauopathy, there are unique advantages in the spreading models. First, there are no confounding effects from the insertion of the exogenous gene and from the overexpression of human mutant tau [4, 5]. Second, the initiation of tau pathology is spatiotemporally regulated, and by strategically choosing the targeted brain region and the time of inoculation, we can specifically recapitulate the developmental stages of different tauopathies. Third, tau spreading models recapitulate the strain-specific features of tau pathology [6–8] in that they faithfully reproduce the characteristic pathological tau associated with specific human tauopathies, after inoculation with tau derived from diseased brains. Despite these advantages, the use of human tau spreading mouse models is limited, because of the availability and quality of tau seeds.

The tau seeding assay has exhibited great potential for increasing the quantity and quality of pathogenic tau seeds [9–13], which, as the primary substrates, are critical for successful amplification. This chapter describes our protocol for in vitro seeding-based amplification of the pathogenic tau seeds derived from diseased, human, postmortem brains, emphasizing the importance of the proper extraction of human tau seeds and the purification of monomeric tau.

2 Materials

2.1 Brain Extraction

1. *4% Paraformaldehyde* (PFA), in phosphate-buffered saline (PBS), freshly made.
2. Glass Dounce homogenizer.
3. 50 mL conical tubes.
4. Plastic centrifugation tubes for ultracentrifuge (depends on the centrifuge type and sample size).
5. *Protease inhibitor cocktail* (PI): 1% w/v pepstatin, 1% w/v leupeptin, 1% w/v N-tosyl-L-phenylalanyl chloromethyl ketone (TPCK), 1% w/v N-α-tosyl-L-lysinyl-chloromethylketone (TLCK), 1% w/v trypsin inhibitor, 0.1 M EDTA, in H_2O, stored at room temperature (RT).
6. 0.5 M phenylmethylsulfonyl fluoride (PMSF), stored at RT.
7. *Phosphatase inhibitor* (PPI): 200 mM imidazole, 100 mM NaF, 100 mM Na orthovanadate, in H_2O, stored at −20 °C.
8. 200 mM 1,4-dithiothreitol (DTT), stored at −20 °C
9. *PHF extraction buffer*: 10 mM Tris base, 10% sucrose, 0.8 M NaCl, 1 mM EDTA, pH 7.4, with 1× PI, 1× PMSF, and 1× PPI, with 2 mM DTT added just before use, stored at 4 °C.
10. 25% (w/v) *sarkosyl*, in H_2O, stored at 4 °C.

11. *PHF extraction buffer* with 0.1% sarkosyl, stored at 4 °C.
12. *Dulbecco's phosphate-buffered saline* (dPBS) without Mg^{2+}/Ca^{2+}, pH = 7.3, stored at RT.
13. Kimwipe® tissues.
14. Plastic powder funnel.
15. 27G 1/2-gauge needle and syringe.
16. Glass beaker (500 mL).
17. Magnetic stir bar.
18. Orbital shaker.
19. Probe-based sonicator.
20. Ultracentrifuge.

2.2 Monomer Expression and Purification

1. Plasmid containing full-length MAPT (*see* **Note 1**).
2. Competent BL21(DE3)RIL bacterial cells.
3. Lysogeny broth (LB) medium without ampicillin, stored at RT.
4. LB medium with 100 μg/mL ampicillin, stored at RT.
5. 10 cm sterile LB agar plate, stored at 4 °C.
6. 500 mL and 2000 mL bacterial culture Erlenmeyer flasks.
7. Fast-protein liquid chromatography (FPLC) device, with fraction collector.
8. 0.5 M isopropyl β-d-1-thiogalactopyranoside (IPTG), stored at −20 °C
9. Glass Dounce homogenizer.
10. *High-salt RAB buffer*: 750 mM NaCl, 20 mM NaF, 100 mM 2-(N-morpholino) ethanesulfonic acid (MES), pH 7.0, 1 mM EGTA, 500 nM $MgSO_4$, with 1× PI, 1× PPI, 1× PMSF, and 1× DTT, stored at 4 °C.
11. *Fast-protein liquid chromatography (FPLC) buffer A*: 20 mM piperazine-N,N′-bis(2-ethanesulfonic acid) (PIPES), pH 6.5 (using NaOH), 10 mM NaCl, 1 mM EGTA, 1 mM $MgSO_4$, 1× DTT, 1× PMSF, stored at 4 °C.
12. *FPLC buffer B*: FPLC buffer A with 1-M NaCl, stored at 4 °C.
13. HiTrap SP HP cation exchange chromatography column (5 mL capacity).
14. Dialysis tubing with MW cutoff (MWCO) <25,000 kDa.
15. 10,000-kDa MWCO protein concentrator.
16. Optical density reader.
17. Cuvet for optical density reader.
18. Microbiological incubator.
19. 100 °C water bath.

20. Ultracentrifuge.
21. Ultracentrifuge tubes.
22. 5 L beaker.
23. Spectrophotometer for testing absorbance at 280 nm (A280).

2.3 In Vitro Seeding Reaction

1. Sterile dPBS without Mg^{2+}/Ca^{2+}, pH 7.3, stored at RT.
2. Bath sonicator (*see* **Note 2**).
3. Thermomixer C (*see* **Note 3**).
4. Thermocycler.
5. 200 mM DTT, stored at −20 °C
6. 0.5-M PSMF, stored at RT.
7. PI (the same as in Subheading 2.1, **item 5**).
8. Low-binding PCR tubes (1 tube per reaction).

3 Methods

3.1 Selection of Postmortem Brain Tissue

1. Choose brain tissues in the late stages of disease, with a high tau burden and short postmortem intervals (PMIs; less than 20 h if possible) (*see* **Note 4**, Fig. 1a–c).
2. Fix a 2-mm-thick piece of brain, adjacent to the brain region of interest, in 4% PFA overnight, for immunohistochemistry (IHC).

3.2 Tau Seeds Extraction

1. Thaw the brain tissue and remove the meninges and blood vessels (*see* **Note 5**).
2. Weigh the brain tissue in a 10 cm Petri dish and transfer the tissue into a Falcon tube.
3. Add 9 volumes (v/w, mL/g) of PHF extraction buffer with 0.1% sarkosyl into the Falcon tube.
4. Homogenize the brain tissue in a glass Dounce homogenizer, on ice, until all the tissue is resuspended, and then reserve a 100 μL sample of this total homogenate fraction to use for quality control (homogenate; *see* Fig. 2a and **Note 6**).
5. Transfer the homogenate to plastic centrifuge tubes and balance the tubes with PHF extraction buffer.
6. Centrifuge the homogenate at 10,000 g for 10 min at 4 °C.
7. Filter the supernatant into labeled 50 mL Falcon tubes by using a plastic powder funnel lined with a piece of Kimwipe® tissue that has been folded into two layers. Do not discard the pellet.
8. Reserve a 100-μL sample of the filtrate for quality control, and label as "Sup 1." Set aside the remainder of the filtered supernatant.

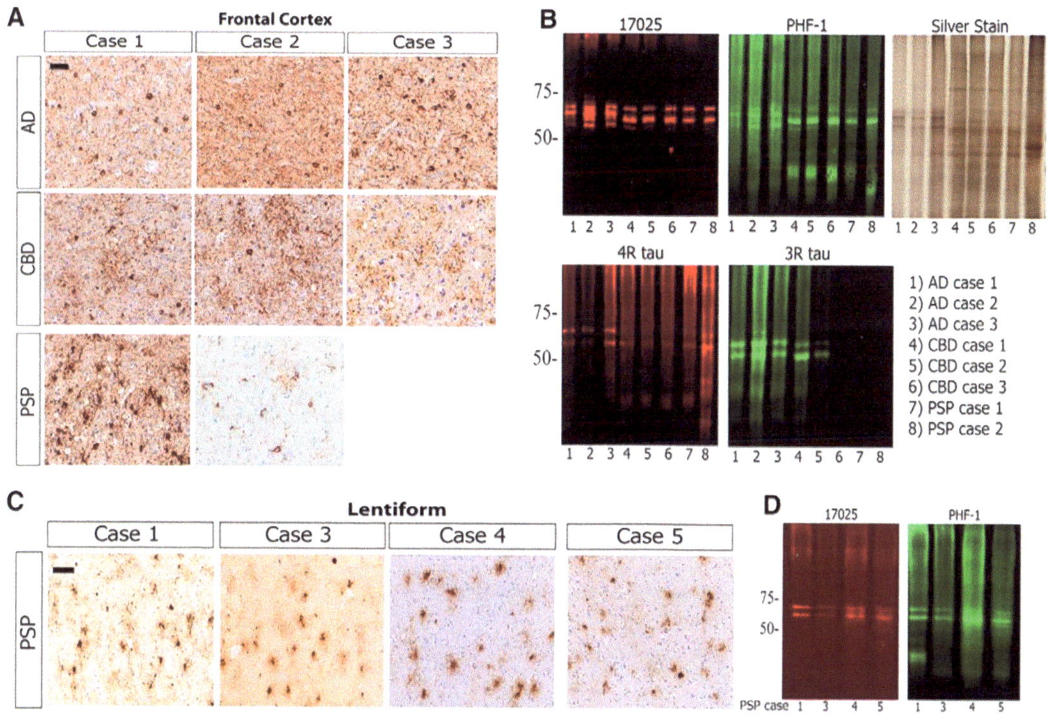

Fig. 1 Biochemical characterization of tau strains from human tauopathy brains. (**a**) Immunohistochemistry (IHC) was conducted on mid-frontal cortex sections with anti-tau MAb PHF-1 to show abundance of tau pathology (on all Alzheimer's disease [AD] and corticobasal degeneration [CBD] cases, and on progressive supranuclear palsy [PSP] Case 1); and with MAb CP13 (in PSP Case 2). *Scale bar*, 50 μm. (**b**) Immunoblots were performed using anti-tau antibodies 17025 (*red*), PHF-1 (*green*), anti-4R tau (CosmoBio) (*red*), and RD3 (*green*), plus silver staining, on the final supernatants from sequential extraction of the three AD, three CBD, and two PSP cases from (**a**). (**c**) IHC in the lentiform nucleus for PHF-1 from PSP cases. *Scale bar*, 50 μm. (**d**) Western blots for 17025 (*red*) and PHF-1 (*green*) on final supernatants from sequential extraction of the lentiform nucleus of the PSP cases from (**c**). (Adapted from [7])

9. Estimate the volume of the pellet produced in **Step** 7 and add 9 volumes of PHF buffer to the pellet.
10. Resuspend the pellet briefly, and then transfer to the glass Dounce homogenizer.
11. Homogenize the pellet until the suspension is completely free of solid brain tissues.
12. Reserve 100 μL of the homogenate for quality control, and label it as "Pel 1."
13. Repeat **Steps 6** and 7 with the remainder of the homogenate to ensure complete extraction.
14. Reserve a 100 μL sample of the filtrate for quality control, and label as "Sup 2." Set aside the remainder.
15. Resuspend the pellet in 9 volumes of PHFs buffer, and save 100 μL as "Pel 2." Discard the rest of the pellet.

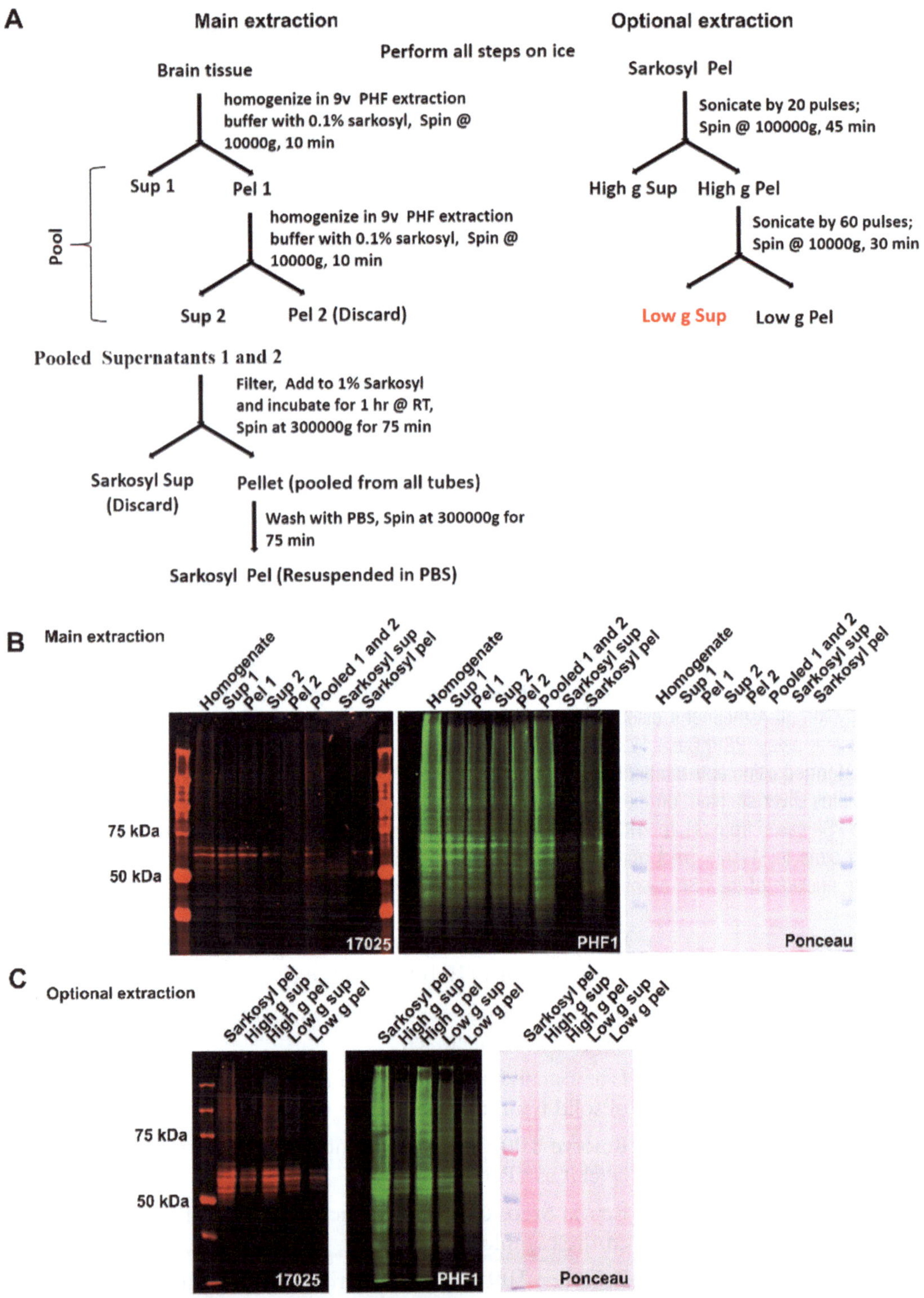

Fig. 2 Schematic workflow of brain extraction and tau quality control in each fraction. (**a**) Schematic workflow of brain extraction. (**b**) and (**c**): Tau species in each fraction extracted from an Alzheimer's disease (AD) case, revealed using 17025 and PHF-1 tau antibodies; the total protein is determined by Ponceau S staining. All samples were loaded using the same relative proportions to their respective fractions

16. Combine the remainder of Supernatants 1 and 2 in a 200 mL glass beaker and roughly estimate the combined volume. Add a sufficient volume of 25% sarkosyl to achieve a final concentration of 1%.
17. Reserve a 100-μL sample for quality control and label it as "Total sup."
18. Using a magnetic stirrer and stir bar, stir the combined supernatant at RT for 1.5 h.
19. Transfer the supernatant to a high-speed centrifuge tube.
20. Balance the tube and spin at 300,000 g for 75 min at 4 °C.
21. Reserve a 100 μL sample of the supernatant for quality control, as "Sark sup." The rest of the supernatant can be discarded.
22. Wash the pellet by adding 6 mL of dPBS, without disturbing the pellet, and then discard the dPBS. Repeat this process two more times.
23. Resuspend the washed pellet in 1 mL of dPBS by repeatedly pipetting with P1000 tips.
24. Spin again at 300,000 g for 30 min. Reserve a 50 μL sample of the supernatant for quality control, and label as "PBS wash." Save the pellet and discard the supernatant.
25. Add to the pellet 0.1 mL of dPBS per gram of the starting brain tissue, and shake the tube overnight at RT.
26. Slowly triturate the homogenate using a 27-G 1/2 gauge needle to break down the pellet.
27. Sonicate the homogenate with a probe sonicator for 20 pulses at 10–20% power. Each pulse should be about a half-second in duration with a one-second interval between pulses.
28. Estimate the volume of the homogenate, for the next steps.
29. Centrifuge the homogenate at 100,000 g for 30 min at 4 °C, then reserve 100 μL sample of the supernatant for quality control, and label it is as "High-g sup."
30. Add one-fifth of the total volume of the homogenate (just before the last centrifugation) to the pellet and sonicate it with a manual probe sonicator for 60 pulses at 10–20% power, on ice, to break the pellet apart thoroughly (*see* **Note** 7).
31. Spin the homogenate again at 10,000 g for 30 min at 4 °C; the resulting supernatant should be labeled as "Low-g sup" and will contain the enriched pathological tau seeds.
32. Save 100 μL of the pellet as "Low-g pellet" for quality control and discard the rest.
33. Determine the tau concentration in the extracts using immunoblots.

3.3 Expression and Purification of Full-Length Monomeric Tau (See Note 8)

All equipment and materials used in this section should be sterile.

1. Transduce 0.1 μg of plasmid pRK172 with MAPT into 20 μL of competent BL21(DE3)RIL bacterial cells, by gently mixing them together.
2. Allow the bacteria to grow overnight on a plate with LB medium containing ampicillin, at 37 °C.
3. Choose a single colony and expand it as a starter culture in 150 mL of LB with ampicillin in a 500 mL bacterial culture Erlenmeyer flask. Place it in a microbiological incubator at 180 RPM at 37 °C, overnight.
4. Add 8 mL of the above starter culture to a 2000 mL Erlenmeyer flask containing 500 mL of LB with ampicillin and shake at 180 RPM at 37 °C, until the optical density at 600 nm (OD_{600nm}) reaches 1.0 (*see* **Note 9**).
5. To induce tau expression, add 1 mL of 0.5-M IPTG to the above flask (1 mM final IPTG concentration), and rotate flask at 180 RPM for 2 h at 37 °C in a microbiological incubator.
6. Centrifuge culture at 5000 g for 15 min at RT and discard the supernatant.
7. Transfer the bacterial pellet into a glass Dounce homogenizer and add 40 mL high-salt RAB buffer to the pellet for every 1 L of bacterial culture.
8. Homogenize the bacterial pellet in high-salt RAB buffer, on ice, until its fully resuspended. Boil the homogenate in a 100 °C water bath for 20 min to coagulate the contaminating non-tau proteins.
9. Transfer the boiled lysate into the centrifuge tube, and balance the tubes with high-salt RAB buffer.
10. Centrifuge the boiled lysate at 186,000 g for 30 min at 37 °C.
11. Dialyze the supernatant with dialysis tubing (MWCO <25,000 kDa) in two beakers of 4 L FPLC buffer A at 4 °C overnight.
12. Purify the dialyzed lysate using the HiTrap SP HP 5 mL column at a 5 mL/min flow rate. Tau protein will bind to the column in FPLC buffer A and be eluted between 10% and 40% of FPLC buffer B/buffer A mixture, within 10 column volumes.
13. Collect the eluted fractions with the fraction collector.
14. Store the fractions at 4 °C, reserving 10 μL of each fraction for the next step.
15. Run 10 μL of each fraction on a 10% sodium dodecyl sulfate–polyacrylamide gel electrophoresis (SDS PAGE) gel to determine which ones contain monomeric tau (*see* **Note 10** and Fig. 3c). The T40 band should be between 55 and 70 kDa.

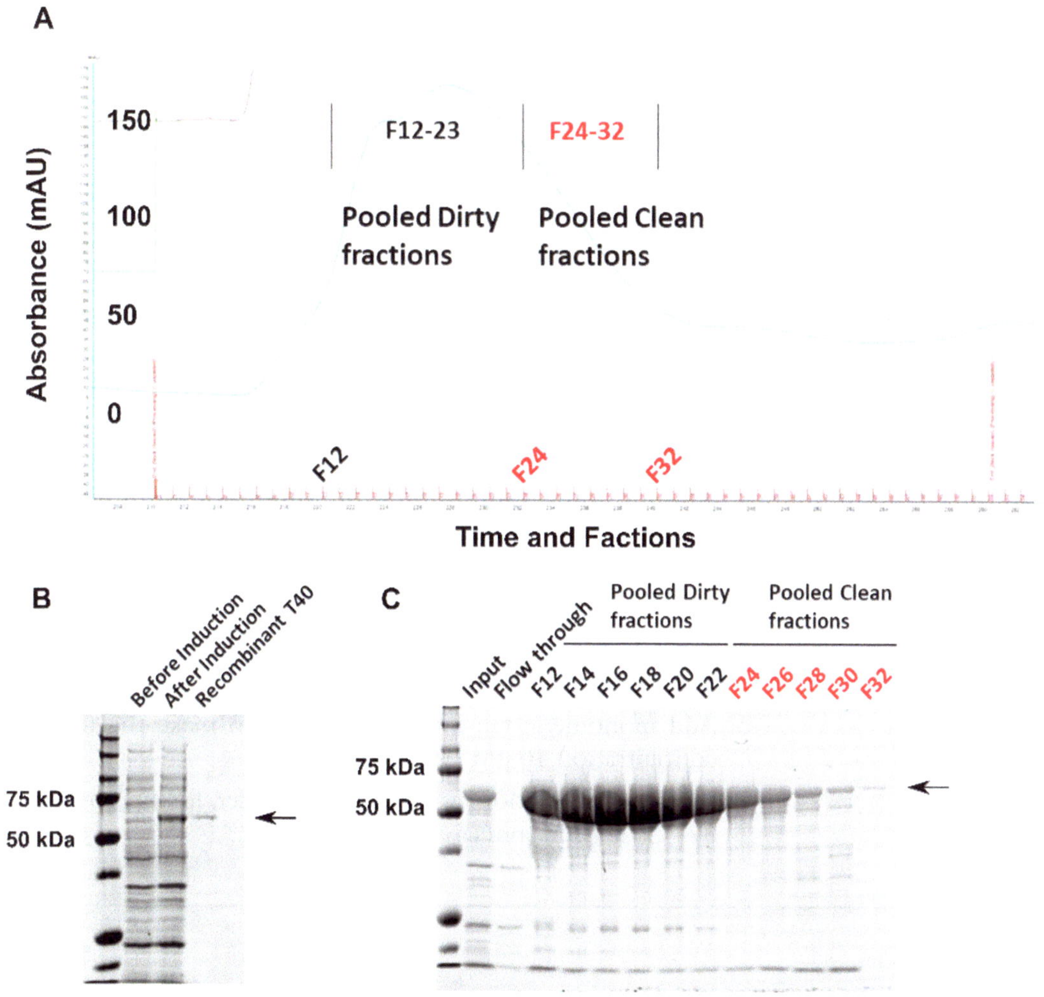

Fig. 3 Analysis of eluted fractions of human wild-type tau (T40) in fast-protein liquid chromatography (FPLC). (**a**) Ultraviolet (UV) absorbance of T40 during the elution with a 10–40% gradient of FPLC buffer B, and samples collected using a fraction collector. (**b**) Bacterial culture before and after isopropyl β-d-1-thiogalactopyranoside (IPTG) induction. (**c**) Eluted fractions were run on a sodium dodecyl sulfate–polyacrylamide gel electrophoresis (SDS PAGE) gel and stained with Coomassie blue. *Arrow* indicates the position of the T40 band (2N4R tau isoform)

16. Pool the cleanest tau fractions (the major band should be the tau band, between the 50 and 75 kDa protein standard, as seen in Fig. 3c).
17. Dialyze the pooled fractions with dialysis tubing (MWCO <25,000 kDa) in 4 L of dPBS at 4 °C overnight (*see* **Note 11**).
18. Concentrate the FPLC-purified tau monomer using a 10,000 MWCO protein concentrator, following the manufacturer's instructions.

19. Add 200-nM DTT to the tube, and reserve 2 μL of the purified tau solution to determine the concentration using a Nanodrop® spectrophotometer.
20. Store the purified tau (in 1 mM DTT) at −80 °C (*see* **Note 12**).

3.4 In Vitro Seeding Reaction

All the materials should be sterile in this section.

1. Calculate the seed–monomer ratio based on the tau concentration (*see* **Notes 13** and **14**). The volume of dPBS = total reaction volume minus the volumes of seeds, monomer, DTT, PSMF, and PI.
2. Mix dPBS with DTT (final concentration = 2 mM), PSMF (final concentration = 50 mM) and PI (1:100) in sterile PCR tubes (*see* **Notes 15** and **16**).
3. Add the appropriate amount of tau seeds into the mixture. Keep the tube on ice when not use.
4. Incubate the mixture at 56 °C for 30 min in a thermocycler (*see* **Note 17**).
5. Sonicate tau seeds in a bath sonicator for 20 min, with 30′ on, 30′ off, at the highest intensity.
6. Add monomeric tau into the solution to make the final tau concentration 40 μM (*see* **Note 18**).
7. Agitate the reaction using a thermomixer for up to 7 d at 37 °C, at 1000 rpm (*see* **Note 19**).

4 Notes

1. Subclone full-length human *tau* (2N4R) DNA into the EcoRI site of M13mpl8. Use site-directed mutagenesis to introduce a NdeI site in the context of the initiator codon. Following cleavage with NdeI and EcoRI, the resulting cDNA fragments should be subcloned downstream of the T7 RNA polymerase promoter into the NdeI/EcoRI cut expression plasmid pRK172 [14].
2. Tau seeds have filamentous structures, which vary dramatically in their sizes after enrichment, because of different strains and patient cases. The elongation of tau filaments occurs at the open ends through their interactions with tau monomers [8, 15]. Intensive sonication is key to homogenizing the different sizes of tau seeds, to generating open ends, and to reducing variations between different disease cases. The average length of ~40-nm tau seeds after sonication can be achieved in our preparations. Alternative sonicators could be used, but the sonication intensity and time should be tested to achieve a homogenous size distribution of the sonicated tau seeds [8].

3. Using a mixer with a temperature-stable chamber can reduce the evaporation of the reaction solution, and using a non-elastic tube rack (or a plastic tube rack) can reduce the potential variations caused by the agitation intensity.
4. It is vital to choose cases and brain regions with high tau burden and no other comorbidities (*see* Fig. 1a, c). Comorbidities will result in increased amounts of other contaminant proteins in the lysates, potentially interfering with the seeding reactions.
5. Tau pathology in an Alzheimer's brain is abundant in gray matter. Separating the gray and white matter can significantly reduce the amount of lipid and myelin interfering with the extraction and increase the purity of tau seeds at the end. Determine the tau burden and look for potential co-pathology, e.g., use staining for phospho-alpha-synuclein for Lewy body pathology and phospho-TDP-43 for TDP-43 pathology. Exclude brain tissue with a high degree of co-pathology.
6. Using a glass Dounce homogenizer will reduce the number of bubbles generated during the homogenization of the brain tissue and preserve the filamentous tau species in the homogenate. We do not recommend sonication-based homogenization for this step since it will break tau filaments into smaller fragments that make them hard to pellet by sedimentation in later steps.
7. These steps are designed to dissociate the tau filaments from the other insoluble contaminating material in the lysate. Whether or not to carry out the steps depends on the abundance of tau filaments in the samples and the amount of insoluble contaminant present. These steps are typically recommended to enrich Alzheimer's disease (AD) tau filaments since they will significantly improve its purity, with a reasonable yield. However, skipping these steps can increase the pathological tau yield for other tau strains while lowering the purity.
8. This approach (Subheading 3.3) could apply to all six isoforms of wild-type tau.
9. We titrated the OD_{600} and found that a value of 1.0 resulted in a better yield of T40, in our protocol, than values of 0.6 and 0.8.
10. Good-quality monomeric tau is essential for the success of seeding reactions. Earlier fractions will result in lower purity (*see* Fig. 3a, c). Monomeric tau with low purity will result in the formation of non-pathogenic tau aggregates during the agitation.

11. Pooled "dirty" fractions could be purified again with the same protocol to increase the yield.
12. Oxidation of the cysteine residue on tau monomers will result in intra- or intermolecular disulfide bonds on tau monomers and significantly impact fibrillization. Storing monomeric tau in 1 mM DTT will inhibit the disulfide bond formation and avoid aberrant tau aggregate formation [16].
13. Although different seed–monomer ratios can be used, the reaction outcome will be more variable with lower ratios. We recommend using 4 μM tau seeds in the 40 μM total tau reactions (seed–monomer ratio = 1:9) for AD–tau and corticobasal degeneration [CBD]–tau. Lower tau concentrations (<20 μM) in the reaction will significantly reduce the rate of fibrillization. Higher tau concentrations (>80 μM) will lead to tau protein precipitating in the solution and/or forming the non-pathogenic tau aggregates.
14. Owing to the distinctive potencies of the different tau strains, titrating different seed–monomer ratios for different tau strains can optimize their amplification.
15. Adding DTT into the reaction significantly reduces the degradation of tau seeds and monomers that is associated with the proteolytic activity of the lysates.
16. Using a low-protein-binding tube will significantly reduce the attachment of substrates and insoluble components to the tube during the reaction.
17. Use of 56 °C heat treatment further reduces the proteolytic activity of the lysates without affecting the seeding capacity of tau seeds. Higher temperatures will denature the tau filaments and compromise the activity of tau seeds.
18. The reaction volume is essential for agitation intensity. We tested 50 to 100 μL reaction volumes and found comparable reaction results. Different reaction volumes should be optimized before use.
19. For a low quantity of tau seeds, a longer reaction time is needed.

5 Discussion

In our approach, the quality of the tau seeds and tau monomers is essential for successful seeding. To achieve faithful amplification of tau strains and avoid formation of aberrant tau filaments, we abandoned polyanionic inducers and instead seed the monomer with human-derived pathogenic tau seeds, in a single reaction. The high purity of tau monomer limits the possible types of seed–monomer interactions and any deviations in the reactions.

Tau monomers can interact with the insoluble components in the human brain lysates. To reduce these aberrant interactions, we need to enrich the human pathogenic tau seeds and improve their purity. The in vitro amplification reactions could faithfully conserve the strain-specific pathogenic properties, thereby expanding the availability of various tau strains. The amplification of pathogenic tau seeds can also improve their relative purity by increasing the ratio of seeds to other components. Moreover, labeled monomers can be generated and integrated into the amplification reaction as tools to elucidate the pathogenic behavior and structure of tau seeds in vitro and in vivo. In sum, the in vitro seeding of human-derived tau opens a new avenue for understanding tau fibrillization in vitro and expands our ability to enrich human pathogenic tau seeds.

Acknowledgments

The Brightfocus Foundation supported this work. We thank Lakshmi Changolkar for providing the brain extraction immunoblot images and Garrett Gibbons and Alex Crowe for providing the tau monomer expression and purification protocol.

References

1. Wang Y, Mandelkow E (2016) Tau in physiology and pathology. Nat Rev Neurosci 17(1): 5–21. https://doi.org/10.1038/nrn.2015.1
2. Iqbal K, Liu F, Gong CX (2016) Tau and neurodegenerative disease: the story so far. Nat Rev Neurol 12(1):15–27. https://doi.org/10.1038/nrneurol.2015.225
3. Leuzy A, Chiotis K, Lemoine L, Gillberg PG, Almkvist O, Rodriguez-Vieitez E, Nordberg A (2019) Tau PET imaging in neurodegenerative tauopathies-still a challenge. Mol Psychiatry 24(8):1112–1134. https://doi.org/10.1038/s41380-018-0342-8
4. Guo JL, Narasimhan S, Changolkar L, He Z, Stieber A, Zhang B, Gathagan RJ, Iba M, McBride JD, Trojanowski JQ, Lee VM (2016) Unique pathological tau conformers from Alzheimer's brains transmit tau pathology in nontransgenic mice. J Exp Med 213(12): 2635–2654. https://doi.org/10.1084/jem.20160833
5. Gamache J, Benzow K, Forster C, Kemper L, Hlynialuk C, Furrow E, Ashe KH, Koob MD (2019) Factors other than hTau overexpression that contribute to tauopathy-like phenotype in rTg4510 mice. Nat Commun 10(1):2479. https://doi.org/10.1038/s41467-019-10428-1
6. He Z, McBride JD, Xu H, Changolkar L, Kim SJ, Zhang B, Narasimhan S, Gibbons GS, Guo JL, Kozak M, Schellenberg GD, Trojanowski JQ, Lee VM (2020) Transmission of tauopathy strains is independent of their isoform composition. Nat Commun 11(1):7. https://doi.org/10.1038/s41467-019-13787-x
7. Narasimhan S, Guo JL, Changolkar L, Stieber A, McBride JD, Silva LV, He Z, Zhang B, Gathagan RJ, Trojanowski JQ, Lee VMY (2017) Pathological tau strains from human brains recapitulate the diversity of tauopathies in nontransgenic mouse brain. J Neurosci 37(47):11406–11423. https://doi.org/10.1523/JNEUROSCI.1230-17.2017
8. Xu H, O'Reilly M, Gibbons GS, Changolkar L, McBride JD, Riddle DM, Zhang B, Stieber A, Nirschl J, Kim SJ, Hoxha KH, Brunden KR, Schellenberg GD, Trojanowski JQ, Lee VM (2021) In vitro amplification of pathogenic tau conserves disease-specific bioactive characteristics. Acta Neuropathol 141(2):193–215. https://doi.org/10.1007/s00401-020-02253-4

9. Kraus A, Saijo E, Metrick MA 2nd, Newell K, Sigurdson CJ, Zanusso G, Ghetti B, Caughey B (2019) Seeding selectivity and ultrasensitive detection of tau aggregate conformers of Alzheimer disease. Acta Neuropathol 137(4): 585–598. https://doi.org/10.1007/s00401-018-1947-3
10. Saijo E, Metrick MA 2nd, Koga S, Parchi P, Litvan I, Spina S, Boxer A, Rojas JC, Galasko D, Kraus A, Rossi M, Newell K, Zanusso G, Grinberg LT, Seeley WW, Ghetti B, Dickson DW, Caughey B (2020) 4-Repeat tau seeds and templating subtypes as brain and CSF biomarkers of frontotemporal lobar degeneration. Acta Neuropathol 139(1): 63–77. https://doi.org/10.1007/s00401-019-02080-2
11. Meyer V, Dinkel PD, Rickman Hager E, Margittai M (2014) Amplification of Tau fibrils from minute quantities of seeds. Biochemistry 53(36):5804–5809. https://doi.org/10.1021/bi501050g
12. Kaufman Sarah K, Sanders David W, Thomas Talitha L, Ruchinskas Allison J, Vaquer-Alicea J, Sharma Apurwa M, Miller Timothy M, Diamond Marc I (2016) Tau prion strains dictate patterns of cell pathology, progression rate, and regional vulnerability in vivo. Neuron 92(4): 796–812. https://doi.org/10.1016/j.neuron.2016.09.055
13. Clavaguera F, Bolmont T, Crowther RA, Abramowski D, Frank S, Probst A, Fraser G, Stalder AK, Beibel M, Staufenbiel M, Jucker M, Goedert M, Tolnay M (2009) Transmission and spreading of tauopathy in transgenic mouse brain. Nat Cell Biol 11(7): 909–913. https://doi.org/10.1038/ncb1901
14. Goedert M, Jakes R (1990) Expression of separate isoforms of human tau protein: correlation with the tau pattern in brain and effects on tubulin polymerization. EMBO J 9(13): 4225–4230
15. Huseby CJ, Bundschuh R, Kuret J (2019) The role of annealing and fragmentation in human tau aggregation dynamics. J Biol Chem 294(13):4728–4737. https://doi.org/10.1074/jbc.RA118.006943
16. Crowe A, James MJ, Lee VM, Smith AB 3rd, Trojanowski JQ, Ballatore C, Brunden KR (2013) Aminothienopyridazines and methylene blue affect Tau fibrillization via cysteine oxidation. J Biol Chem 288(16): 11024–11037. https://doi.org/10.1074/jbc.M112.436006

Chapter 16

Aβ and Tau Prions Causing Alzheimer's Disease

Carlo Condello, Gregory E. Merz, Atsushi Aoyagi, William F. DeGrado, and Stanley B. Prusiner

Abstract

Studies show that patients with Alzheimer's disease (AD) have both Aβ and tau prions, and thus, AD is a double-prion disease. AD patients with the greatest longevity exhibited low levels of both Aβ and tau prions; tau prions were nearly absent in the brains of almost half of the patients who lived beyond 80 years of age. Using cellular bioassays for prions in postmortem samples, we found that both Aβ and tau proteins misfold into prions leading to AD, which is either a sporadic or familial dementing disorder. Although AD is transmissible experimentally, there is no evidence that AD is either communicable or contagious. Since the progression of AD correlates poorly with insoluble Aβ in the central nervous system (CNS), it was difficult to distinguish between inert amyloids and Aβ prions. To measure the progression of AD, we devised rapid bioassays to measure the abundance of isoform-specific Aβ prions in the brains of transgenic (Tg) mice and in postmortem human CNS samples from AD victims and people who died of other neurodegenerative diseases (NDs). We found significant correlations between the longevity of individuals with AD, sex, and genetic background, despite the fact that all postmortem brain tissue had essentially the same confirmed neuropathology.

Although brains from all AD patients had measurable levels of Aβ prions at death, the oldest individuals had lower Aβ prion levels than the younger ones. Additionally, the long-lived individuals had low tau prion levels that correlated with the extent of phosphorylated tau (p-tau). Unexpectedly, a longevity-dependent decrease in tau prions was found in spite of increasing amounts of total insoluble tau. When corrected for the abundance of insoluble tau, the tau prion levels decreased exponentially with respect to the age at death with a half-time of approximately one decade, and this correlated with the abundance of phosphorylated tau.

Even though our findings with tau prions were not unexpected, they were counterintuitive; thus, tau phosphorylation and tau prion activity decreased exponentially with longevity in patients with AD ranging from ages 37 to 99 years. Our findings demonstrated an inverse correlation between longevity in AD patients and the abundance of neurotoxic tau prions. Moreover, our discovery may have profound implications for the selection of phenotypically distinct patient populations and the development of diagnostics and effective therapeutics for AD.

Key words Neurodegenerative disease, Prion, Amyloid, Tauopathy, Alzheimer's disease, Parkinson's disease, Multiple system atrophy, Aβ, Tau, Synuclein, Cellular bioassay, Cryo-EM

Carlo Condello and Atsushi Aoyagi contributed equally with all other contributors.

Jerold Chun (ed.), *Alzheimer's Disease: Methods and Protocols*,
Methods in Molecular Biology, vol. 2561, https://doi.org/10.1007/978-1-0716-2655-9_16,

1 Introduction

In 1911, Alois Alzheimer described the plaques of the senile dementia that bear his name [1]; soon thereafter, Oskar Fisher is thought to have described the neurofibrillary tangles (NFTs) of AD [2]. In 1984, the Aβ peptide was identified [3], and tau protein in tangles was reported soon afterward [4–7]. Over the past four decades, studies have shown repeatedly that both Aβ and tau proteins adopt pathogenic, self-propagating conformations characteristic of prions. The structural features of Aβ and tau causing AD are similar to PrP^{Sc}, which is the prototypic prion protein (PrP) causing Creutzfeldt–Jakob disease, Gerstmann–Sträussler–Scheinker (GSS) disease, fatal familial insomnia, and kuru [8]. For clarity, accuracy, and fidelity, we refer to the self-propagating conformers of Aβ and tau as "Aβ prions" and "tau prions," respectively [9–13]. Using cellular bioassays for prions in postmortem brain specimens, we found that both the Aβ and tau proteins misfold into prions leading to AD, which is either a sporadic or familial dementing disorder. Over 90% of these prion diseases are sporadic disorders, while <10% are inherited. Early studies on familial transmissible NDs including GSS [14] presented a conundrum, which remained unresolved until 1989 when one of us reported genetic linkage of the P102L mutation in the PrP open reading frame (ORF) that causes GSS [15]. Two years later, familial AD was linked to the V717I mutation in the amyloid precursor protein (APP) ORF [16]. The importance of Aβ was further elucidated in investigations of AD pathogenesis based on genetic linkage studies between inherited AD and mutations in either APP or its processing enzymes [17, 18]. The role of tau in AD was further clarified in 1998 when three different point mutations in the *MAPT* gene were shown to cause familial Pick's disease (PiD) but not familial AD [19–21].

In familial cases with mutations in the Aβ coding region causing cerebral amyloid angiopathy (CAA), death can occur relatively early (in the fourth or fifth decade of life) and in the absence of substantial accumulation of insoluble tau [22, 23]. In contrast to CAA, tau prions spread through many brain regions in sporadic AD (sAD) resulting in cognitive decline. Hypotheses about the role of Aβ abound: At one end of the AD spectrum, Aβ deposition is considered either inconsequential [24, 25] or Aβ transiently initiates tau polymerization into NFTs [26–29]. Aβ might also have transient toxicity due to the accumulation of Aβ oligomers, which peak in the early prodromal phase of disease progression [30, 31]. At the other end of the AD spectrum, an early, steady progression of Aβ oligomerization, prion formation, and deposition are envisioned. The latter hypothesis is supported by the correlation of increasing AD severity with the spread of Aβ prions [32, 33] and decreased Aβ42 in cerebrospinal fluid (CSF) [34]. In this view, early and progressive formation of Aβ prions leads to CNS dysfunction accompanied by the inability to clear misfolded tau prions.

To address these hypotheses, we needed rapid bioassays to measure and rapidly compare Aβ and tau prions in the brains of deceased AD patients and animal models [13]. Prior to our cell culture bioassay for Aβ prions, measurement of Aβ prions relied on either time-consuming mouse models or in vitro biophysical methods that had to be performed at super-physiological peptide concentrations. Furthermore, clinical imaging ligands measure insoluble amyloids, whose role in AD pathogenesis is uncertain, rather than active, neuroinvasive Aβ and tau prions. In contrast, highly reproducible and rapid cell-based methods have been devised to quantify the ability of tau and α-synuclein prions to induce misfolding and aggregation of the corresponding soluble proteins expressed in mammalian cells [35–38]. Such cells were transfected with mutant tau fused to yellow fluorescent protein (YFP). The ability of tau prions to enter reporter cell lines could then be quantitatively assessed by measuring the fraction of infected cells using fluorescence confocal microscopy. Here, we describe an analogous cellular bioassay for Aβ, allowing Aβ and tau prions to be compared in frozen postmortem AD brain samples. Strikingly, Aβ prions were detected in all patients, even at the time of death, and they decreased significantly in the most long-lived patients. Even more striking, tau prion abundance decreased with longevity despite the presence of increasing NFTs. This decrease in tau prions occurred in parallel with similar decreases in tau phosphorylation. Thus, the greatest tau prion levels were found in individuals who died at relatively young ages, despite having the lowest abundance of total insoluble tau. Indeed, it is the relative abundance of active phosphorylated tau prions—and not the total amount of inert insoluble tau—that correlates with longevity.

2 Development of Isoform- and Sequence-Specific Aβ Prion Assays

To measure Aβ prion levels in postmortem brains from AD victims, we created Aβ prion bioassays that are exquisitely selective for the length and sequence of Aβ isoforms [13]. Sequential cleavage of APP predominately yields Aβ40 and Aβ42 as the major isoforms, which can polymerize into oligomers or fibrils and ultimately form amyloid deposits in the brain [18]. Here, we describe cell lines capable of detecting both Aβ40 and Aβ42 prions and a cell line that is specific for Aβ40 prions, which proved useful for studying isotype-specific prion propagation (*see* Figs. 1 and 2). Following earlier studies [39–41], we built constructs in which Aβ40 and Aβ42 were either N-terminally or C-terminally fused to YFP (*see* Fig. 2a). Several structures of Aβ42 show that its C-terminal

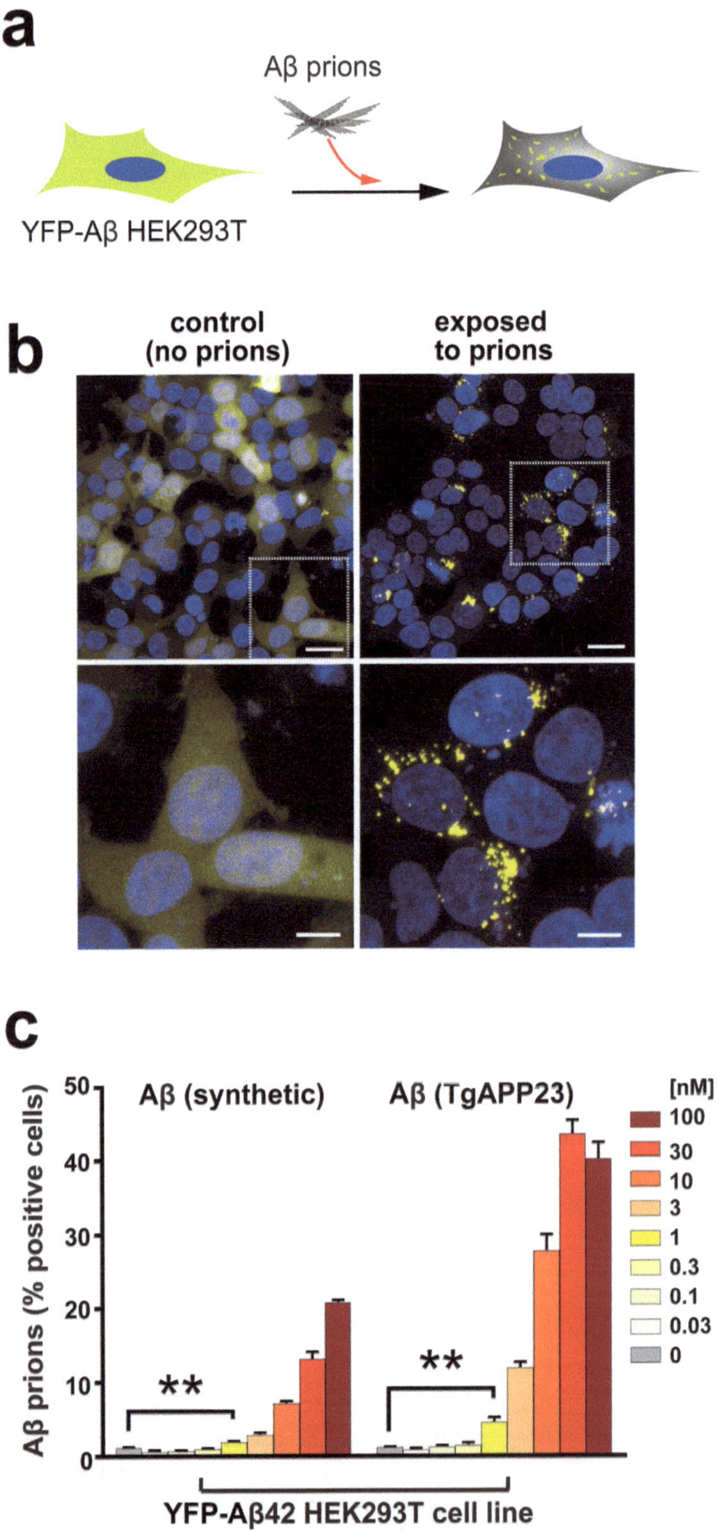

Fig. 1 HEK293T cells expressing Aβ fusion proteins are susceptible to aggregation of synthetic and brain-derived Aβ prions. (**a**) Schematic illustration of stably

residues are buried in an inaccessible core [42–44], while the N-terminus is largely disordered, suggesting that the fusion of a fluorescent protein might be better tolerated at the N-terminus. We, therefore, constructed N- and C-terminal YFP fusions to Aβ40 and Aβ42 and examined the ability of synthetic fibrils to induce aggregation of these constructs in the cytoplasm of mammalian cells (*see* Figs. 1 and 2). For these studies, we used a preparation of synthetic fibrils, which had previously been extensively characterized biophysically and shown to initiate prion propagation in a strain-dependent manner in Tg mice [45, 46]. As expected, Aβ40 and Aβ42 fibrils were able to infect and induce aggregation of YFP–Aβ42 fusions in a dose-dependent manner, and this was most efficient when the YFP was fused to the N-terminus (*see* Figs. 1 and 2). Thus, this cell line, designated YFP–Aβ42, was found to be an efficient, broad-spectrum reporter of Aβ prions.

Our Aβ prion assays also proved useful for examining isotype specificity and infectivity in what is intrinsically a heterogeneous system. Interestingly, Aβ40 fibrils—but not Aβ42 fibrils—were able to infect YFP–Aβ40 cells, indicating that this cell line is isoform specific (*see* Fig. 2). To investigate this phenomenon further, we introduced several familial AD sequence mutations (*see* Fig. 3), which had previously been shown to increase the propensity of Aβ peptides to aggregate in vitro and in vivo [47, 48], in YFP–Aβ40 cells. Expression of the E22G and E22Q mutants increased the sensitivity of the cells to wild-type (WT) Aβ40 fibrils but not to WT Aβ42 fibrils (*see* Fig. 3). In contrast, WT Aβ40 was unable to infect YFP–Aβ40(ΔE22), a highly aggregation-prone deletion mutant (*see* Fig. 3). Our findings are consistent with an earlier biophysical study showing that the Aβ40(ΔE22) mutant could stimulate assembly of WT Aβ40 fibrils but not vice versa [49]. In summary, the availability of our battery of cell lines provides tools for evaluating the sequence and structural specificity of Aβ prion infection and the kinetics of propagation over days (*see* Fig. 4).

Fig. 1 (continued) transfected cells expressing a YFP–Aβ fusion construct before and after lipofectamine-based transduction with synthetic Aβ prions. (**b**) Representative confocal images of HEK293T cells expressing Aβ42 fused to YFP at the N-terminus (Clone #1) treated with phosphate-buffered saline (PBS; *left*, unexposed) or exposed to Aβ40 prions (*right*, initial monomeric concentration of 1 μM). *Lower panels* show delineated areas in upper images at higher magnification. *Scale bars* represent 20 μm (*upper*) or 5 μm (*lower*). (**c**) YFP–Aβ42 cells were treated with two different types of Aβ prions ranging from 0.03 to 100 nM (initial monomeric concentration): synthetic Aβ40 prions (*left*) or Aβ prions purified from Tg(APP23) mouse brains (*right*). Cell prion activity was quantified 2 days after the initial exposure to various prion preparations. Data shown as mean ± SEM were determined from four images per well in four wells

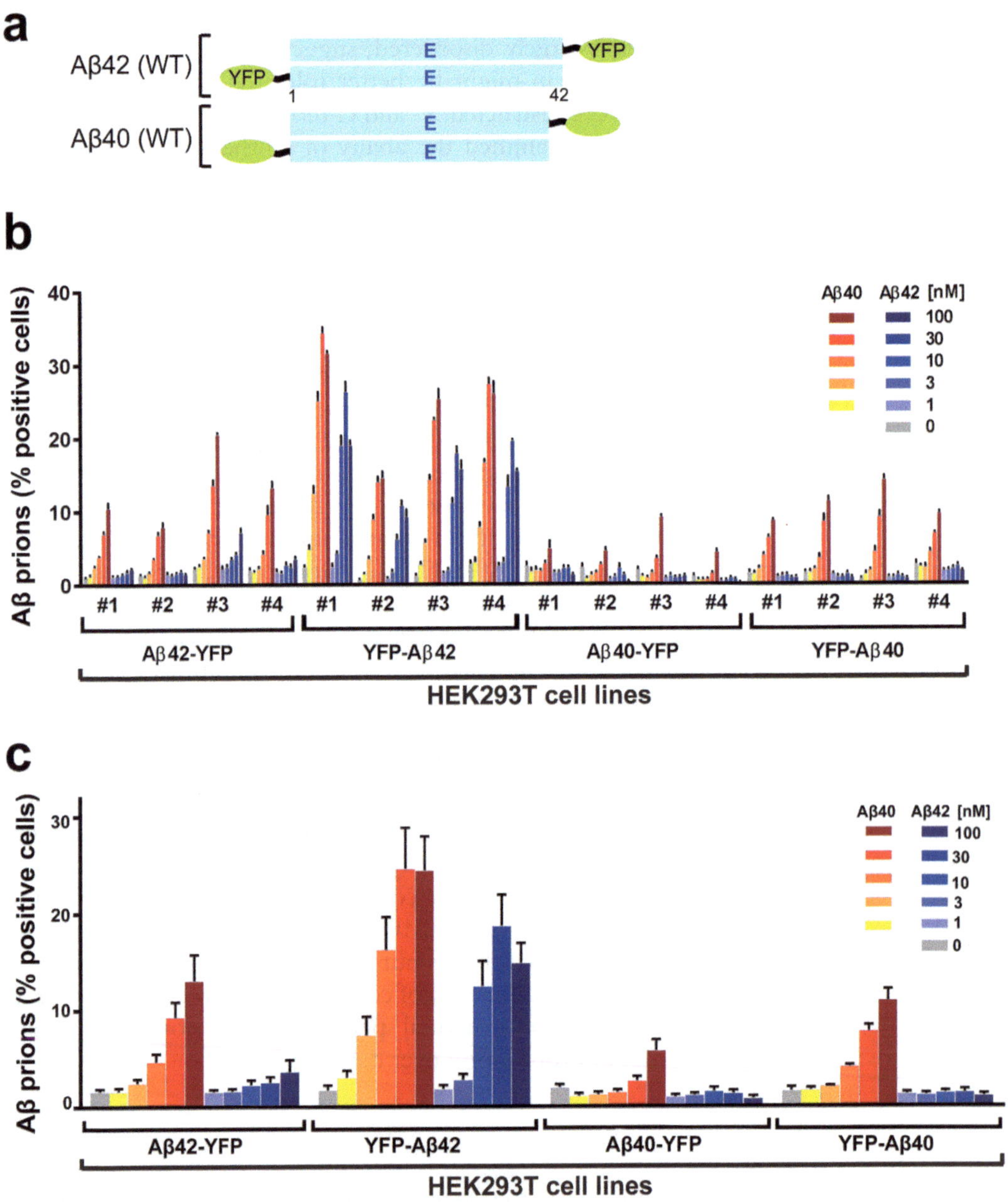

Fig. 2 Synthetic and brain-derived Aβ prions propagate in HEK293T cells expressing wild-type (WT) Aβ fusion proteins. (**a**) Diagram illustrating WT Aβ constructs used in this study. (**b**) Cell lines stably expressing four different WT Aβ constructs shown in (**a**) were developed, and prion propagation was compared with both synthetic Aβ40 and Aβ42 prions. Quantification of prion loads in 16 monoclonal cell lines (four randomly chosen clones from each construct) to increasing concentrations of synthetic Aβ40 or Aβ42 prions (1–100 nM) at day 2 post exposure. Data shown as mean ± SEM as determined from four images per well in four wells. (**c**) Cell lines stably expressing four different WT Aβ constructs shown in (**a**) were developed, and prion loads were compared with both synthetic Aβ40 and Aβ42 prions. Quantification of Aβ prions in 16 monoclonal cell lines (four randomly chosen clones from each construct; *see* Fig. 2a) 2 days after exposure to increasing concentrations of synthetic Aβ40 or Aβ42 prions (1–100 nM; *see* Fig. 4). Data shown as mean ± SEM as determined from four images per well in four wells

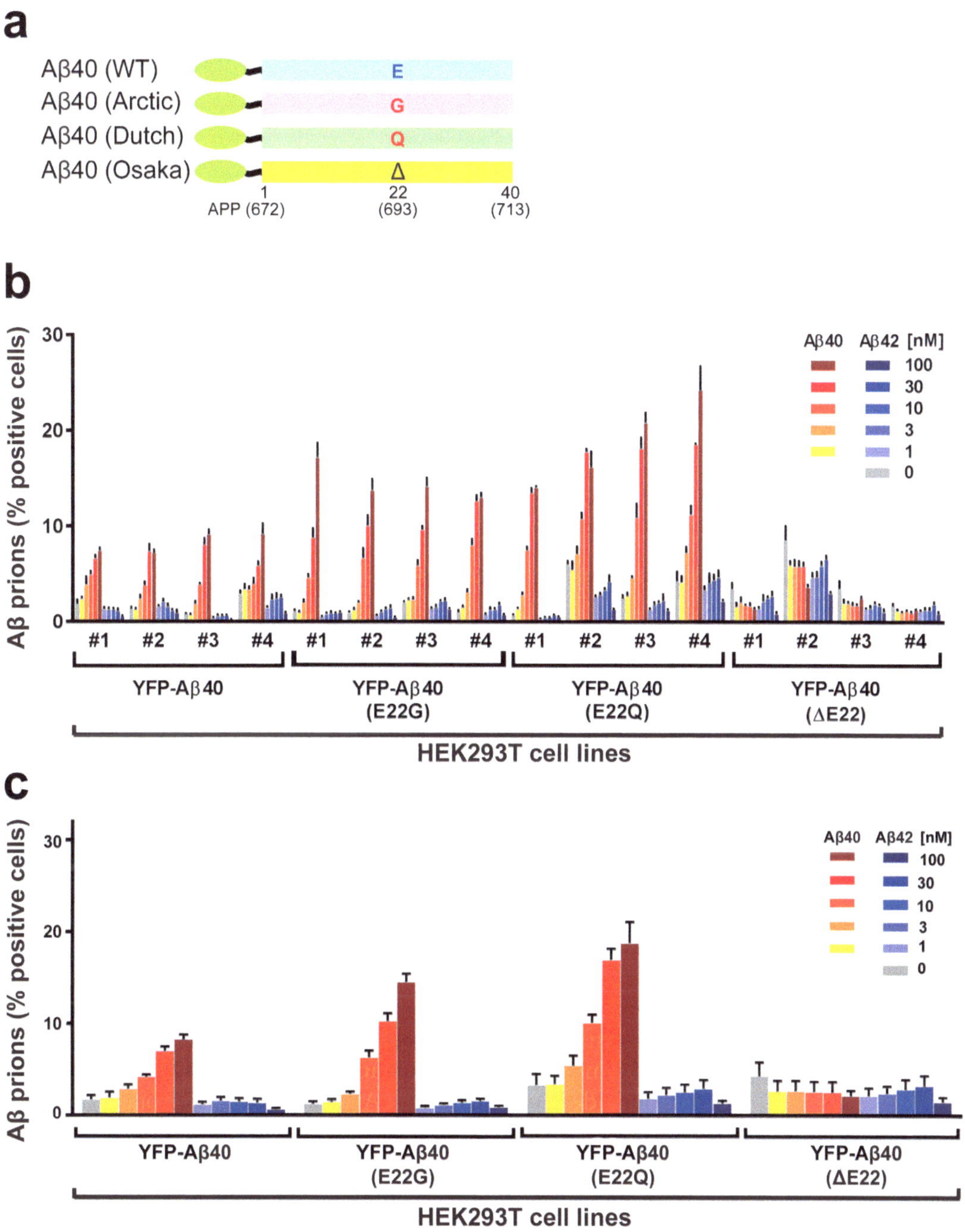

Fig. 3 Synthetic and brain-derived Aβ prions propagate in HEK293T cells expressing familial mutant Aβ fusion proteins. (**a**) Diagram illustrating mutant Aβ40 constructs used in this study. (**b**) Cell lines stably expressing four different Aβ40 constructs fused to yellow fluorescent protein (YFP) at the N-terminus were developed (shown in **a**), and prion loads were compared with both synthetic Aβ40 and Aβ42 prions. Quantification of prions in 16 monoclonal cell lines (four randomly chosen clones from each construct) to increasing concentrations of synthetic Aβ40 or Aβ42 prions (1–100 nM) at day 2 post-exposure. Data shown as mean ± SEM as determined from four images per well in four wells. (**c**) Cell lines stably expressing four different Aβ40 constructs fused to YFP at the N-terminus shown in (**a**) were developed, and prion loads were compared with both synthetic Aβ40 and Aβ42 prions. Quantification of Aβ prions in 16 monoclonal cell lines (four randomly chosen clones from each construct; *see* Fig. 2b) exposed to increasing concentrations of synthetic Aβ40 or Aβ42 prions (1–100 nM) at day 2 after exposure (*see* Fig. 4). Data shown as mean ± SEM as determined from four images per well in four wells

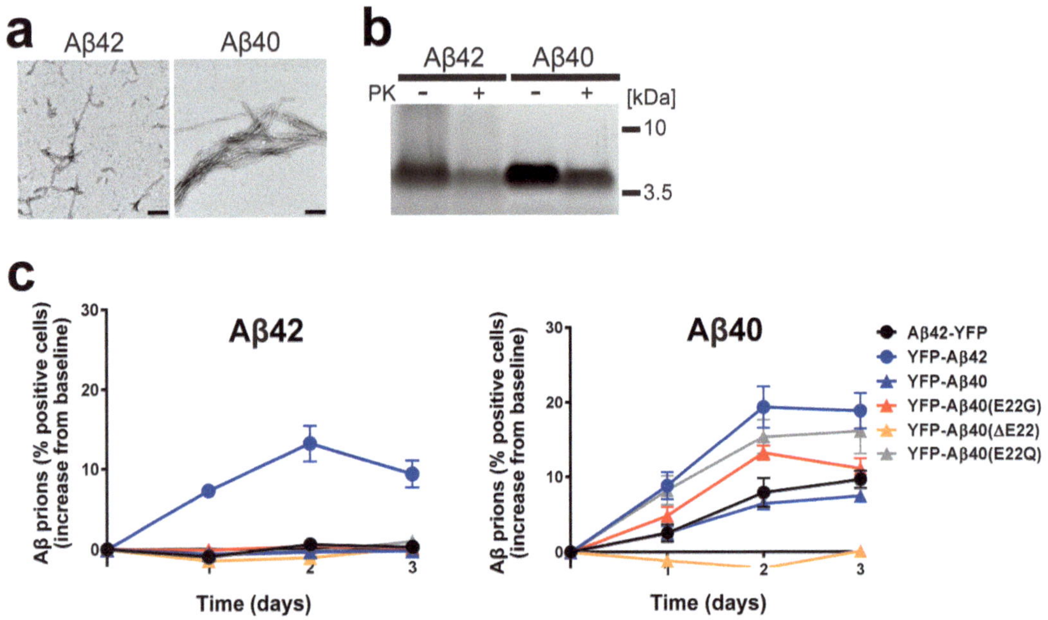

Fig. 4 Synthetic Aβ prion formation and kinetics of propagation in HEK293T cells expressing Aβ fusion proteins. (**a**) Electron micrographs of synthetic Aβ42 and Aβ40 prions. *Scale bar* represents 100 nm. (**b**) Coomassie-stained sodium dodecyl-sulfate polyacrylamide gel electrophoresis (SDS-PAGE) of Aβ42 and Aβ40 amyloid shows a proteinase K (PK)-resistant fraction (PK+). (**c**) Kinetics of prion propagation following exposure to Aβ42 (*left*) or Aβ40 (*right*). Quantification of Aβ prions in five representative clones expressing different constructs every 24 h during incubation with Aβ prions (100 nM). Percentage of cells with prions normalized using PBS-treated controls were shown as the mean ± SEM from the same regions of four wells. Cell lines include Aβ42–yellow fluorescent protein (YFP; *black circle*), YFP–Aβ42 (*blue circle*), YFP–Aβ40 (*blue triangle*), YFP–Aβ40(E22G) (*red triangle*), YFP–Aβ40(ΔE22) (*orange triangle*), and YFP–Aβ40(E22Q) (*gray triangle*)

Next, we examined homogenates of mouse brain–derived Aβ prions extracted using sodium phosphotungstic acid (PTA) [50] from Tg mice expressing human mutant *APP* [51] that had spontaneously developed extensive Aβ deposition. The concentration of total Aβ—determined by enzyme-linked immunosorbent assay (ELISA)—required to induce infection of cells expressing YFP–Aβ42 was approximately tenfold lower than for synthetic fibrils (*see* Fig. 1). Notably, we observed a similar difference in transmission efficiency between synthetic versus Tg mouse brain–derived Aβ prions in recipient mouse models [45]. To demonstrate the specificity of the assay, we also examined prions from mouse models for tau and α-synuclein pathology, *MAPT* [52] and *SNCA* [53], respectively, each of which failed to infect the YFP–Aβ42 cells (*see* Fig. 5). Similarly, cell lines expressing YFP fusions of the corresponding α-synuclein and tau mutations were highly specific for their cognate prions (*see* Fig. 5). These results demonstrate the homophilicity of prion propagation, which is consistent with

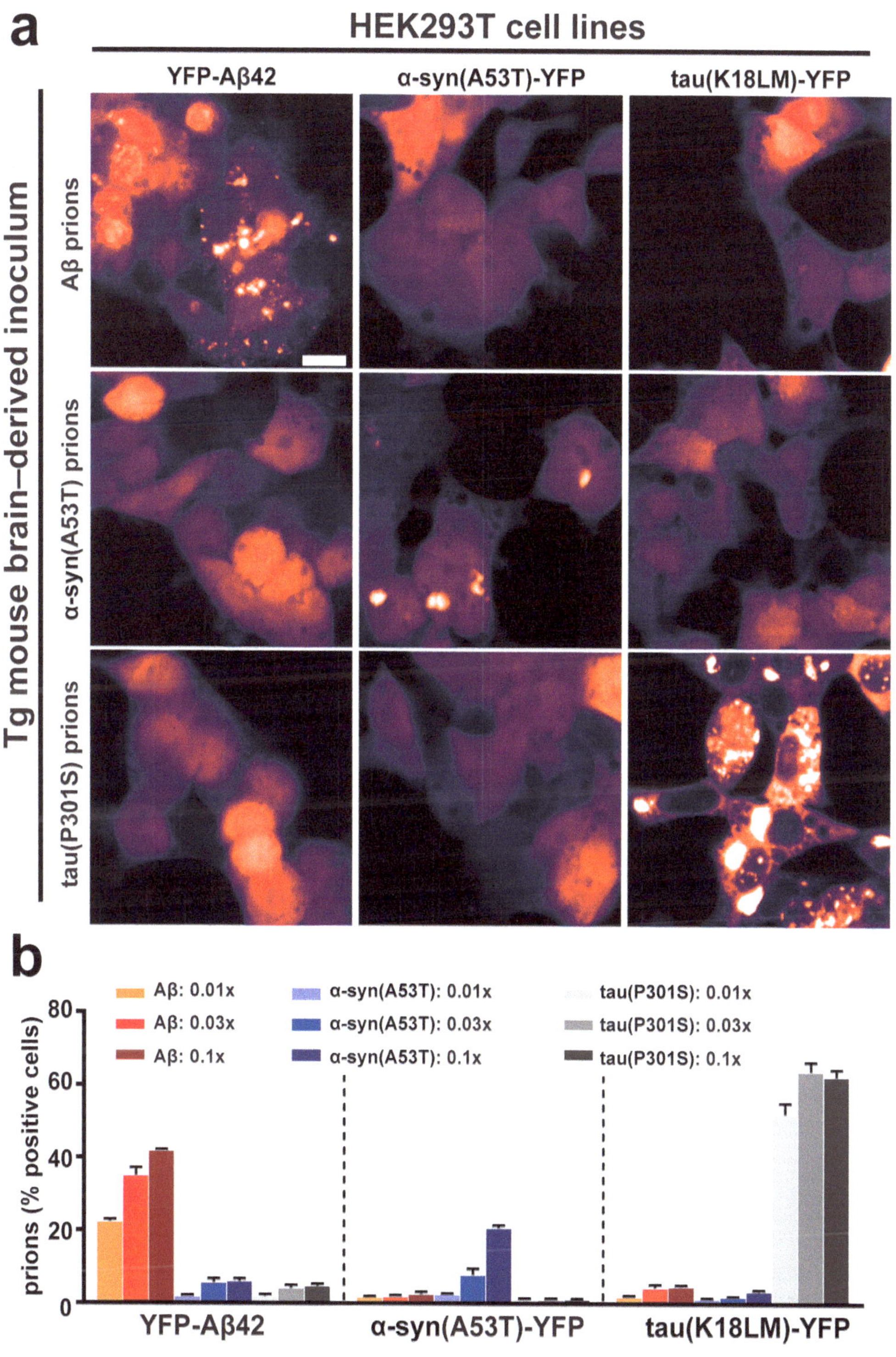

Fig. 5 Specificity of the Aβ cell assay: Aβ, α-synuclein, and tau cell lines display homotypic prion formation. (**a**) Representative confocal images of HEK293T cells stably expressing yellow fluorescent protein (YFP)–Aβ42 (*left column*), α-syn(A53T)–YFP (*middle column*), or tau(K18LM)–YFP (*right column*) treated with brain-derived prions from amyloid precursor protein (APP) Tg(APP23) mice (0.1× phosphotungstic acid [PTA] sample, *top row*), Tg[α-Syn*A53T] mice (0.1× PTA sample, *middle row*), or homozygous Tg[(0N4R)tau*P301S] mice

previous studies using cell culture assays [37, 54]. Importantly, the ability of our YFP–Aβ cell culture system to detect only Aβ prions is a critical attribute given the coexistence of tau prions [36, 38, 55] and the occasional presence of Lewy bodies composed of α-synuclein in the brains of patients with AD [56, 57].

3 Aβ Prions in Tg Mice Increased with Disease Progression

Using our bioassays to measure Aβ prion infectivity, we found that Aβ prion infectivity increases with time and correlates with measures of neuroinflammation and pathology in a well-studied model of AD using mice expressing a human mutant APP transgene [Tg (APP23)]. These Tg mice exhibited progressive Aβ deposition with age (*see* Fig. 6), including glial inflammation, neuronal dysfunction, and cognitive deficits [51]. To determine the time course of Aβ prion accumulation, we collected brain samples over a ~2-year span in a slowly progressing model. Tg(APP23) mice display the first neuropathological changes between 6 and 9 months of age [51], with overt Aβ deposition appearing after ~1 year (390 days) (*see* Fig. 6a). Using our bioassay, we found that the first detectable Aβ prion infectivity appeared between 200 and 300 days of age (*see* Fig. 6c) in parallel with the initial appearance of plaques (*see* Fig. 6a) and astrocytic gliosis measured using a bioluminescent reporter gene driven by the glial fibrillary acidic protein (GFAP) promoter (*see* Fig. 6b) [58]. Infectivity continued to increase as the animals reached ~2 years of age (732 days) (*see* Fig. 6c). Collectively, our bioassay readily detected Aβ prions in Tg mice, consistent with a recent study [59], thus indicating its potential utility for detecting Aβ prions in human AD brain tissue.

4 Correlation of Aβ with Tau Prion Abundance as a Function of Age, Sex, APOE ε4, and Disease Phenotype in Deceased People

Although the molecular basis of prion strains was once enigmatic, a wealth of evidence argues that prion strain–specific information is enciphered in the conformations of prion proteins in fungi and Tg mice [60–62]. The identification of Aβ prion strains in AD brains

Fig. 5 (continued) (0.1× PTA sample, *bottom row*). Only homotypic prions were formed. *Scale bar* represents 20μm. (**b**) Quantification of the responses of YFP–Aβ42, α-syn(A53T)–YFP, and tau(K18LM)–YFP cells to increasing concentrations of Aβ prions (0.01×–0.1× PTA sample, *orange to dark red*), α-syn(A53T) prions (0.01×–0.1× PTA sample, *light blue to dark blue*), and tau(P301S) prions (0.01×–0.1× PTA sample, *light gray to dark gray*) at day 2 after exposure. Data shown as mean ± SEM as determined from four images per well in four wells

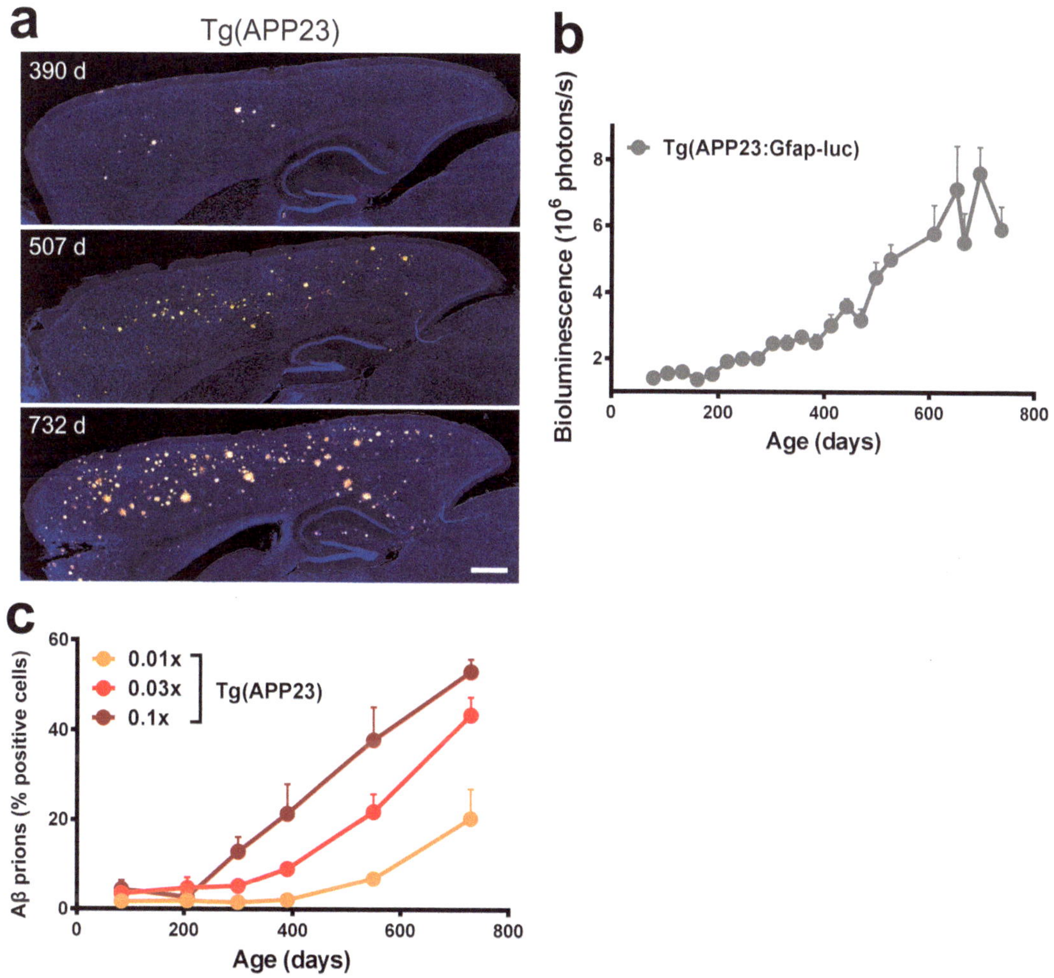

Fig. 6 Aβ amyloid plaque pathology, astrogliosis, and Aβ prion activities during disease progression in Tg (APP23). (**a**) Age-dependent progressive increase in Aβ amyloid plaque pathology in Tg(APP23) mice (*top*, 390 days old; *middle*, 507 days old; *bottom*, 732 days old). Sagittal sections (10μm thick) labeled by propidium iodide (cell nuclei, *blue*) and FSB (amyloid, *orange*). *Scale bar* represents 100μm. (**b**) A spontaneous upregulation of the bioluminescence imaging (BLI) signal in Tg(APP23:*Gfap*-luc) mice was observed in vivo. Data shown as mean ± SEM obtained from six to eight animals per time point. (**c**) Age-dependent progressive increase in prion content as quantified by the activity of YFP–Aβ42 expressing cells to phosphotungstic acid (PTA)-precipitated brain samples (0.01×–0.1× dilution) from Tg(APP23) mice at different ages (84–732 days; 4 animals per time point). Data shown as mean ± SEM at day 2 post prion exposure as determined from four images per well in four wells

has been particularly informative [63–66]. Prior to studies reported here, quantification of infectious Aβ prion measurements required time-consuming and expensive animal studies. Our Aβ prion cellular bioassay in combination with analogous tau assays [36, 38] allowed for comparison of Aβ and tau prion infectivity in numerous human samples. We examined postmortem human specimens from brain banks (*see* Table 1) in which the reported cause of

Table 1
Human brain samples used for analysis

Class	Patient ID#	Sex	Age	NPDX	Mutation	Thal phase	CERAD score	Braak stage	Prion propagation mean ± SEM (%)			APOE genotype	Brain region	Source
									Aβ	tau	syn			
sAD	BB#06-01	F	83	AD		–	C3	V–VI	18.9 ± 0.8	13.9 ± 0.3	1.0 ± 0.3	–	Frontal cortex	UCSF VAMC
sAD	N1415	F	85	AD		1	C2	V–VI	20.3 ± 0.7	10.9 ± 1.3	1.1 ± 0.1	–	Temporal cortex	Uppsala University
sAD	P2217	M	63	AD		5	C3	VI	25.1 ± 1.1	22.8 ± 1.0	1.1 ± 0.3	3/4	Temporal cortex	UCSF NDBB
sAD	P2273	M	63	AD		5	C3	VI	31.4 ± 0.2	24.0 ± 1.6	2.5 ± 0.8	3/3	Temporal cortex	UCSF NDBB
sAD	P2287	F	69	AD		5	C3	VI	21.1 ± 0.3	18.4 ± 0.3	0.7 ± 0.2	3/3	Temporal cortex	UCSF NDBB
sAD	P2312.10	M	59	AD		5	C3	VI	24.4 ± 0.4	26.4 ± 0.9	0.9 ± 0.1	4/4	Temporal cortex	UCSF NDBB
sAD	P2330.11	M	70	AD		5	C3	VI	24.4 ± 1.2	29.3 ± 1.5	0.7 ± 0.2	3/4	Temporal cortex	UCSF NDBB
sAD	P2350.11	F	65	AD		5	C3	VI	17.8 ± 0.4	10.8 ± 0.9	0.5 ± 0.1	2/3	Temporal cortex	UCSF NDBB
sAD	BBN_6072	M	78	AD, CAA, SVD		5	C2	III	19.4 ± 1.1	21.4 ± 2.0	0.8 ± 0.1	3/4	Temporal cortex	MRC UK (Manchester)
sAD	BBN_3416	M	87	AD, CAA, SVD		4	C2	IV	14.9 ± 0.4	6.2 ± 0.4	0.8 ± 0.1	3/4	Temporal cortex	MRC UK (Manchester)
sAD	BBN_15592	M	62	AD, LBD		5	C3	–	17.4 ± 1.8	8.4 ± 0.7	1.2 ± 0.04	3/4	Temporal cortex	MRC UK (Manchester)
sAD	20242958	F	88	AD		3	C3	V	12.8 ± 0.8	1.6 ± 0.1	0.7 ± 0.2	3/3	Frontal cortex	Rush University
sAD	20611993	F	90	AD		–	C3	V	10.0 ± 0.6	0.7 ± 0.2	0.9 ± 0.1	3/3	Frontal cortex	Rush University

sAD	21123474	F	98	AD	–	C3	IV	19.7 ± 1.0	12.4 ± 5.1	2.2 ± 0.4	2/3	Frontal cortex	Rush University
sAD	20160019	F	94	AD	5	C3	V	18.5 ± 0.3	7.2 ± 0.4	1.8 ± 0.3	3/4	Frontal cortex	Rush University
sAD	20185860	F	95	AD	–	C3	IV	17.2 ± 0.8	0.7 ± 0.1	1.4 ± 0.1	3/3	Frontal cortex	Rush University
sAD	20260284	F	99	AD	–	C3	V	7.8 ± 0.8	6.0 ± 0.6	0.5 ± 0.02	3/3	Frontal cortex	Rush University
sAD	20964295	F	86	AD	5	C3	V	20.0 ± 1.1	6.3 ± 0.9	0.7 ± 0.2	3/4	Frontal cortex	Rush University
sAD	20207437	F	87	AD	3	C3	VI	11.5 ± 0.4	7.6 ± 0.7	0.7 ± 0.3	3/3	Frontal cortex	Rush University
sAD	20208992	F	95	AD	5	C3	V	9.8 ± 0.5	1.6 ± 0.4	0.7 ± 0.2	3/3	Frontal cortex	Rush University
sAD	20242958	F	88	AD	3	C3	V	12.1 ± 0.7	4.1 ± 0.5	0.7 ± 0.2	3/3	Temporal cortex	Rush University
sAD	20611993	F	90	AD	–	C3	V	18.1 ± 0.7	6.7 ± 1.6	0.5 ± 0.1	3/3	Temporal cortex	Rush University
sAD	21123474	F	98	AD	–	C3	IV	15.7 ± 0.7	18.0 ± 0.6	0.7 ± 0.1	2/3	Temporal cortex	Rush University
sAD	20160019	F	94	AD	5	C3	V	13.6 ± 0.5	16.2 ± 0.4	0.7 ± 0.02	3/4	Temporal cortex	Rush University
sAD	20185860	F	95	AD	–	C3	IV	10.0 ± 0.6	2.9 ± 0.3	1.8 ± 0.3	3/3	Temporal cortex	Rush University
sAD	20260284	F	99	AD	–	C3	V	10.7 ± 0.3	2.9 ± 1.0	1.4 ± 0.2	3/3	Temporal cortex	Rush University
sAD	20964295	F	86	AD	5	C3	V	10.6 ± 0.6	9.0 ± 0.7	1.5 ± 0.3	3/4	Temporal cortex	Rush University
sAD	20204538	F	88	AD	–	C3	V	12.4 ± 0.4	6.9 ± 0.6	0.7 ± 0.1	3/3	Temporal cortex	Rush University

(continued)

Table 1 (continued)

Class	Patient ID#	Sex	Age	NPDX	Mutation	Thal phase	CERAD score	Braak stage	Prion propagation mean ± SEM (%)			APOE genotype	Brain region	Source
									Aβ	tau	syn			
sAD	20207437	F	87	AD		3	C3	VI	14.2 ± 0.7	22.1 ± 1.1	1.1 ± 0.1	3/3	Temporal cortex	Rush University
sAD	20208992	F	95	AD		5	C3	V	15.3 ± 0.6	14.5 ± 1.9	0.7 ± 0.02	3/3	Temporal cortex	Rush University
sAD	20946545	F	97	AD		1	C3	V	19.0 ± 0.8	24.3 ± 3.2	1.0 ± 0.1	3/4	Temporal cortex	Rush University
sAD	12-1181	M	62	AD		5	C3	VI	36.4 ± 1.0	27.0 ± 1.4	0.9 ± 0.1	–	Frontal cortex	University of Washington (Seattle)
sAD	07-0997	F	87	AD		5	C3	VI	25.4 ± 2.3	33.4 ± 2.3	0.6 ± 0.1	3/3	Occipital cortex	University of Washington (Seattle)
sAD	2412.12	M	88	AD		5	C3	VI	31.6 ± 0.9	26.0 ± 2.0	1.7 ± 0.7	3/4	Parietal cortex	UCSF NDBB
sAD	2376.11	M	80	AD		4	C3	VI	17.2 ± 0.6	18.4 ± 0.4	0.9 ± 0.1	3/3	Parietal cortex	UCSF NDBB
sAD	BB#06-04	M	62	AD		–	C3	V–VI	45.4 ± 2.3	28.0 ± 1.3	0.7 ± 0.1	–	Frontal cortex	UCSF VAMC
sAD	P2478	M	78	AD		4	C3	V	18.3 ± 0.7	23.5 ± 1.4	0.8 ± 0.1	–	Temporal cortex	UCSF NDBB
fAD APP	0276-05	F	61	AD, CAA	KM670/671NL APP (Swedish)	3	C3	V–VI	25.7 ± 0.8	28.4 ± 1.2	1.1 ± 0.4	3/4	Temporal cortex	Uppsala Unversity (Sweden)
fAD APP	6901	M	62	AD	KM670/671NL APP (Swedish)	–	C3	VI	30.0 ± 1.4	25.3 ± 0.7	0.9 ± 0.1	3/3	Frontal cortex	Karolinska Institutet (Sweden)
fAD APP	493	M	68	AD	KM670/671NL APP (Swedish)	–	C3	–	27.3 ± 1.2	16.0 ± 1.4	0.7 ± 0.1	2/3	Parietal cortex	Karolinska Institutet (Sweden)

fAD APP	39794	M	66	AD	KM670/671NL APP (Swedish)	–	C3	VI	15.9 ± 0.5	13.7 ± 1.9	0.7 ± 0.1	2/3	Frontal cortex	Karolinska Institutet (Sweden)
fAD APP	7795	M	56	AD	KM670/671NL APP (Swedish)	–	C3	VI	25.7 ± 0.8	21.8 ± 0.9	0.9 ± 0.2	4/4	Frontal cortex	Karolinska Institutet (Sweden)
fAD APP	14096	F	62	AD	KM670/671NL APP (Swedish)	–	C3	VI	16.9 ± 0.3	17.0 ± 2.1	0.9 ± 0.1	3/3	Frontal cortex	Karolinska Institutet (Sweden)
fAD APP	BBN_13890	M	61	AD	V717I APP (London)	5	C3	VI	16.3 ± 0.6	14.1 ± 0.6	0.9 ± 0.1	3/3	Temporal cortex	MRC UK (London)
fAD APP	BBN_9608	F	62	AD	V717I APP (London)	4	C3	VI	22.9 ± 1.2	22.2 ± 1.5	0.6 ± 0.1	3/4	Temporal cortex	MRC UK (London)
fAD APP	BBN_13861	F	69	AD	V717G APP	5	C3	VI	18.4 ± 1.7	19.0 ± 0.8	1.1 ± 0.1	3/3	Temporal cortex	MRC UK (London)
fAD APP	BBN_9637	F	72	AD	V717I APP (London)	–	C3	VI	17.3 ± 1.4	17.2 ± 1.5	2.7 ± 0.3	3/3	Temporal cortex	MRC UK (London)
fAD APP	6214	M	55	AD	V717I APP (London)	–	C3	V–VI	23.2 ± 1.7	18.1 ± 1.6	0.6 ± 0.1	3/3	Frontal cortex	Karolinska Institutet (Sweden)
fAD APP	N1084	M	64	AD	E693G APP (Arctic)	5	–	V–VI	21.5 ± 0.5	16.6 ± 1.1	0.6 ± 0.6	3/3	Temporal cortex	Uppsala University
fAD APP	7612	F	66	AD	E693G APP (Arctic)	–	–	VI	27.3 ± 0.5	16.3 ± 1.5	0.6 ± 0.1	3/3	Frontal cortex	Karolinska Institutet (Sweden)
fAD PSEN	BBN_13829	M	42	AD	Δ4 PSEN1	4	C3	VI	22.4 ± 1.0	18.0 ± 0.9	6.6 ± 1.1	3/3	Temporal cortex	MRC UK (London)
fAD PSEN	BBN_3246	F	57	AD	M139V PSEN1	5	C3	VI	18.3 ± 0.9	13.8 ± 0.5	0.7 ± 0.2	3/3	Temporal cortex	MRC UK (Manchester)
fAD PSEN	3206	M	45	AD	I143T PSEN1	–	C3	VI	22.4 ± 0.7	19.0 ± 1.1	0.8 ± 0.1	3/3	Frontal cortex	Karolinska Institutet (Sweden)
fAD PSEN	12908	F	43	AD	I143T PSEN1	–	C3	VI	25.6 ± 0.6	25.0 ± 1.5	0.8 ± 0.1	3/4	Frontal cortex	Karolinska Institutet (Sweden)
fAD PSEN	00-0700	M	37	AD	I143T PSEN1	5	C3	VI	19.8 ± 1.5	24.7 ± 1.9	0.8 ± 0.04	3/3	Frontal cortex	University of Washington (Seattle)

(continued)

Table 1 (continued)

Class	Patient ID#	Sex	Age	NPDX	Mutation	Thal phase	CERAD score	Braak stage	Prion propagation mean ± SEM (%)			APOE genotype	Brain region	Source
									Aβ	tau	syn			
fAD PSEN	4602	F	50	AD	I143T PSEN1	–	C3	VI	29.8 ± 1.2	24.8 ± 2.3	0.9 ± 0.05	3/4	Frontal cortex	Karolinska Institutet (Sweden)
fAD PSEN	2598	F	48	AD	I143T PSEN1	–	C3	VI	22.1 ± 0.7	28.8 ± 2.3	0.7 ± 0.1	3/4	Frontal cortex	Karolinska Institutet (Sweden)
fAD PSEN	1104	M	68	AD	H163Y PSEN1	–	C3	V–VI	25.4 ± 1.4	20.7 ± 1.1	0.7 ± 0.2	3/4	Frontal cortex	Karolinska Institutet (Sweden)
fAD PSEN	8715	M	63	AD	H163Y PSEN1	–	C3	V–VI	11.9 ± 0.6	15.1 ± 1.5	0.8 ± 0.2	2/4	Frontal cortex	Karolinska Institutet (Sweden)
fAD PSEN	12829	M	45	AD	G209V PSEN1	5	C3	VI	16.4 ± 0.6	19.1 ± 1.3	0.9 ± 0.1	3/3	Frontal cortex	University of Washington (Seattle)
fAD PSEN	25044	F	51	AD	G209V PSEN1	5	C3	VI	13.5 ± 0.6	15.9 ± 0.4	1.1 ± 0.2	2/3	Frontal cortex	University of Washington (Seattle)
fAD PSEN	01-0164	M	58	AD	G209V PSEN1	5	C3	V	13.3 ± 0.6	13.3 ± 1.4	1.2 ± 0.1	2/3	Frontal cortex	University of Washington (Seattle)
fAD PSEN	25212	F	65	AD	A260V PSEN1	5	C3	V	18.3 ± 1.0	7.6 ± 0.6	1.0 ± 0.2	3/3	Frontal cortex	University of Washington (Seattle)
fAD PSEN	02-0621	F	44	AD	A260V PSEN1	5	C3	VI	22.7 ± 1.2	27.1 ± 0.7	0.6 ± 0.2	3/3	Frontal cortex	University of Washington (Seattle)
fAD PSEN	BBN_13813	F	65	AD	E280G PSEN1	5	C3	VI	12.1 ± 0.7	16.5 ± 1.8	1.2 ± 0.3	3/4	Temporal cortex	MRC UK (London)

fAD PSEN	P2197	M	51	AD	A431E PSEN1	5	C3	VI	33.4 ± 2.8	17.8 ± 1.5	0.5 ± 0.05	2/3	Temporal cortex	UCSF NDBB
fAD PSEN	00-0281	M	61	AD	A431E PSEN1	5	C3	VI	17.9 ± 1.4	14.6 ± 0.8	0.5 ± 0.1	3/3	Frontal cortex	University of Washington (Seattle)
fAD PSEN	940	M	55	AD	L435F PSEN1	–	C3	V	23.3 ± 1.3	14.0 ± 1.3	0.6 ± 0.1	–	Frontal cortex	MGH Harvard
fAD PSEN	935	F	59	AD	L435F PSEN1	–	C3	V	23.5 ± 0.7	10.7 ± 1.2	0.7 ± 0.05	–	Frontal cortex	MGH Harvard
fAD PSEN	1831	M	51	AD	L435F PSEN1	–	C3	VI	22.2 ± 1.6	22.4 ± 2.6	0.8 ± 0.1	–	Frontal cortex	MGH Harvard
fAD PSEN	12540	M	72	AD	N141I PSEN2 (Volga German)	5	C3	VI	13.0 ± 0.7	15.3 ± 1.4	0.8 ± 0.2	3/4	Frontal cortex	University of Washington (Seattle)
fAD PSEN	23156	F	78	AD	N141I PSEN2 (Volga German)	5	C3	V	16.8 ± 0.4	5.0 ± 1.8	1.3 ± 0.1	2/3	Frontal cortex	University of Washington (Seattle)
fAD PSEN	26111	F	53	AD	N141I PSEN2 (Volga German)	5	C3	V	12.2 ± 0.8	16.8 ± 1.7	1.0 ± 0.2	3/4	Frontal cortex	University of Washington (Seattle)
fAD PSEN	11-0929	M	58	AD	N141I PSEN2 (Volga German)	5	C3	VI	15.4 ± 0.6	12.6 ± 2.0	0.7 ± 0.1	3/3	Frontal cortex	University of Washington (Seattle)
fAD PSEN	12-0370	M	73	AD	N141I PSEN2 (Volga German)	5	C3	VI	13.3 ± 0.9	9.5 ± 0.9	2.0 ± 0.4	–	Frontal cortex	University of Washington (Seattle)
sCAA	BBN_24306	F	83	CAA		–	–	III	23.2 ± 1.6	15.6 ± 2.8	1.1 ± 0.1	–	Occipital cortex	MRC UK (Edinburgh)
sCAA	BBN_24302	F	90	CAA		–	–	II	18.7 ± 0.9	5.8 ± 0.9	1.0 ± 0.3	–	Occipital cortex	MRC UK (Edinburgh)
sCAA	BBN_1082	F	92	CAA		–	C1	I–II	19.0 ± 1.0	4.9 ± 1.5	0.8 ± 0.2	3/4	Temporal cortex	MRC UK (London)

(continued)

Table 1
(continued)

Class	Patient ID#	Sex	Age	NPDX	Mutation	Thal phase	CERAD score	Braak stage	Prion propagation mean ± SEM (%)			APOE genotype	Brain region	Source
									Aβ	tau	syn			
fCAA	BBN_3310	M	63	AD	A692G APP (Flemish)	5	C3	IV	23.5 ± 0.2	27.4 ± 1.6	0.8 ± 0.2	3/4	Temporal cortex	MRC UK (Manchester)
fCAA	NP2043	M	60	AD	A692G APP (Flemish)	–	C3	IV	30.7 ± 1.7	15.7 ± 1.4	0.9 ± 0.3	3/3	Occipital cortex	Sydney Brain Bank (Australia)
fCAA	E14-69-4J	F	52	CAA	E693Q APP (Dutch)	–	C0	0	14.0 ± 0.8	0.4 ± 0.03	0.5 ± 0.1	–	Frontal cortex	Leiden Unversity (Netherlands)
fCAA	E13-99-4B	M	59	CAA	E693Q APP (Dutch)	–	C0	0	15.7 ± 1.0	1.4 ± 0.9	0.8 ± 0.3	–	Frontal cortex	Leiden Unversity (Netherlands)
fCAA	E06-0004	M	57	CAA	E693Q APP (Dutch)	–	C0	0	13.6 ± 0.8	0.5 ± 0.1	0.9 ± 0.1	–	Frontal cortex	Leiden Unversity (Netherlands)
fCAA	E12-82-4D	F	55	CAA	E693Q APP (Dutch)	–	C0	0	24.9 ± 0.4	0.4 ± 0.04	0.7 ± 0.2	–	Frontal cortex	Leiden Unversity (Netherlands)
fCAA	1193	M	66	CAA	D694N APP (Iowa)	–	C0	0	16.6 ± 0.5	3.2 ± 0.5	0.8 ± 0.1	–	Frontal cortex	MGH Harvard (Boston)
fCAA	1569	M	45	CAA	D694N APP (Iowa)	–	C0	0	18.6 ± 0.3	2.5 ± 0.2	0.8 ± 0.1	–	Frontal cortex	MGH Harvard (Boston)
PSP	P2326	M	66	PSP		0	C0	0	4.8 ± 0.3	35.6 ± 0.5	1.5 ± 0.5	3/3	Frontal cortex	UCSF NDBB
PSP	P2326	M	66	PSP		0	C0	0	4.7 ± 0.6	25.3 ± 1.1	0.9 ± 0.3	3/3	Occipital cortex	UCSF NDBB
PSP	P2113	F	80	PSP		1	C0	III–IV	5.4 ± 0.4	39.0 ± 0.9	0.8 ± 0.04	3/3	Frontal cortex	UCSF NDBB
PSP	P2397.12	F	84	PSP		0	C0	II	5.5 ± 0.3	28.4 ± 1.9	0.8 ± 0.1	2/2	Parietal cortex	UCSF NDBB

PSP	P2418.12	F	78	PSP	0	C0	I	5.4 ± 0.3	36.0 ± 1.1	0.9 ± 0.2	2/2	Parietal cortex	UCSF NDBB
PSP	P2461.13	F	73	PSP	3	C0	0	7.4 ± 0.4	30.6 ± 1.4	0.9 ± 0.1	2/3	Frontal cortex	UCSF NDBB
PSP	P2461.13	F	73	PSP	3	C0	0	5.6 ± 0.5	11.9 ± 0.2	0.7 ± 0.1	2/3	Parietal cortex	UCSF NDBB
CBD	P2431.12	M	71	CBD	0	C0	0	4.6 ± 0.3	20.5 ± 1.9	0.6 ± 0.1	2/3	Parietal cortex	UCSF NDBB
CBD	P2444.12	F	70	CBD	2	C0	III	5.3 ± 0.6	23.9 ± 0.4	0.8 ± 0.2	3/3	Parietal cortex	UCSF NDBB
CBD	P1838	F	80	CBD	–	–	–	5.1 ± 0.2	27.1 ± 2.0	0.7 ± 0.1	–	Parietal cortex	UCSF NDBB
MSA	NP22365	M	60	MSA-P	0	C0	0	4.1 ± 0.1	0.6 ± 0.2	21.3 ± 1.6	–	Pons	Sydney Brain Bank (Australia)
MSA	NP22494	M	82	MSA-P	0	C0	0	5.2 ± 0.8	0.2 ± 0.04	6.7 ± 0.3	–	Pons	Sydney Brain Bank (Australia)
MSA	NP22514	M	64	MSA-P	0	C0	0	5.4 ± 0.3	0.3 ± 0.1	17.9 ± 2.0	–	Pons	Sydney Brain Bank (Australia)
MSA	NP22527	M	61	MSA-P	1	C1	0	5.8 ± 0.4	0.5 ± 0.1	8.1 ± 0.9	–	Pons	Sydney Brain Bank (Australia)
MSA	P43/12	F	67	MSA	–	–	–	7.1 ± 0.7	1.3 ± 0.2	22.6 ± 2.2	–	Basal ganglia	University College London
MSA	P91/10	M	78	MSA	–	–	–	6.0 ± 0.5	2.7 ± 0.2	23.9 ± 0.7	–	Basal ganglia	University College London
MSA	P40/12	F	59	MSA	–	–	–	5.7 ± 0.3	3.9 ± 0.8	28.9 ± 2.0	–	Midbrain	University College London
Control	#23	F	88	NCI	0	C0	II	5.3 ± 0.4	1.1 ± 0.4	0.9 ± 0.1	–	Frontal cortex	Uppsala University
Control	#24	M	63	NCI	0	C0	0	5.4 ± 0.4	0.8 ± 0.1	0.8 ± 0.1	–	Frontal cortex	Uppsala University

(continued)

Table 1 (continued)

Class	Patient ID#	Sex	Age	NPDX	Mutation	Thal phase	CERAD score	Braak stage	Prion propagation mean ± SEM (%) Aβ	tau	syn	APOE genotype	Brain region	Source
Control	HCTMI_15_001	M	62	NCI		–	–	–	6.3 ± 0.3	0.9 ± 0.3	1.0 ± 0.2	–	Putamen	U. of Miami Brain Bank
Control	HCTQJ_15_0002	F	67	NCI		–	–	–	6.3 ± 0.6	1.0 ± 0.2	0.7 ± 0.1	–	Putamen	U. of Miami Brain Bank
Control	HCTNU_15_0008	M	59	NCI		–	–	–	4.9 ± 0.6	1.1 ± 0.2	1.0 ± 0.1	–	Putamen	U. of Miami Brain Bank
Control	HCTLT_15_0009	M	65	NCI		–	–	–	5.0 ± 0.3	0.4 ± 0.1	0.8 ± 0.2	–	Putamen	U. of Miami Brain Bank
Control	P78/06	F	68	NCI		–	–	–	5.4 ± 0.2	0.3 ± 0.1	0.6 ± 0.2	–	Frontal cortex	University College London
Control	P2207	F	81	NCI		1	C0	II	4.3 ± 0.7	0.2 ± 0.04	0.7 ± 0.1	3/3	Temporal cortex	UCSF NDBB
Control	PDC023	–	–	NCI		–	–	–	5.2 ± 0.4	0.4 ± 0.1	1.0 ± 0.1	–	Frontal cortex	UKPD Brain Bank
Control	P2594	F	86	NCI		0	C0	II	6.1 ± 0.7	0.7 ± 0.1	0.9 ± 0.2	–	Temporal cortex	UCSF NDBB

Abbreviations: *AD* Alzheimer's disease, *CAA* cerebral amyloid angiopathy, *CBD* corticobasal degeneration, *CERAD* Consortium to Establish a Registry for Alzheimer's Disease, *CVD* cerebrovascular disease, *DLB* dementia with Lewy bodies, *LBD* Lewy body dementia, *MGH* Massachusetts General Hospital, *MSA* multiple system atrophy, *MRC UK* Medical Research Council United Kingdom, *NCI* no cognitive impairment, *NPDX* neuropathological diagnosis, *PSP* progressive supranuclear palsy, *SVD* small vessel disease, *UCSF NDBB*, University of California, San Francisco, Neurodegenerative Disease Brain Bank, *UKPD* United Kingdom Parkinson's disease. (−) denotes missing information for some samples where NPDX scores were not available at the time of autopsy or could not be determined for this study because the samples were exhausted. Only samples with Braak stage V and higher were used for data presented in Figs. 7, 8, 9, 10, 11, 12, and 13 death included a spectrum of diseases ranging from "tau-only" dementias (e.g., frontotemporal lobar degenerations with tau-immunoreactive inclusions [FTLD-tau]) to Aβ-centric familial cerebral amyloid angiopathies (fCAA). More specifically, we examined the levels of infectious prions in (1) 2 types of FTLD-tau samples (7 cases of progressive supranuclear palsy [PSP] and 3 cases of corticobasal degeneration [CBD]) in addition to 10 aged controls; (2) 37 sporadic AD samples; (3) 47 samples of familial AD or fCAA-bearing disease-causing mutations in *APP*, *PSEN1*, or *PSEN2*; and (4) 3 samples of sporadic CAA. In each case, we used PTA to nonspecifically enrich all prions from a given brain homogenate, and we then carried out bioassays for Aβ, tau, and α-synuclein prions using YFP–Aβ42, tau(K18LM)–YFP, or α-syn(A53T)–YFP cells (*see* Fig. 7).

Aβ prions were detected in all of the AD and CAA samples (*see* Fig. 7c, d), arguing that the YFP–Aβ42 bioassay is sensitive to different Aβ prion strains [63, 66]. In each case, the abundance of Aβ prions was at least 20 standard errors above the mean for the control samples, indicating that infectious Aβ prions remained active at the time of death and could be preserved by freezing brains from patients with each of these diseases. Additionally, all AD and CAA samples tested were devoid of α-synuclein prions as compared with samples from eight multiple system atrophy (MSA) cases (*see* Fig. 7e), further demonstrating the specificity of our bioassays. Although tau prions were detectable in all FTLD-tau brains, no Aβ prions were found in these specimens (*see* Fig. 7b); our findings are consistent with the classification of PSP and CBD as primary tauopathies.

In fCAA cases, Aβ prions were dominant and tau prions were either low or absent. In Dutch (E22Q in Aβ or E693Q in APP) or Iowa (D22N) mutants (*see* Fig. 7d), we found high amounts of Aβ prions and insignificant tau prion load in these patients. Our findings are consistent with the absence of NFTs in the brains of these individuals [22, 23]. Interestingly, both Aβ and tau prions were detected in two patients carrying the Flemish (A21G) mutation (*see* Fig. 7d). This finding confirms earlier reports of widespread NFTs and hyperphosphorylated tau observed in dystrophic neurites in such patients [67, 68], similar to the samples we examined histologically. The two patients carrying the Flemish mutation (*see* Fig. 7d) were older at death (63 and 60 years old) than the corresponding Iowa and Dutch patients (55.7 ± 2.9, standard error).

A plot of Aβ versus tau prion infectivity is instructive when considered in relation to disease (*see* Fig. 8). The FTLD-tau cases lie near the *y*-axis in a well-separated region at a very low Aβ and, for these cases, there is no correlation between the titer of the two types of prions. The Dutch and Iowa fCAA cases lie near the *x*-axis at a very low tau prion infectivity, which is not significantly different

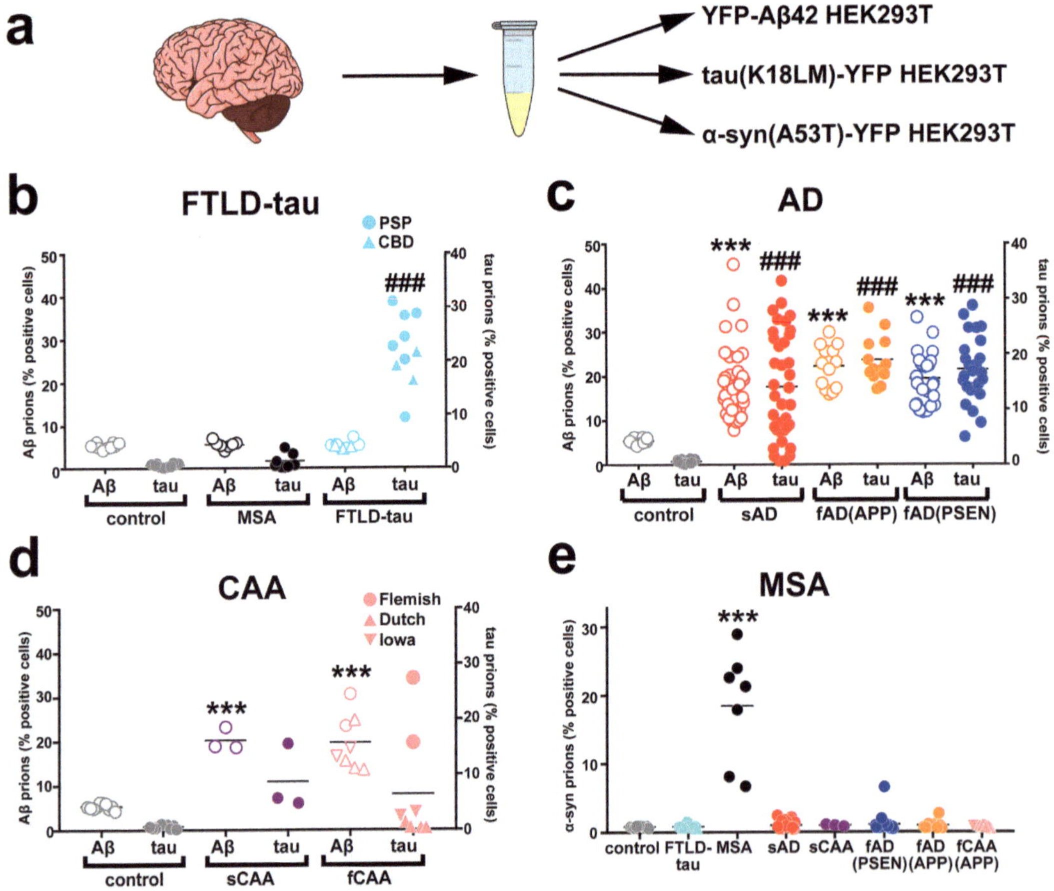

Fig. 7 Quantitative comparison of human brain–derived Aβ prions and tau prions in parallel cellular assays. (**a**) Schematic diagram showing treatment of the same brain samples with three different cell reporter systems detecting Aβ, tau, and α-synuclein prions, respectively. (**b–d**) Aβ prion and tau prion abundance as quantified by percent of cells expressing yellow fluorescent protein (YFP)–Aβ42 or tau(K18LM)–YFP. Cells designated prion positive after exposure to a 0.03× dilution of phosphotungstic acid (PTA)-precipitated brain homogenates from 86 patient samples of sporadic and familial Alzheimer's disease (AD) and cerebral amyloid angiopathy (CAA), 10 aged controls, and 10 frontotemporal lobar degenerations with tau-immunoreactive inclusion (FTLD-tau; 7 progressive supranuclear palsy [PSP] and 3 corticobasal degeneration [CBD]) cases. Data shown as mean ± SEM as determined from four images per well in four wells per sample. Statistical significance indicated as *** ($P < 0.0001$) for Aβ prion activity compared to control, or ### ($P < 0.0001$) for tau prion abundance compared to control. (**e**) α-Synuclein prion abundance as quantified by percent of α-syn (A53T)–YFP expressing cells that were prion positive after exposure to a 0.03× dilution of PTA-precipitated brain homogenates from all patient samples. Data shown as mean ± SEM as determined from four images per well in four wells per sample

from the control samples. The remaining cases are broadly distributed along the diagonal with a modest but significant ($P < 0.0001$) linear correlation. This plot should be useful for typing the molecular origins of diverse disease types and illustrates the coexistence and correlation between Aβ and tau prions.

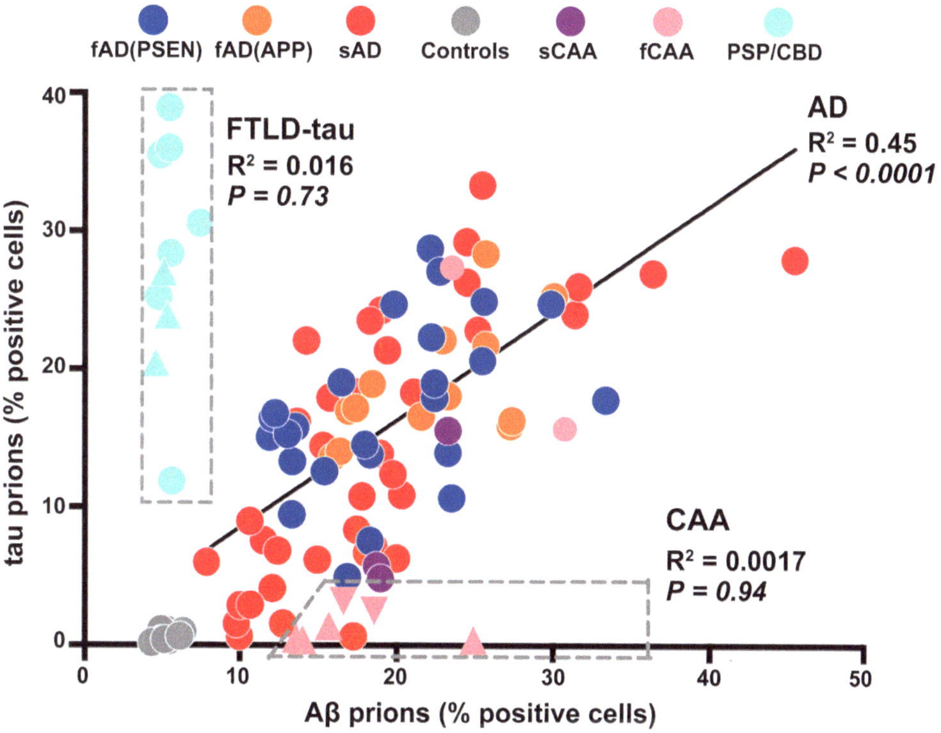

Fig. 8 Correlation analyses between Aβ prion load (*x*-axis) and tau prion load (*y*-axis): summary plot from 75 Alzheimer's disease (AD) cases (37 sporadic AD [sAD], *red*; 25 fAD [PSEN], *blue*; 13 fAD [APP], *orange*) compared with aged controls (*gray*), and cerebral amyloid angiopathy (CAA; sCAA, *purple*; fCAA, *pink*) and frontotemporal lobar degenerations with tau-immunoreactive inclusions (FTLD-tau; progressive supranuclear palsy [PSP]/corticobasal degeneration [CBD], *cyan*) samples. One sAD patient lacked a Braak stage score and had comorbid Lewy body dementia, and two sAD patients had Braak stage III/IV, but all three data points fell well within the range of other close-lying points and thus were not removed

We next asked whether the extent of Aβ and tau prion infectivity observed at the time of death correlated with variables relating to the genetic background of the donor or sample collection. While no significant correlation was observed with respect to variables such as brain bank and preservation method, significant correlations were observed with the longevity of the donor, sex, and genetic background for AD patients with confirmed neuropathology (CERAD neuritic plaque, C3; Braak stage V to VI). All samples, including familial AD patients, had Aβ prion infectivity well above those of healthy controls at the time of death (*see* Fig. 9a, b). The Aβ prion load decreased modestly in patients with the greatest longevity (*see* Fig. 9a, b). The tau prion infectivity showed a similar linear decrease with age (*see* Fig. 9c). Most interestingly, by age 80, approximately half of the patients exhibited low tau prion infectivity, despite having reached Braak stage V or VI and a CERAD score of 3 (*see* Fig. 9d). A preponderance of these elderly AD cases was female (*see* Fig. 10a, b), although we cannot rule out

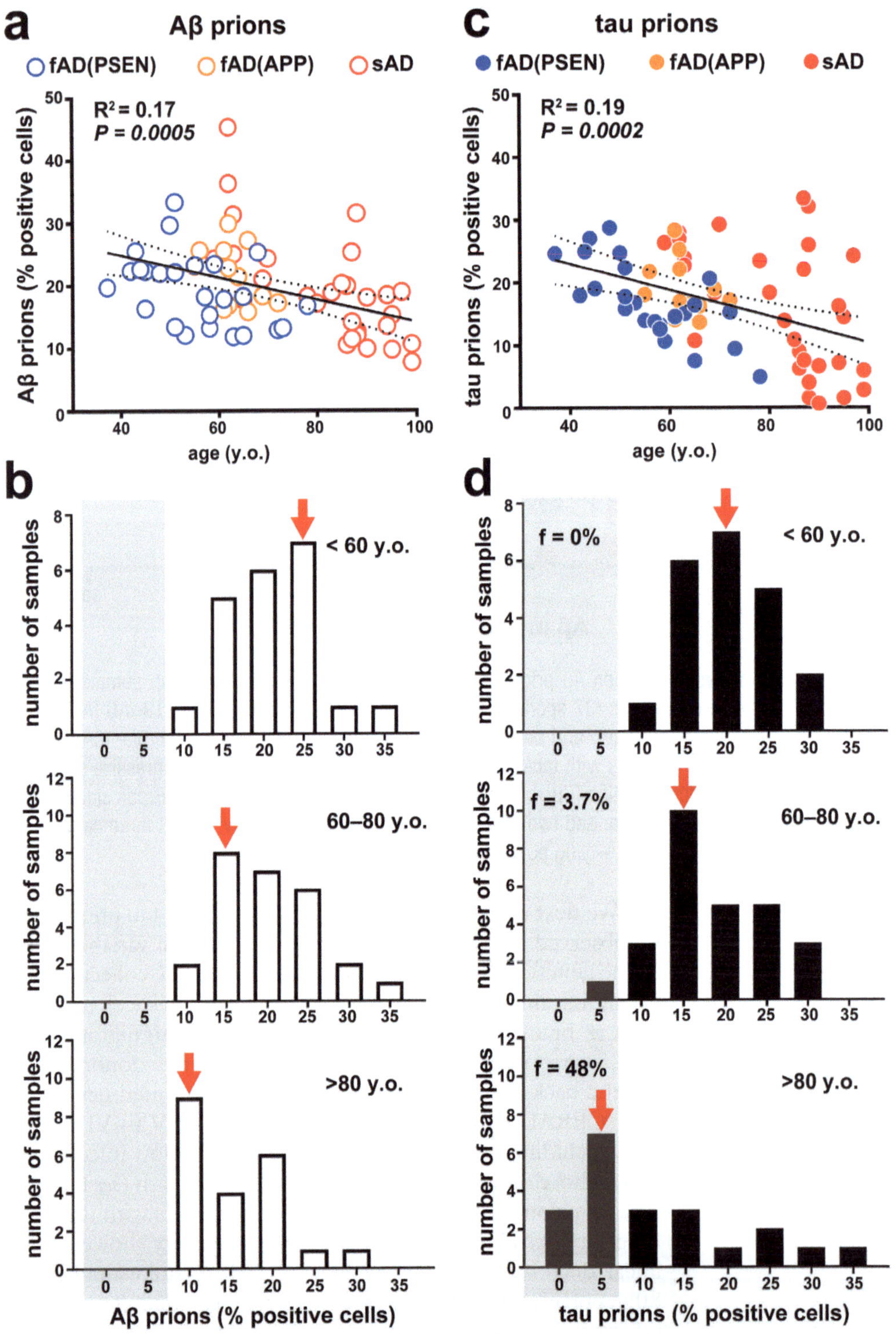

Fig. 9 Inverse correlation of Aβ prion and tau prion abundance with Alzheimer's disease (AD) patient age at death. (**a**, **b**) Aβ prion load of sporadic and familial AD samples (Braak stage V to VI) plotted as a function of patient age at death (*open circles*; fAD [PSEN], *blue*; fAD [APP], *orange*; sporadic AD [sAD], *red*). Statistical values for correlation, linear regression, and 95% confidence interval reported (**a**). Aβ prion load histogram for

collection bias as the great majority of the elderly patients were female. Overall, these linear trends were observed over a wide range of sporadic and familial forms of AD. Thus, what had appeared to be a set of disparate disorders can now be seen as a continuous spectrum, with the defining feature being the presence of both Aβ and tau prions.

The presence of the ε4 allele of the gene that encodes apolipoprotein E is the major AD risk factor [69, 70]. We, therefore, compared the effect of the *APOE* ε4 allele on Aβ and tau prion infectivity (*see* Fig. 10c, d). Correlation analysis showed that patients who were *APOE* ε4 non-carriers had lower tau prion infectivity than *APOE* ε4 carriers, and this effect became particularly pronounced in the population of patients over the age of 80 (*see* Fig. 10d). Similar trends were observed for Aβ prion levels, although they fell slightly short of 95% statistical confidence ($P = 0.057$) for the given sample size (*see* Fig. 10c). It will be interesting to examine the differential effects of *APOE* genotypes on Aβ versus tau prion activity in future studies with larger patient cohorts. Finally, a weak correlation of the tau prion level was observed between samples taken from the frontal and temporal lobes (*see* Fig. 10f).

5 Relationships Between Infectivity Versus the Abundance of Various Forms of Tau and Aβ

It is possible that the intersubject variations of Aβ and tau prion infectivity simply reflect differences in the amount of expression and/or the concentration of misfolded forms of these proteins. Alternatively, individuals with higher Aβ and tau prion activity might have biochemically or physically distinct forms with greater intrinsic potency. To differentiate between these possibilities, we measured the total abundance of various forms of soluble and insoluble Aβ and tau proteins using ELISA. Interestingly, the abundance of APP, Aβ40, and Aβ42 showed a significant trend ($P < 0.005$ in all cases) toward lower values (*see* Fig. 11a–c), matching the lower Aβ prion activity seen in Fig. 9a. The linear

Fig. 9 (continued) the same data set binned into three age groups (**b**): <60 years old (*top*), 60–80 years old (*middle*), and >80 years old (*bottom*). Red arrows indicate the infectivity bin with the greatest number of samples. (**c**, **d**) Tau prion load of sporadic and familial AD samples plotted as a function of patient age at death (filled circles; fAD [PSEN], *blue*; fAD [APP], *orange*; sAD: *red*). Statistical values for correlation, linear regression, and 95% confidence interval reported (**c**). Tau prion abundance histogram for the same data set binned into three age groups (**d**): <60 years old (*top*), 60–80 years old (*middle*), and >80 years old (*bottom*). Fraction of total sample number in each age bin with 0–5% tau prion-positive cells. The standard deviations of individual data points are similar to those in Fig. 1 and are much smaller than the deviation from the regression line, indicating that measurement error did not contribute significantly to the deviations from the trend line

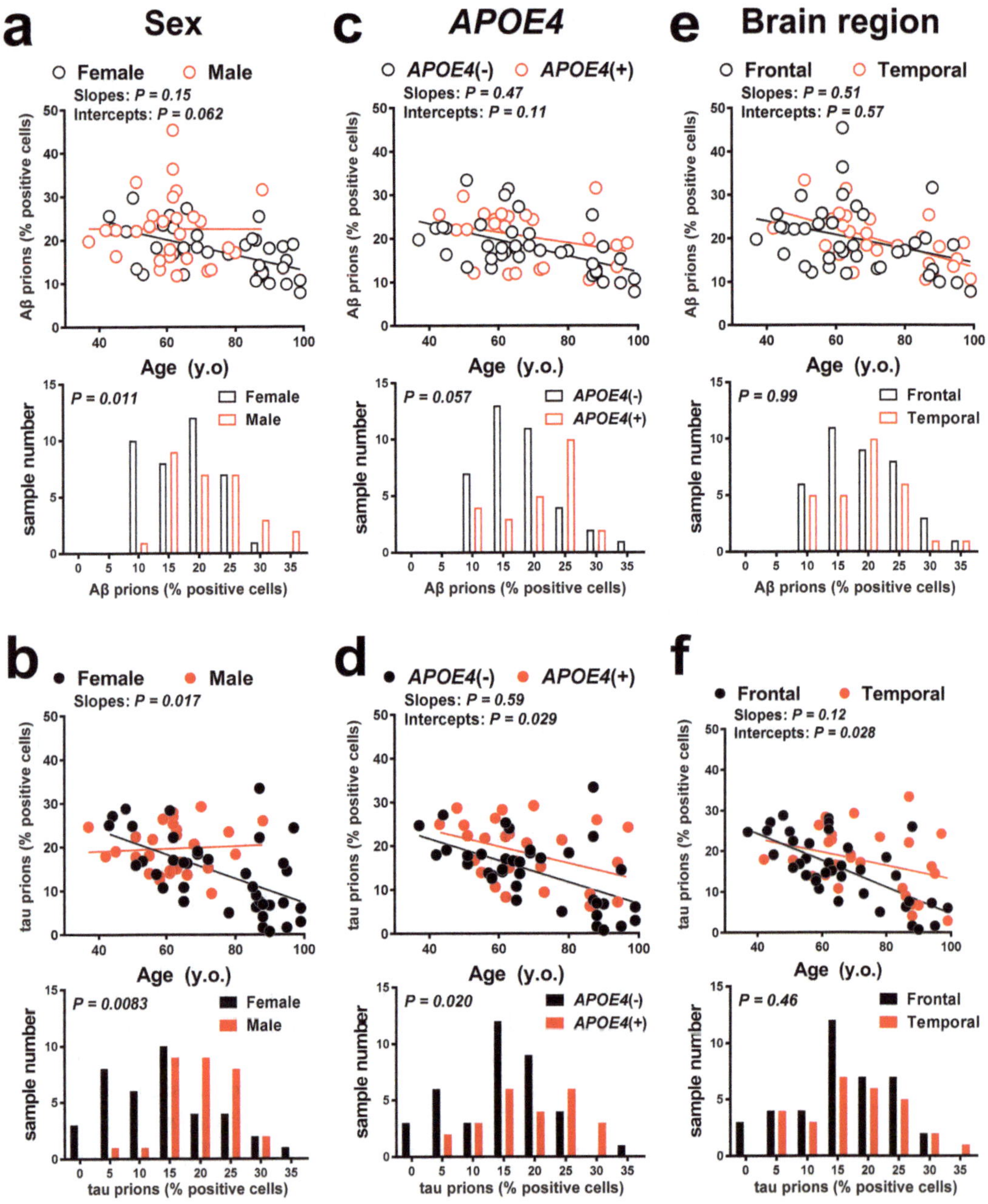

Fig. 10 Host factors influence the extent of Aβ and tau prion infectivity in Alzheimer's disease (AD): sex, *APOE* ε4 status, and brain region. Aβ prion infectivity (*open circles* and *open bars*) and tau prion infectivity (*filled circles* and *filled bars*) of sporadic and familial AD samples (Braak stage V to VI) separated by (**a**, **b**) sex (female, *black*; male, *red*), (**c**, **d**) *APOE* ε4 allele status (noncarrier, *black*; carrier of at least one allele, *red*), or (**e**, **f**) brain region of the donor sample (frontal cortex, *black*; temporal cortex, *red*). In each panel, cell infectivity data are plotted as a function of patient age at death (*top*) or presented as a histogram (*bottom*). A statistical comparison by linear regression (slopes and *y*-intercepts) of each group is reported (*top panels*). An infectivity histogram is displayed for the same data set (*bottom panels*). The frequency distribution of each group is statistically compared by the nonparametric Mann–Whitney *U* test (*bottom panels*)

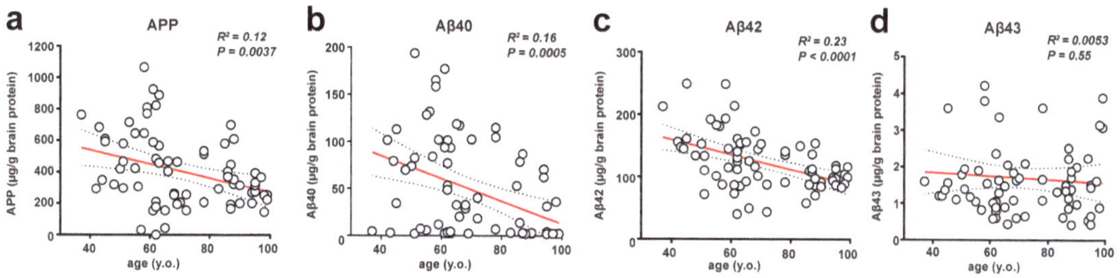

Fig. 11 Correlation of different Aβ species with Alzheimer's disease (AD) patient age at death: older individuals have lower amyloid precursor protein (APP) and Aβ levels. Enzyme-linked immunosorbent assay (ELISA) measurements in familial and sporadic AD brain samples (Braak stage V to VI) for the following proteins: (**a**) APP in the clarified brain homogenate (phosphate-buffered saline [PBS] soluble) and (**b–d**) formic acid–soluble Aβ40, Aβ42, and Aβ43. All data plotted as a function of patient age at death (years old). Statistical values for correlation, linear regression, and 95% confidence interval reported. The measurements were made in duplicate and are much smaller than the deviation from the regression line, indicating that measurement error did not contribute significantly to the deviations from the trend line

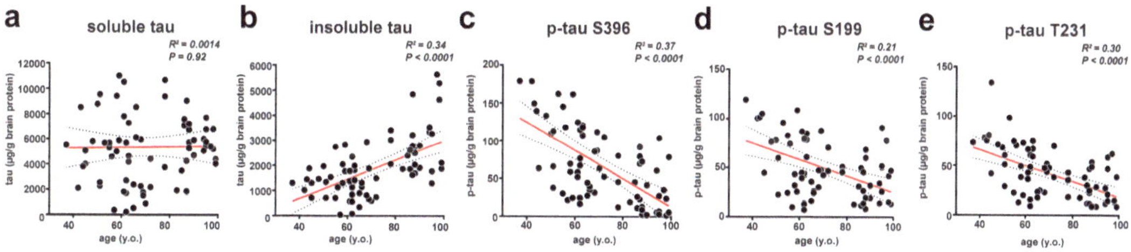

Fig. 12 Correlation of different tau species with Alzheimer's disease (AD) patient age at death: divergence of total insoluble tau and phosphorylated tau accumulation in the longest-lived individuals. Enzyme-linked immunosorbent assay (ELISA) measurements in familial and sporadic AD brain samples (Braak stage V to VI) for the following proteins: (**a**) total tau in the clarified brain homogenate (phosphate-buffered saline [PBS] soluble), (**b**) formic acid–soluble total tau, and (**c–e**) formic acid–soluble phosphorylated tau (phosphorylation epitope serine 396, serine 199, or threonine 231). All data plotted as a function of patient age at death (years old). Statistical values for correlation, linear regression, and 95% confidence interval reported. The measurements were made in duplicate and are much smaller than the deviation from the regression line, indicating that measurement error did not contribute significantly to the deviations from the trend line

decrease in APP is particularly interesting, as this represents the precursor to Aβ peptides. If this trend were true early in life, it might indicate that low APP expression contributes to longevity.

Very different trends were observed for the abundance of various forms of tau protein (*see* Fig. 12). There was no significant relationship between age and total concentration of soluble tau, as measured using an antibody that recognized all splice forms of the protein (*see* Fig. 12a). Furthermore, the amount of total insoluble tau increased with age (*see* Fig. 12b), contrasting with the decrease in infectivity seen in Fig. 9c. We, therefore, examined the abundance of phosphorylated tau (p-tau) because the amount of p-tau is known to accompany clinical severity [71, 72]. Antibodies for three different phosphorylated tau epitopes (S396, S199, and T231) gave very similar results (*see* Fig. 12c–e). Each showed that unlike total insoluble tau (*see* Fig. 12b), the extent of insoluble p-tau decreased linearly with age (*see* Fig. 12c–e), matching the result for tau infectivity (*see* Fig. 9c).

We next normalized the tau prion infectivity, relative to the total insoluble tau concentration, to provide a measure of the specific activity of the insoluble tau within a given sample. The data are extremely well described by an exponential decay ($R^2 = 0.79$) over five half-lives, a remarkable correlation given the stochastic nature of the underlying biological processes. This excellent correlation is, in part, a result of the data that span a wide range of ages from 37 to 99 years, resulting in a large range of the dependent variable relative to sampling and other errors. If a smaller age range were considered, the overall range of specific activities would be smaller, while experimental and other errors would remain approximately the same, leading to a lower correlation coefficient. Thus, the exponential nature of the process is most clearly revealed by including as wide a range of ages as available.

The extent of phosphorylation of tau as a function of age was similarly evaluated by calculating the amount of a given p-tau epitope in the insoluble fraction relative to the total insoluble tau. Again, a single exponential decay was observed over five half-lives, with R^2 values ranging from 0.75 to 0.78 ($P < 0.0001$), depending on the epitope used to quantify insoluble p-tau (*see* Fig. 13b–d). The half-life obtained from the normalized infectivity assay data (*see* Fig. 13a) was 12 years, while the corresponding half-lives for p-tau ranged from 7 to 10 years. The 95% confidence intervals for each of the half-lives overlap (*see* Fig. 13e), so it is not currently possible to tell whether the differences in half-life are meaningful. At present, we can conclude, however, that both tau prion infectivity and the extent of phosphorylation decrease about twofold for each decade of longevity. Thus, for example, an individual who died at age 40 would have about $2^5 = 32$-fold higher infectivity than an individual who lived five decades longer and died at age 90 (when normalized for total insoluble tau). Furthermore, the same holds true for the degree of tau phosphorylation. These data clearly show that the biophysical and biochemical changes that accompany high tau phosphorylation correlate strongly with disease progression at a young age.

6 Commentary and Analysis

Although a wealth of evidence argues that both Aβ and tau adopt prion-specific conformations leading to AD pathogenesis, the findings described here and first reported elsewhere [13] establish the presence of both Aβ and tau prions in the brains of patients who died of either sporadic or inherited AD. Notably, findings by others

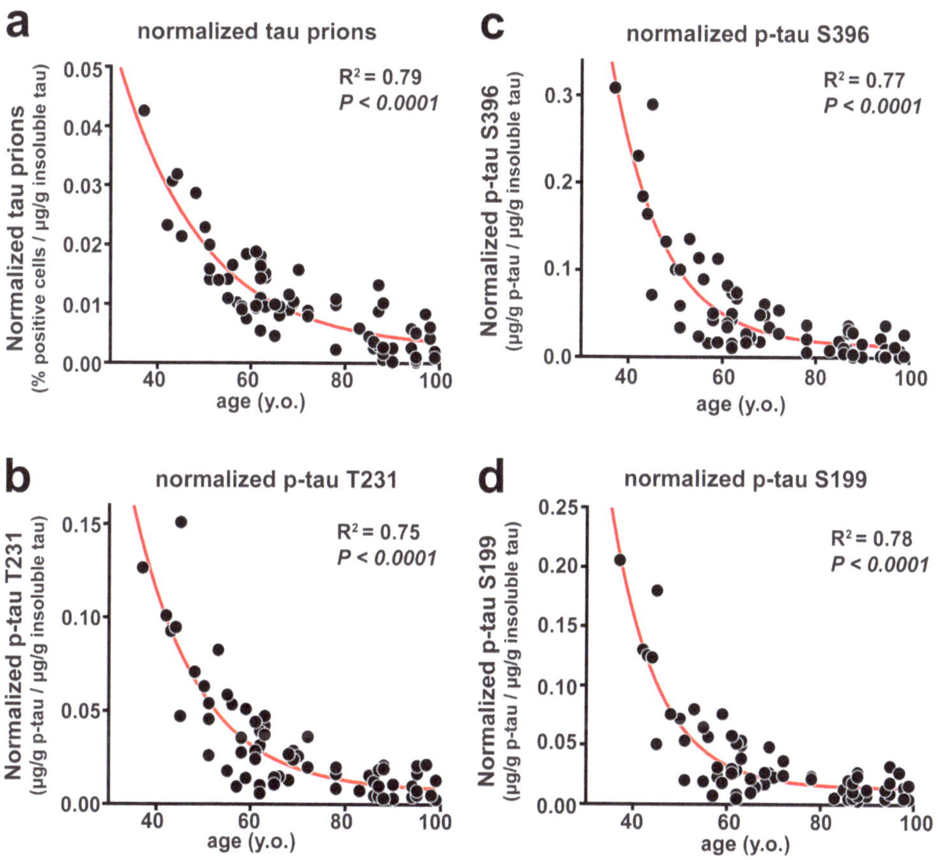

e

Normalized to insoluble tau	k	k (95% CI)	$T_{1/2}$	$T_{1/2}$ (95% CI)
tau prions	0.06	0.04–0.07	12.4	9.3–17.8
p-tau T231	0.07	0.05–0.09	9.5	7.4–12.6
p-tau S396	0.09	0.07–0.11	7.7	6.1–9.8
p-tau S199	0.1	0.08–0.12	6.9	5.6–8.7

Fig. 13 Tau prion abundance and phosphorylated tau normalized to insoluble tau: exponential decay model. (**a**) Tau prion abundance measured in Alzheimer's disease (AD) samples (Braak stage V to VI) normalized to the adjusted value of insoluble tau as measured by enzyme-linked immunosorbent assay (ELISA). Normalized data plotted as a function of AD patient age at death and fitted using an exponential decay model equation (one-phase decay). (**b**) Phosphorylated tau threonine 231 measured in AD samples normalized to the adjusted value of insoluble tau as measured by ELISA. Normalized data plotted as a function of AD patient age at death and fitted using an exponential decay model equation (one-phase decay). To normalize the data, the prion values from Fig. 9 or the p-tau concentration from Fig. 10 were divided by concentration of total insoluble tau obtained from the regression line for the total tau versus age shown in Fig. 10e. (**c**) Phosphorylated tau serine 396 measured in AD samples normalized to the adjusted value of insoluble tau as measured by ELISA. (**d**) Phosphorylated tau serine 199 measured in AD samples normalized to the adjusted value of insoluble tau as measured by ELISA. (**e**) Statistical values for correlation, decay constant (k), half-life ($T_{1/2}$), and their respective 95% confidence intervals (CI)

using cellular and Tg mouse models previously demonstrated tau prions in the brains of patients who died of FTLD [36, 73, 74]. Our results extend those earlier studies by demonstrating that the brains of FTLD patients are devoid of both Aβ and α-synuclein prions.

Earlier molecular genetic studies by others and our recent findings are beginning to provide a more complete view of the chemical processes that feature in the pathogenesis of AD. While a few samples from the brains of AD patients demonstrated transmission of Aβ prions to marmosets [75] and, more recently, Tg mice [46, 63, 76], our rapid Aβ prion infectivity cell assay allowed measurements from the brains of large numbers of people who died of AD and other NDs [13]. The studies reported here argue persuasively that AD patients have two prions in their brains: Aβ and tau prions. By combining the Aβ, α-synuclein, and tau prion bioassays, we are not only able to demonstrate the specificity of prions featuring in NDs, but we can also begin to dissect the molecular basis of the prion biology underlying each of the neurodegenerative processes.

6.1 Kinetics of Aβ and Tau Prion Formation in AD

Both human and animal studies argue that Aβ prions arise initially in AD and that these prions initiate the subsequent formation of tau prions [16, 26, 77–80]. Presumably, the formation of Aβ prions begins in one or more CNS cells and then propagates to other cells. The movement of PrP, α-synuclein, tau, and possibly Aβ prions from one CNS region to another argues for trans-synaptic spread [81–86]. The apparent spread of prions in the CNS is reflected by their regional distribution and has been well documented in neuropathological studies [87, 88].

Aβ peptides assemble into aggregates, which are called oligomers when the aggregate size is less than about 50 peptides [89]. A multitude of studies on human brain samples have reported the existence of soluble Aβ oligomers ranging in size from dimers [90] to dodecamers [91]. It has been reported that oligomer size correlates inversely with cellular toxicity [92, 93]. Moreover, the abundance of Aβ oligomers correlates well with the progression of cognitive deficits [94–96] and can differentiate AD patients from non-demented subjects with comparable amyloid plaque burden [97]. Whether Aβ multimer size correlates with prion infectivity and pathological deposition of Aβ and/or tau prions remains to be established. The ability to measure Aβ and tau prion infectivity has allowed us and others to begin to answer these critical questions. Additionally, by carefully quantifying the oligomeric distribution and concentration, it may be possible to establish the relationship between the number of Aβ peptides and Aβ prion infectivity (i.e., the particle-to-infectivity ratio [P/I]). For PrP^{Sc} prions, the P/I is ~5000 [8].

Recent studies report that the minimal size of an infectious tau prion may range from a monomer [98] to linear aggregates of ~200 nm in length [99]. Pentameric or smaller tau aggregates were unable to support prion infectivity [100]. Additionally, tau phosphorylation may also contribute to the tau prion conformation as shown in previous studies where immunodepletion of p-tau in brain extracts used as inocula abolished tau infectivity in recipient cells [55, 101] or animals [100]. These findings are consistent with our data demonstrating a significant relationship between human tau prion infectivity and the extent of tau phosphorylation in AD samples from people who died at various ages (*see* Figs. 12 and 13).

Further studies are needed to elucidate the molecular mechanism in AD by which Aβ prions stimulate tau prion formation in contrast to Aβ prions in CAA that do not stimulate tau prion formation. The mechanism responsible for selective tau prion formation is unknown but might reside in the multiple conformations that Aβ can adopt. Aβ is known to adopt a large number of fibrillar conformations and morphologies, but it remains to be determined which of these forms are related to distinct disease phenotypes. Using the assays reported here, it should be possible to correlate the presence of a given conformational form with its Aβ and tau prion infectivities across a variety of situations including (1) cognitively normal subjects with AD pathology, (2) CAA-specific patients, (3) FTLD-tau, and (4) early- versus late-onset familial and sporadic AD.

6.2 Aβ and Tau Prion Infectivity Inversely Correlates with Longevity in Older People

Development of a bioassay for Aβ prions allowed for comparison of prion titers from synthetic preparations, which is useful for examining cross-infection between Aβ subtypes and alleles in familial NDs. It also has enabled parallel quantification of Aβ and tau prion infectivity, providing the first direct quantitative comparison of the active propagating species rather than inert protein deposits [13]. The data described here indicate that AD patients with the greatest longevity have lower levels of both Aβ and tau prions at the time of death compared with patients who die at younger ages from AD-related symptoms. Previous studies show that NFTs composed of tau protein correlate well with the extent of brain atrophy and cognitive decline in AD [34, 102]. Notably, those studies focused only on total insoluble tau. By looking at the age at death as a variable, we found that low tau prion infectivity, which is a direct measure of replication-competent tau protein, is associated with greater longevity.

It is notable that both the extent of Aβ infectivity and the abundance of APP, Aβ40, and Aβ42 decrease with longevity in a roughly synchronous manner. This finding is consistent with the hypothesis that Aβ prions feature early during the formation of tau prions. Moreover, measurable Aβ prion activity was found in the oldest patients, arguing that it continues to participate throughout the pathogenesis of AD. However, the R^2 values that we found range from 0.12 to 0.2, indicating that many factors ranging from sample collection methods to genetic factors appear to have a sizable influence on the observed correlation. Clearly, genetic factors such as the *APOE* ε4 allele and *TREM2* variants, which have been implicated in Aβ metabolism and Aβ plaque passivation, can strongly increase the risk of AD [69, 103]. Although we found interesting trends with respect to the *APOE* ε4 genotype and sex, we need to perform larger studies that carefully sample all of these different patient populations.

The even stronger associations among longevity, tau prion infectivity, and phosphorylation are particularly intriguing; these are consistent with the greater contribution of tau versus Aβ protein misfolding to AD pathology and neurological dysfunction. One particularly striking result is the accumulation of insoluble tau that increases with age at death, while the converse is true for phosphorylated insoluble tau (*see* Fig. 12). Although these relationships were clear from examining the extent of infectivity and phosphorylation per gram of tissue, they became more striking when the data were normalized according to the abundance of insoluble tau (*see* Fig. 13). Our findings suggest that all insoluble tau is not equally neurotoxic or virulent and that biochemical events such as phosphorylation strongly influence the formation of tau prions and/or modulate their toxicity. Thus, future work aimed at the development of diagnostic reagents and effective therapeutics will need to focus on infectious tau and its associated posttranslational modifications rather than total insoluble tau.

7 New Frontiers in Developing Effective Therapeutics for AD

While our approach and that of others to measuring Aβ and tau prions from postmortem brain samples has illuminated new insights into the human biology of disease [13, 36], these bioassays are likely to have additional applications. Antemortem detection of biologically active Aβ and tau prions in the CSF or blood of patients, as shown for PrP prions [104–106], may prove to be a more informative diagnostic tool for staging disease. Moreover, the bioassays described here may complement the use of clinical imaging ligands to study the interdependence of insoluble Aβ and tau deposition patterns with respect to brain atrophy and cognitive decline [107–109].

Our dual cell approach provides a new quantitative tool for measuring AD progression in Tg mice following genetic manipulation or therapeutic intervention. Most important, our findings may illuminate both the successes and failures of pharmaceutical approaches that target Aβ and the Aβ–tau axis. The availability of paired, cellular prion infectivity bioassays should greatly enable future drug discovery programs targeted at stopping the progressive dementia of AD.

In addition to the novel quantitative approach of our bioassays, another technological advancement is enabling molecular studies of AD prions not possible previously. Extraordinary developments in cryogenic electron microscopy (cryo-EM) have led to informative high-resolution structures of three different mammalian prions.

8 Cryo-EM Structures of Tau, α-Synuclein, and Aβ Prions

Beginning more than 50 years ago with the observation of paired helical filaments (PHFs) in the brain tissue of patients with AD [110], transmission electron microscopes have been instrumental in the study of prions. While we were able to discern the general filament structural and morphological characteristics of prions, our ability to understand their structures at the atomic level remained elusive until recently. However, with improvements in sample preparation and microscope technology, the advent of direct electron detectors, and advances in computational methods [111, 112], generating an atomic model of prions via cryo-EM has become possible. In the space of a few years, our understanding of prions has leapt forward; overall folds and individual atomic interactions of multiple NDs across several different prions have been elucidated. Importantly, this information will greatly aid our ability to diagnose and treat these diseases through the potential for rational design and optimization of PET probes and therapeutics.

Despite differences in the causative proteins and diseases, all prion filamentous structures share certain characteristics. The filaments consist of stacked β-sheets, stabilized by a network of intersheet hydrogen bonding, where spacing between each sheet is approximately 4.7 Å. These stacked β-sheets form the "ordered core" of the protein (i.e., those residues that are observable at atomic resolution by cryo-EM). The remainder of the protein—generally the residues toward each terminus—is known as the "fuzzy coat." Finally, for those diseases in which cryo-EM structures have been solved from multiple patients, the molecular structures are consistent across individuals. This lends credence to the strain hypothesis, which proposes that a unique molecular structure of the misfolded protein determines each prion disease [113]. Here, we delve into the molecular structures of three different prions solved from patient-derived tissue by cryo-EM: tau, α-synuclein, and Aβ.

8.1 Tau

Tau is a soluble protein with six different isoforms expressed depending upon the number of N-terminal inserts (0, 1, or 2) and microtubule-binding repeat domains toward the C-terminus (3 or 4 referred to as 3R or 4R tau). Tauopathies can be classified according to the isoforms found in the NFTs of patients. The tau inclusions found in AD and chronic traumatic encephalopathy (CTE) consist of all six isoforms and are referred to as 3R + 4R tauopathies. In CBD, only 4R tau is found in the insoluble inclusions, and in PiD, only 3R tau is present. Tauopathies in which the inclusions are formed by the same isoforms show clear structural similarities.

In 2017, tau from AD inclusions became the first prion to be characterized at atomic resolution by cryo-EM [114]. These tau filaments are comprised of two ultrastructural motifs: PHFs, which constitute a majority of the total filaments, and straight filaments (SFs). Each ultrastructure is comprised of a homodimer of the ordered core region; the monomers comprising the homodimer are called protofilaments (*see* Fig. 14a). While the protofilaments that form PHFs and SFs are almost identical, their arrangement relative to each other differs: the PHF protofilaments pack with $C2_1$

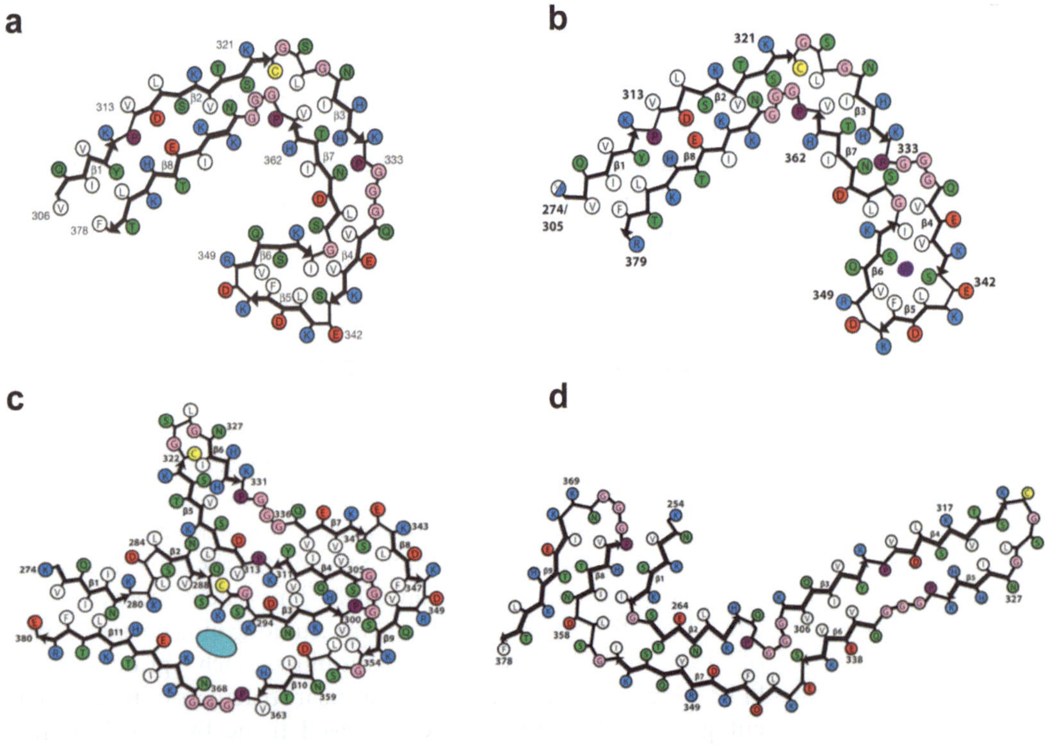

Fig. 14 Diagram showing the protofilament folds of (**a**) Alzheimer's disease (AD), (**b**) chronic traumatic encephalopathy (CTE), (**c**) corticobasal degeneration (CBD), and (**d**) Pick's disease (PiD). Adapted by permission from Springer Nature: refs. [114] (a), [116] (b), [117] (c), and [118] (d)

symmetry, and the SF protofilaments form an asymmetric dimer. Therefore, the protofilaments in PHFs are offset by half the rise of SFs along the helical axis (2.36 versus 4.72 Å) with a twist of 179.4° and −1.05°, respectively. The protofilaments themselves have a C-shaped arrangement of primarily β-strands consisting mostly of the R3 and R4 repeat regions. Later, structures solved from five additional patients confirmed that the ultrastructural motifs and conformation of the protofilament are identical across multiple AD patients [115], further supporting the strain hypothesis. Despite this structural homogeneity, it remains to be understood why certain AD patients have widely varying levels of tau, phosphorylated tau, and infectivity as measured in cellular prion assays (*see* Figs. 7, 8, 9, 10, 12, and 13). Structural characterization of the tau from a subset of these patients could confirm or disprove whether alternative tau ultrastructures play a role in these differences.

The other 3R + 4R tauopathy to be structurally characterized is CTE [116], a disease caused by repetitive head trauma. As with AD, two types of filaments were observed in purified CTE tissue: type I and type II filaments. Type I filaments are wider than AD SFs and have a more distinct twist; type II filaments resemble the AD PHFs in their ultrastructure but are considerably wider. Like the AD PHFs, both type I and type II CTE filaments are homodimers with $C2_1$ symmetry and have almost identical rise and twist parameters as the PHFs. Moreover, the interface between the protofilaments in type II filaments is identical to the interface in AD PHFs: residues 331–338 (KPGGGQVE). Type I filaments form a novel interface, which includes residue S324 as part of an anti-parallel steric zipper. This serine is known to be phosphorylated in AD, which may explain why this dimer arrangement is found only in CTE. The CTE protofilament (*see* Fig. 14b), while sharing the exact residues and general C-shape of the AD ordered core, has several unique structural differences. The most striking of these is the nonproteinaceous density in a hydrophobic cavity of the protofilament. This density is thought to represent a small hydrophobic molecule, perhaps a metabolite, which is present in near-stoichiometric amounts to tau. This density may explain why the CTE protofilaments have a more open C-shape than their AD counterparts and other minor side chain packing differences.

CBD is the only 4R-only tauopathy for which there is a cryo-EM filament structure [117]. Purified CBD tissue contains two types of filament polymorphs: narrow, with a width ranging from 80 to 130 Å, and wide, with a width ranging from 80 to 260 Å. Narrow CBD filaments are simply a single protofilament, while the wide filaments are a C2-symmetric homodimer formed by two protofilaments. The protofilament itself consists of 11 β-strands spanning residues 274–380, which includes the entirety of the R2, R3, and R4 regions of tau (*see* Fig. 14c). The ordered core of CBD filaments is much longer than those of CTE or AD

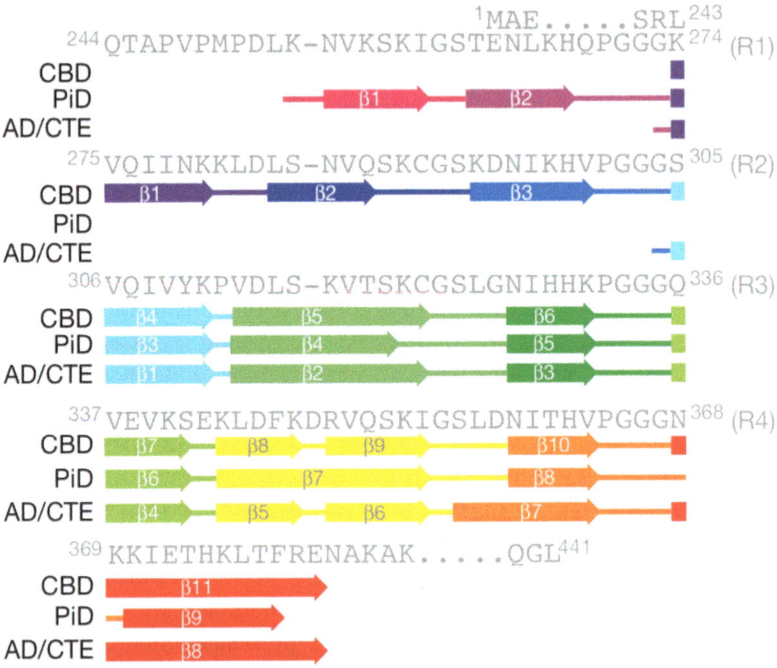

Fig. 15 Diagram showing residues comprising each β-strand of the ordered core of the structurally characterized tauopathy filaments. Adapted by permission from Springer Nature: ref. [117]

(73 residues; *see* Fig. 15), perhaps being due to structural homogeneity conferred by the 4R-only composition of these filaments. The structure of the CBD protofilament is completely different than the AD and CTE arrangements, with multiple folds back upon adjacent β-strands. Like CTE, CBD filaments also contain additional density in a pocket of the protofilament. However, in CBD, this density is situated in a hydrophilic pocket surrounded by three lysine residues, indicating it is likely a polyanion.

The final known tau filament structure is from the tissue of a patient with PiD, a 3R-only tauopathy [118]. As with the other tauopathies, there are two types of PiD filaments: narrow Pick filaments (NPFs) and wide Pick filaments (WPFs). Like CBD filaments, the NPFs are a single protofilament (*see* Fig. 14d), and the WPFs are a C2-symmetric protofilament homodimer. However, the vast majority of PiD filaments (~93%) are NPFs, as opposed to CBD in which the two filament types exist in an approximately equal ratio. This is likely due to the very small, three-residue interface between the two protofilaments, which is less stable than the interfaces of the other tauopathy filament dimers. The ordered core of PiD filaments spans three repeat domains as with CBD, but R2 in CBD is replaced by R1 in PiD. The PiD protofilament is narrow and extended, and it is incompatible with 4R tau because of would-be steric clashes of a lysine for threonine substitution at position 294/263 and a glutamate for valine substitution at position 300/269.

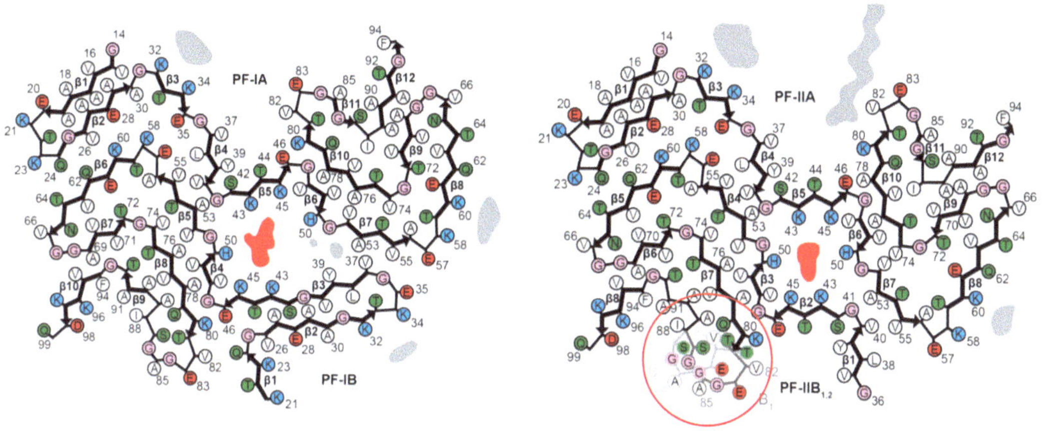

Fig. 16 Schematic diagram of multiple system atrophy (MSA) type I filaments (*left*) and type II filaments (*right*). The type II_1 trace is in light gray, and the type II_2 trace is in dark gray (*red circle*). Adapted by permission from Springer Nature: ref. [119]

8.2 α-Synuclein

α-Synuclein is a cytoplasmic protein comprised of 140 amino acids with a single isoform expressed. α-Synuclein inclusions are found in the brain tissue of patients with several diseases, including Parkinson's disease, MSA, and dementia with Lewy bodies. To date, only glial cytoplasmic inclusion filaments from MSA patient tissue have been purified and structurally characterized by cryo-EM [119]. Two molecular conformations of filaments (type I and type II) were found, despite all filaments being ultrastructurally indistinguishable. Type II filaments have two subtypes, II_1 and II_2, with slight differences at the C-terminal end of the B protofilament (*see* Fig. 16). Both types exist as dimers of two slightly different protofilaments, and each type is found in varying ratios across the five patients from which MSA structures have been solved. The ordered core consists of 45–60% of the overall length of the protein, giving a much more structural insight into MSA than any of the characterized tauopathies in which the ordered core is no more than one-third of the tau sequence. Both conformations have an unidentified nonproteinaceous density between the two protofilaments, which appears to be polyanionic as the side chains facing inward are those of positively charged histidine and lysine residues. These structures also provide strong evidence that the structure of Parkinson's disease filaments differs from these MSA filaments. E46K, a well-studied familial Parkinson's disease mutation, is not compatible with any of the MSA filament structures as this mutation would break a salt bridge between E46 and K80, which is found in all MSA protofilaments.

8.3 *Aβ*

Aβ is a short peptide formed by the sequential cleavage of APP and is thought to be the other causative prion (along with tau) in AD [13]. This cleavage results in peptides of various lengths, the vast majority of which are 36–43 residues. Aβ40 and Aβ42 are the two most commonly found peptides and also the best studied. In AD, Aβ42 aggregates accumulate in the brain, and Aβ42 is thought to be the pathogenic species. Aβ filaments purified from the brain tissue of a patient with AD showed multiple ultrastructures, with three of them comprising roughly 75% of all purified filaments (morphologies I, II, and III); each of these has a unique width and crossover distance [120]. Morphology I was the only structure determined at sufficient resolution by cryo-EM to place side chains and generate an accurate model (*see* Fig. 17). It consists of two elongated protofilaments related by $C2_1$ symmetry, with a lengthy hydrophobic protofilament interface. Interestingly, all 40 amino acids of the Aβ40 peptide are able to be placed into this map, yielding a near total understanding of the filament structure. The lower-resolution maps for morphologies II and III are consistent with side-on packing of multiple protofilament dimers (two for morphology II and three for morphology III). These larger morphologies are stabilized by two salt bridges between E3 and R5.

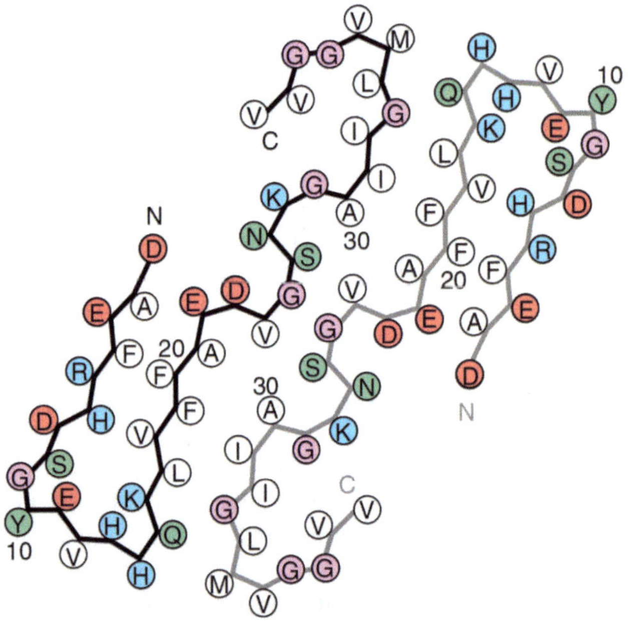

Fig. 17 Schematic of the Aβ protofilament, which is also the morphology I structure. Adapted by permission from Springer Nature: ref. [120]

9 Closing Remarks

Prions were initially identified by copurification of scrapie infectivity and enrichment of a novel protein that we named the scrapie prion protein (PrP^{Sc}). The discovery of mutations in prion proteins that were genetically linked to specific neurodegenerative diseases (NDs) led to the discovery of the inherited prion diseases [15, 16]. With this discovery, it became possible to reconcile how a single disease could be manifested by three different phenotypes: (1) infectious, (2) sporadic, and (3) inherited [121]. Atomic resolution structures of prion filaments solved by cryo-EM represent a major step forward in our understanding of NDs. Each tauopathy appears to be caused by discrete tau protofilament conformations that are consistent across patients; this finding lends credence to the hypothesis that different protein conformations encipher distinct prion strains. A corollary proposes that different prion strains manifest as distinct prion phenotypes.

Whether clinical and neuropathological differences among patients with the same disease can be explained by structural heterogeneity is an important question that cryo-EM is in a unique position to answer. Additionally, the filament conformations that form inside the brain may be influenced by nonproteinaceous small molecules, which might be able to distinguish similar strains from each other (i.e., CTE versus AD). There is also great potential for use of these structures in the diagnosis and treatment of NDs. Small molecules could be designed to fluoresce upon binding to one of these prion structures as a diagnostic method in living patients. Furthermore, prion therapeutics could be designed to break important interactions in the protofilament, which may slow down or eliminate disease progression. As structures continue to be solved at an increasing resolution, our knowledge of prion disease pathology will only continue to grow.

Acknowledgments

This work was supported by grants from the National Institutes of Health (NIH) (AG002132 and AG031220) and by support from the Dana Foundation, the Rainwater Charitable Foundation, and the Sherman Fairchild Foundation. Human brain tissue was received from the UCSF Neurodegenerative Disease Brain Bank, which is supported by the NIH (AG023501 and AG19724 to W. W.S.), the Tau Consortium, and the Consortium for Frontotemporal Dementia Research. Human brain tissue provided by the Brain Bank at Karolinska Institutet (KI), Stockholm, Sweden, received financial support from StratNeuro at KI, Swedish Brain Power, and Stockholm County Council. Human brain tissue provided by the Massachusetts Alzheimer's Disease Research Center (Director: Matthew P. Frosch) received financial support from the NIH (P50 AG005134). Autopsy tissue obtained from the University of Washington Neuropathology Core is supported by the Alzheimer's Disease Research Center (AG05136), the Adult Changes in Thought Study (AG006781), and the Morris K. Udall Center of Excellence for Parkinson's Disease Research (NS062684). Autopsy tissue was also supplied by the King's College London (Department of Clinical Neuroscience), the University of Edinburgh (Department of Neuropathology), and the Manchester Brain Bank (University of Manchester), which is part of the Brains for Dementia Research Initiative, jointly funded by Alzheimer's Society and Alzheimer's Research UK. We thank the Queen Square Brain Bank for Neurological Disorders (supported by the Reta Lila Weston Trust for Medical Research, the Progressive Supranuclear Palsy [Europe] Association, and the Medical Research Council) at the UCL Institute of Neurology, University College London, for provision of the UK human brain tissue samples. The Sydney Brain Bank is supported by Neuroscience Research Australia and the University of New South Wales. Glenda M. Halliday is a National Health and Medical Research Council of Australia Senior Principal Research Fellow (1079679). Human brain tissue provided by the Rush Alzheimer's Disease Center, Rush University Medical Center (Director: David A. Bennett) [122], received financial support from the NIH (P30AG10161 and R01AG15819).

Competing Interests S.B.P. is the founder of Prio-Pharma, which has not contributed financial or any other support to these studies. W.F.D. is a member of the scientific advisory boards of Alzheon Inc., Pliant, Longevity, CyteGen, Amai, and ADRx Inc., none of which have contributed financial or any other support to these studies.

References

1. Alzheimer A (1911) Über eigenartige Krankheitsfälle des späteren Alters. Zentralbl Gesamte Neurol Psychiatr 4:356–385
2. Goedert M (2009) Oskar Fischer and the study of dementia. Brain 132:1102–1111
3. Glenner GG, Wong CW (1984) Alzheimer's disease: initial report of the purification and characterization of a novel cerebrovascular amyloid protein. Biochem Biophys Res Commun 120:885–890
4. Brion JP, Couck AM, Passareiro E, Flament-Durand J (1985) Neurofibrillary tangles of Alzheimer's disease: an immunohistochemical study. J Submicrosc Cytol 17:89–96
5. Kosik KS, Joachim CL, Selkoe DJ (1986) Microtubule-associated protein tau is a major antigenic component of paired helical filaments in Alzheimer disease. Proc Natl Acad Sci U S A 83:4044–4048
6. Grundke-Iqbal I, Iqbal K, Tung Y-C, Quinlan M, Wisniewski HM, Binder LI (1986) Abnormal phosphorylation of the microtubule-associated protein (tau) in Alzheimer cytoskeletal pathology. Proc Natl Acad Sci U S A 83:4913–4917

7. Pollock NJ, Mirra SS, Binder LI, Hansen LA, Wood JG (1986) Filamentous aggregates in Pick's disease, progressive supranuclear palsy, and Alzheimer's disease share antigenic determinants with microtubule-associated protein, tau. Lancet 2:1211
8. Prusiner SB, McKinley MP, Bowman KA, Bolton DC, Bendheim PE, Groth DF et al (1983) Scrapie prions aggregate to form amyloid-like birefringent rods. Cell 35: 349–358
9. Prusiner SB (2012) A unifying role for prions in neurodegenerative diseases. Science 336: 1511–1513
10. Rasmussen J, Jucker M, Walker LC (2017) Aβ seeds and prions: how close the fit? Prion 11: 215–225
11. Condello C, Stöhr J (2018) Aβ propagation and strains: implications for the phenotypic diversity in Alzheimer's disease. Neurobiol Dis 109:191–200
12. Ayers JI, Giasson BI, Borchelt DR (2018) Prion-like spreading in tauopathies. Biol Psychiatry 83:337–346
13. Aoyagi A, Condello C, Stöhr J, Yue W, Lee JC, Rivera BM et al (2019) Aβ and tau prion-like activities decline with longevity in the Alzheimer's disease human brain. Sci Transl Med 11(eaat8462):1–13
14. Masters CL, Gajdusek DC, Gibbs CJ Jr (1981) Creutzfeldt-Jakob disease virus isolations from the Gerstmann-Sträussler syndrome. Brain 104:559–588
15. Hsiao K, Baker HF, Crow TJ, Poulter M, Owen F, Terwilliger JD et al (1989) Linkage of a prion protein missense variant to Gerstmann-Sträussler syndrome. Nature 338: 342–345
16. Goate A, Chartier-Harlin M-C, Mullan M, Brown J, Crawford F, Fidani L et al (1991) Segregation of a missense mutation in the amyloid precursor protein gene with familial Alzheimer's disease. Nature 349:704–706
17. Goate A, Hardy J (2012) Twenty years of Alzheimer's disease-causing mutations. J Neurochem 120(Suppl. 1):3–8
18. TCW J, Goate AM (2017) Genetics of β-amyloid precursor protein in Alzheimer's disease. In: Prusiner SB (ed) Prion diseases. Cold spring harb. perspect. med. Cold Spring Harbor Laboratory Press, Cold Spring Harbor, NY, pp 203–213
19. Hutton M, Lendon CL, Rizzu P, Baker M, Froelich S, Houlden H et al (1998) Association of missense and 5′-splice-site mutations in tau with the inherited dementia FTDP-17. Nature 393:702–705
20. Hong M, Zhukareva V, Vogelsberg-Ragaglia-V, Wszolek Z, Reed L, Miller BI et al (1998) Mutation-specific functional impairments in distinct tau isoforms of hereditary FTDP-17. Science 282:1914–1917
21. Poorkaj P, Bird TD, Wijsman E, Nemens E, Garruto RM, Anderson L et al (1998) Tau is a candidate gene for chromosome 17 frontotemporal dementia. Ann Neurol 43(6): 815–825
22. Maat-Schieman MLC, Yamaguchi H, Hegeman-Kleinn IM, Welling-Graafland C, Natté R, Roos RAC et al (2004) Glial reactions and the clearance of amyloid β protein in the brains of patients with hereditary cerebral hemorrhage with amyloidosis-Dutch type. Acta Neuropathol 107:389–398
23. Van Duinen SG, Castano EM, Prelli F, Bots GTAB, Luyendij KW, Frangione B (1987) Hereditary cerebral haemorrhage with amyloidosis in patients of Dutch origin is related to Alzheimer's disease. Proc Natl Acad Sci U S A 84:5991–5994
24. Karran E, De Strooper B (2016) The amyloid cascade hypothesis: are we poised for success or failure? J Neurochem 139(Suppl. 2): 237–252
25. Viola KL, Klein WL (2015) Amyloid β oligomers in Alzheimer's disease pathogenesis, treatment, and diagnosis. Acta Neuropathol 129:183–206
26. Gotz J, Chen F, van Dorpe J, Nitsch RM (2001) Formation of neurofibrillary tangles in P301L tau transgenic mice induced by Aβ42 fibrils. Science 293:1491–1495
27. Lewis J, Dickson DW, Lin WL, Chisholm L, Corral A, Jones G et al (2001) Enhanced neurofibrillary degeneration in transgenic mice expressing mutant tau and APP. Science 293:1487–1491
28. Bennett RE, DeVos SL, Dujardin S, Corjuc B, Gor R, Gonzalez J et al (2017) Enhanced tau aggregation in the presence of amyloid β. Am J Pathol 187:1601–1612
29. He Z, Guo JL, McBride JD, Narasimhan S, Kim H, Changolkar L et al (2018) Amyloid-β plaques enhance Alzheimer's brain tau-seeded pathologies by facilitating neuritic plaque tau aggregation. Nat Med 24:29–38
30. Lesne SE (2014) Toxic oligomer species of amyloid-β in Alzheimer's disease, a timing issue. Swiss Med Wkly 144:w14021. https://doi.org/10.4414/smw.2014.14021
31. Musiek ES, Holtzman DM (2015) Three dimensions of the amyloid hypothesis: time, space and 'wingmen'. Nat Neurosci 18: 800–806

32. Thal DR, Rüb U, Orantes M, Braak H (2002) Phases of Aβ-deposition in the human brain and its relevance for the development of AD. Neurology 58:1791–1800
33. Murray ME, Lowe VJ, Graff-Radford NR, Liesinger AM, Cannon A, Przybelski SA et al (2015) Clinicopathologic and ^{11}C-Pittsburgh compound B implications of Thal amyloid phase across the Alzheimer's disease spectrum. Brain 138:1370–1381
34. Jack CR Jr, Knopman DS, Jagust WJ, Petersen RC, Weiner MW, Aisen PS et al (2013) Tracking pathophysiological processes in Alzheimer's disease: an updated hypothetical model of dynamic biomarkers. Lancet Neurol 12:207–216
35. Kfoury N, Holmes BB, Jiang H, Holtzman DM, Diamond MI (2012) Trans-cellular propagation of tau aggregation by fibrillar species. J Biol Chem 287:19440–19451
36. Sanders DW, Kaufman SK, DeVos SL, Sharma AM, Mirbaha H, Li A et al (2014) Distinct tau prion strains propagate in cells and mice and define different tauopathies. Neuron 82:1271–1288
37. Woerman AL, Stöhr J, Aoyagi A, Rampersaud R, Krejciova Z, Watts JC et al (2015) Propagation of prions causing synucleinopathies in cultured cells. Proc Natl Acad Sci U S A 112:E4949–E4958
38. Woerman AL, Aoyagi A, Patel S, Kazmi SA, Lobach I, Grinberg LT et al (2016) Tau prions from Alzheimer's disease and chronic traumatic encephalopathy patients propagate in cultured cells. Proc Natl Acad Sci U S A 113:E8187–E8196
39. Wurth C, Guimard NK, Hecht MH (2002) Mutations that reduce aggregation of the Alzheimer's Aβ42 peptide: an unbiased search for the sequence determinants of Aβ amyloidogenesis. J Mol Biol 319:1279–1290
40. Kim W, Kim Y, Min J, Kim DJ, Chang YT, Hecht MH (2006) A high-throughput screen for compounds that inhibit aggregation of the Alzheimer's peptide. ACS Chem Biol 1:461–469
41. Ochiishi T, Doi M, Yamasaki K, Hirose K, Kitamura A, Urabe T et al (2016) Development of new fusion proteins for visualizing amyloid-β oligomers in vivo. Sci Rep 6:22712
42. Schmidt M, Rohou A, Lasker K, Yadav JK, Schiene-Fischer C, Fändrich M et al (2015) Peptide dimer structure in an Aβ(1–42) fibril visualized with cryo-EM. Proc Natl Acad Sci U S A 112:11858–11863
43. Lu JX, Qiang W, Yau WM, Schwieters CD, Meredith SC, Tycko R (2013) Molecular structure of β-amyloid fibrils in Alzheimer's disease brain tissue. Cell 154:1257–1268
44. Gremer L, Schölzel D, Schenk C, Reinartz E, Labahn J, Ravelli RBG et al (2017) Fibril structure of amyloid-β(1-42) by cryo-electron microscopy. Science 358:116–119
45. Stöhr J, Watts JC, Mensinger ZL, Oehler A, Grillo SK, DeArmond SJ et al (2012) Purified and synthetic Alzheimer's amyloid beta (Aβ) prions. Proc Natl Acad Sci U S A 109:11025–11030
46. Stöhr J, Condello C, Watts JC, Bloch L, Oehler A, Nick M et al (2014) Distinct synthetic Aβ prion strains producing different amyloid deposits in bigenic mice. Proc Natl Acad Sci U S A 111:10329–10334
47. Nilsberth C, Westlind-Danielsson A, Eckman CB, Condron MM, Axelman K, Forsell C et al (2001) The 'Arctic' APP mutation (E693G) causes Alzheimer's disease by enhanced Aβ protofibril formation. Nat Neurosci 4:887–893
48. Elkins MR, Wang T, Nick M, Jo H, Lemmin T, Prusiner SB et al (2016) Structural polymorphism of Alzheimer's β-amyloid fibrils as controlled by an E22 switch: a solid-state NMR study. J Am Chem Soc 138:9840–9852
49. Cloe AL, Orgel JPRO, Sachleben JR, Tycko R, Meredith SC (2011) The Japanese mutant Aβ (ΔE22-Aβ(1-39)) forms fibrils instantaneously, with low-thioflavin T fluorescence: seeding of wild-type Aβ(1-40) into atypical fibrils by ΔE22-Aβ(1-39). Biochemistry 50:2026–2039
50. Levine DJ, Stöhr J, Falese LE, Ollesch J, Wille H, Prusiner SB et al (2015) Mechanism of scrapie prion precipitation with phosphotungstate anions. ACS Chem Biol 10:1269–1277
51. Sturchler-Pierrat C, Abramowski D, Duke M, Wiederhold KH, Mistl C, Rothacher S et al (1997) Two amyloid precursor protein transgenic mouse models with Alzheimer disease-like pathology. Proc Natl Acad Sci U S A 94:13287–13292
52. Allen B, Ingram E, Takao M, Smith MJ, Jakes R, Virdee K et al (2002) Abundant tau filaments and nonapoptotic neurodegeneration in transgenic mice expressing human P301S tau protein. J Neurosci 22:9340–9351
53. Giasson BI, Duda JE, Quinn SM, Zhang B, Trojanowski JQ, Lee VM (2002) Neuronal α-synucleinopathy with severe movement disorder in mice expressing A53T human α-synuclein. Neuron 34:521–533

54. Holmes BB, Furman JL, Mahan TE, Yamasaki TR, Mirbaha H, Eades WC et al (2014) Proteopathic tau seeding predicts tauopathy in vivo. Proc Natl Acad Sci U S A 111: E4376–E4385
55. Furman JL, Vaquer-Alicea J, White CL III, Cairns NJ, Nelson PT, Diamond MI (2017) Widespread tau seeding activity at early Braak stages. Acta Neuropathol 133:91–100
56. Toledo JB, Gopal P, Raible K, Irwin DJ, Brettschneider J, Sedor S et al (2016) Pathological α-synuclein distribution in subjects with coincident Alzheimer's and Lewy body pathology. Acta Neuropathol 131:393–409
57. Jellinger KA (2003) α-Synuclein pathology in Parkinson's and Alzheimer's disease brain: incidence and topographic distribution—a pilot study. Acta Neuropathol 106:191–201
58. Watts JC, Giles K, Grillo SK, Lemus A, DeArmond SJ, Prusiner SB (2011) Bioluminescence imaging of Aβ deposition in bigenic mouse models of Alzheimer's disease. Proc Natl Acad Sci U S A 108:2528–2533
59. Olsson TT, Klementieva O, Gouras GK (2018) Prion-like seeding and nucleation of intracellular amyloid-β. Neurobiol Dis 113: 1–10
60. Toyama BH, Kelly MJ, Gross JD, Weissman JS (2007) The structural basis of yeast prion strain variants. Nature 449:233–237
61. Legname G, Nguyen H-OB, Baskakov IV, Cohen FE, DeArmond SJ, Prusiner SB (2005) Strain-specified characteristics of mouse synthetic prions. Proc Natl Acad Sci U S A 102:2168–2173
62. Legname G, Nguyen H-OB, Peretz D, Cohen FE, DeArmond SJ, Prusiner SB (2006) Continuum of prion protein structures enciphers a multitude of prion isolate-specified phenotypes. Proc Natl Acad Sci U S A 103:19105–19110
63. Watts JC, Condello C, Stöhr J, Oehler A, Lee J, DeArmond SJ et al (2014) Serial propagation of distinct strains of Aβ prions from Alzheimer's disease patients. Proc Natl Acad Sci U S A 111:10323–10328
64. Qiang W, Yau WM, Lu JX, Collinge J, Tycko R (2017) Structural variation in amyloid-β fibrils from Alzheimer's disease clinical subtypes. Nature 541:217–221
65. Rasmussen J, Mahler J, Beschorner N, Kaeser SA, Häsler LM, Baumann F et al (2017) Amyloid polymorphisms constitute distinct clouds of conformational variants in different etiological subtypes of Alzheimer's disease. Proc Natl Acad Sci U S A 114:13018–13023
66. Condello C, Lemmin T, Stöhr J, Nick M, Wu Y, Watts JC et al (2018) Structural heterogeneity and intersubject variability of Aβ in familial and sporadic Alzheimer's disease. Proc Natl Acad Sci U S A 115: E782–E791
67. Cras P, van Harskamp F, Hendriks L, Ceuterick C, van Duijn CM, Stefanko SZ et al (1998) Presenile Alzheimer dementia characterized by amyloid angiopathy and large amyloid core type senile plaques in the APP 692Ala-->Gly mutation. Acta Neuropathol 96:253–260
68. Kumar-Singh S, Cras P, Wang R, Kros JM, van Swieten J, Lübke U et al (2002) Dense-core senile plaques in the Flemish variant of Alzheimer's disease are vasocentric. Am J Pathol 161:507–520
69. Zhao N, Liu C-C, Qiao W, Bu G (2018) Apolipoprotein E, receptors, and modulation of Alzheimer's disease. Biol Psychiatry 83: 347–357
70. Verghese PB, Castellano JM, Holtzman DM (2011) Apolipoprotein E in Alzheimer's disease and other neurological disorders. Lancet Neurol 10:241–252
71. Augustinack JC, Schneider A, Mandelkow EM, Hyman BT (2002) Specific tau phosphorylation sites correlate with severity of neuronal cytopathology in Alzheimer's disease. Acta Neuropathol 103:26–35
72. Braak H, Alafuzoff I, Arzberger T, Kretzschmar H, Del Tredici K (2006) Staging of Alzheimer disease-associated neurofibrillary pathology using paraffin sections and immunocytochemistry. Acta Neuropathol 112:389–404
73. Clavaguera F, Bolmont T, Crowther RA, Abramowski D, Frank S, Probst A et al (2009) Transmission and spreading of tauopathy in transgenic mouse brain. Nat Cell Biol 11:909–913
74. Frost B, Jacks RL, Diamond MI (2009) Propagation of tau misfolding from the outside to the inside of a cell. J Biol Chem 284: 12845–12852
75. Ridley RM, Baker HF, Windle CP, Cummings RM (2006) Very long term studies of the seeding of beta-amyloidosis in primates. J Neural Transm 113:1243–1251
76. Meyer-Luehmann M, Coomaraswamy J, Bolmont T, Kaeser S, Schaefer C, Kilger E et al (2006) Exogenous induction of cerebral beta-amyloidogenesis is governed by agent and host. Science 313:1781–1784
77. Bolmont T, Clavaguera F, Meyer-Luehmann-M, Herzig MC, Radde R, Staufenbiel M et al (2007) Induction of tau pathology by intracerebral infusion of amyloid-beta-containing brain extract and by amyloid-beta deposition

in APP × Tau transgenic mice. Am J Pathol 171:2012–2020

78. Vasconcelos B, Stancu IC, Buist A, Bird M, Wang P, Vanoosthuyse A et al (2016) Heterotypic seeding of Tau fibrillization by pre-aggregated Abeta provides potent seeds for prion-like seeding and propagation of Tau-pathology in vivo. Acta Neuropathol 131:549–569
79. Griner SL, Seidler P, Bowler J, Murray KA, Yang TP, Sahay S et al (2019) Structure-based inhibitors of amyloid beta core suggest a common interface with tau. eLife 8:1–28
80. Shin WS, Di J, Cao Q, Li B, Seidler PM, Murray KA et al (2019) Amyloid β-protein oligomers promote the uptake of tau fibril seeds potentiating intracellular tau aggregation. Alzheimers Res Ther 11(86):1–13
81. Kordower JH, Chu Y, Hauser RA, Freeman TB, Olanow CW (2008) Lewy body-like pathology in long-term embryonic nigral transplants in Parkinson's disease. Nat Med 14:504–506
82. Li JY, Englund E, Holton JL, Soulet D, Hagell P, Lees AJ et al (2008) Lewy bodies in grafted neurons in subjects with Parkinson's disease suggest host-to-graft disease propagation. Nat Med 14:501–503
83. Desplats P, Lee HJ, Bae EJ, Patrick C, Rockenstein E, Crews L et al (2009) Inclusion formation and neuronal cell death through neuron-to-neuron transmission of alpha-synuclein. Proc Natl Acad Sci U S A 106: 13010–13015
84. Iba M, McBride JD, Guo JL, Zhang B, Trojanowski JQ, Lee VM-Y (2015) Tau pathology spread in PS19 tau transgenic mice following locus coeruleus (LC) injections of synthetic tau fibrils is determined by the LC's afferent and efferent connections. Acta Neuropathol 130:349–362
85. Ye L, Hamaguchi T, Fritschi SK, Eisele YS, Obermuller U, Jucker M et al (2015) Progression of seed-induced Aβ deposition within the limbic connectome. Brain Pathol 25:743–752
86. Wu JW, Hussaini SA, Bastille IM, Rodriguez GA, Mrejeru A, Rilett K et al (2016) Neuronal activity enhances tau propagation and tau pathology in vivo. Nat Neurosci 19: 1085–1092
87. Braak H, Braak E, Yilmazer D, de Vos RA, Jansen EN, Bohl J (1996) Pattern of brain destruction in Parkinson's and Alzheimer's diseases. J Neural Transm 103:455–490
88. Braak H, Del Tredici K (2017) Potential pathways of abnormal tau and α-synuclein dissemination in sporadic Alzheimer's and Parkinson's diseases. In: Prusiner SB (ed) Prion biology. Cold spring harb. perspect. biol. Cold Spring Harbor Laboratory Press, Cold Spring Harbor, NY, pp 377–399
89. Hashimoto T, Adams KW, Fan Z, McLean PJ, Hyman BT (2011) Characterization of oligomer formation of amyloid-β peptide using a split-luciferase complementation assay. J Biol Chem 286:27081–27091
90. Walsh DM, Tseng BP, Rydel RE, Podlisny MB, Selkoe DJ (2000) The oligomerization of amyloid β-protein begins intracellularly in cells derived from human brain. Biochemistry 39:10831–10839
91. Lesne S, Koh MT, Kotilinek L, Kayed R, Glabe CG, Yang A et al (2006) A specific amyloid-beta protein assembly in the brain impairs memory. Nature 440:352–357
92. Shankar GM, Li S, Mehta TH, Garcia-Munoz A, Shepardson NE, Smith I et al (2008) Amyloid-beta protein dimers isolated directly from Alzheimer's brains impair synaptic plasticity and memory. Nat Med 14: 837–842
93. Yang T, Li S, Xu H, Walsh DM, Selkoe DJ (2017) Large soluble oligomers of amyloid β-protein from Alzheimer brain are far less neuroactive than the smaller oligomers to which they dissociate. J Neurosci 37:152–163
94. Cleary JP, Walsh DM, Hofmeister JJ, Shankar GM, Kuskowski MA, Selkoe DJ et al (2005) Natural oligomers of the amyloid-β protein specifically disrupt cognitive function. Nat Neurosci 8:79–84
95. Chiang ACA, Fowler SW, Reddy R, Pletnikova O, Troncoso JC, Sherman MA et al (2018) Discrete pools of oligomeric amyloid-β track with spatial learning deficits in a mouse model of Alzheimer amyloidosis. Am J Pathol 188:739–756
96. Zahs KR, Ashe KH (2013) β-Amyloid oligomers in aging and Alzheimer's disease. Front Aging Neurosci 5:28
97. Esparza TJ, Zhao H, Cirrito JR, Cairns NJ, Bateman RJ, Holtzman DM et al (2013) Amyloid-β oligomerization in Alzheimer dementia versus high-pathology controls. Ann Neurol 73:104–119
98. Mirbaha H, Chen D, Morazova OA, Ruff KM, Sharma AM, Liu X et al (2018) Inert and seed-competent tau monomers suggest structural origins of aggregation. eLife 7: e36584
99. Falcon B, Cavallini A, Angers R, Glover S, Murray TK, Barnham L et al (2015)

Conformation determines the seeding potencies of native and recombinant tau aggregates. J Biol Chem 290:1049–1065

100. Jackson SJ, Kerridge C, Cooper J, Cavallini A, Falcon B, Cella CV et al (2016) Short fibrils constitute the major species of seed-competent tau in the brains of mice transgenic for human P301S tau. J Neurosci 36: 762–772

101. Johnson NR, Condello C, Guan S, Oehler A, Becker J, Gavidia M et al (2017) Evidence for sortilin modulating regional accumulation of human tau prions in transgenic mice. Proc Natl Acad Sci U S A 114:E11029–E11036

102. Xia C, Makaretz SJ, Caso C, McGinnis S, Gomperts SN, Sepulcre J et al (2017) Association of in vivo [^{18}F]AV-1451 tau PET imaging results with cortical atrophy and symptoms in typical and atypical Alzheimer disease. JAMA Neurol 74:427–436

103. Condello C, Yuan P, Grutzendler J (2018) Microglia-mediated neuroprotection, TREM2, and Alzheimer's disease: evidence from optical imaging. Biol Psychiatry 83:377–387

104. Bongianni M, Orrù C, Groveman BR, Sacchetto L, Fiorini M, Tonoli G et al (2017) Diagnosis of human prion disease using real-time quaking-induced conversion testing of olfactory mucosa and cerebrospinal fluid samples. JAMA Neurol 74:155–162

105. Concha-Marambio L, Pritzkow S, Moda F, Tagliavini F, Ironside JW, Schulz PE et al (2016) Detection of prions in blood from patients with variant Creutzfeldt-Jakob disease. Sci Transl Med 8:370ra183

106. Edgeworth JA, Farmer M, Sicilia A, Tavares P, Beck J, Campbell T et al (2011) Detection of prion infection in variant Creutzfeldt-Jakob disease: a blood-based assay. Lancet 377: 487–493

107. Pontecorvo MJ, Devous MD Sr, Navitsky M, Lu M, Salloway S, Schaerf FW et al (2017) Relationships between flortaucipir PET tau binding and amyloid burden, clinical diagnosis, age and cognition. Brain 140:748–763

108. Tosun D, Landau S, Aisen PS, Petersen RC, Mintun M, Jagust W et al (2017) Association between tau deposition and antecedent amyloid-β accumulation rates in normal and early symptomatic individuals. Brain 140: 1499–1512

109. Iaccarino L, Tammewar G, Ayakta N, Baker SL, Bejanin A, Boxer AL et al (2017) Local and distant relationships between amyloid, tau and neurodegeneration in Alzheimer's disease. Neuroimage Clin 17:452–464

110. Kidd M (1963) Paired helical filaments in electron microscopy in Alzheimer's disease. Nature 197:192–193

111. Cheng Y (2015) Single-particle cryo-EM at crystallographic resolution. Cell 161: 450–457

112. He S, Scheres SHW (2017) Helical reconstruction in RELION. J Struct Biol 198: 163–176

113. Bartz JC (2017) Prion strain diversity. In: Prusiner SB (ed) Prion diseases. Cold spring harb. Perspect. Med. Cold Spring Harbor Laboratory Press, Cold Spring Harbor, NY, pp 31–44

114. Fitzpatrick AWP, Falcon B, He S, Murzin AG, Murshudov G, Garringer HJ et al (2017) Cryo-EM structures of tau filaments from Alzheimer's disease. Nature 547:185–190

115. Falcon B, Zhang W, Schweighauser M, Murzin AG, Vidal R, Garringer HJ et al (2018) Tau filaments from multiple cases of sporadic and inherited Alzheimer's disease adopt a common fold. Acta Neuropathol 136: 699–708

116. Falcon B, Zivanov J, Zhang W, Murzin AG, Garringer HJ, Vidal R et al (2019) Novel tau filament fold in chronic traumatic encephalopathy encloses hydrophobic molecules. Nature 568:420–423

117. Zhang W, Tarutani A, Newell KL, Murzin AG, Matsubara T, Falcon B et al (2020) Novel tau filament fold in corticobasal degeneration. Nature 580:283–287

118. Falcon B, Zhang W, Murzin AG, Murshudov G, Garringer HJ, Vidal R et al (2018) Structures of filaments from Pick's disease reveal a novel tau protein fold. Nature 561:137–140

119. Schweighauser M, Shi Y, Tarutani A, Kametani F, Murzin AG, Ghetti B et al (2020) Structures of α-synuclein filaments from multiple system atrophy. Nature 585: 464–469

120. Kollmer M, Close W, Funk L, Rasmussen J, Bsoul A, Schierhorn A et al (2019) Cryo-EM structure and polymorphism of Aβ amyloid fibrils purified from Alzheimer's brain tissue. Nat Commun 10:4760

121. Prusiner SB (1994) Biology and genetics of prion diseases. Annu Rev Microbiol 48: 655–686

122. Bennett DA, Buchman AS, Boyle PA, Barnes LL, Wilson RS, Schneider JA (2018) Religious orders study and rush memory and aging project. J Alzheimers Dis 64: S161–S189

Index

Jerold Chun (ed.), *Alzheimer's Disease: Methods and Protocols*,
Methods in Molecular Biology, vol. 2561, https://doi.org/10.1007/978-1-0716-2655-9,

MIX
Papier aus verantwortungsvollen Quellen
Paper from responsible sources
FSC® C105338

If you have any concerns about our products,
you can contact us on
ProductSafety@springernature.com

In case Publisher is established outside the EU,
the EU authorized representative is:
Springer Nature Customer Service Center GmbH
Europaplatz 3, 69115 Heidelberg, Germany

Printed by Libri Plureos GmbH
in Hamburg, Germany